Beyond the Standard Model of Particle Physics

Beyond the Standard Model of Particle Physics provides a comprehensive overview of the current state of research in the field, as well as the theories and models that are being developed to explain the unexplained phenomena in the universe. It also discusses the experiments and observations that are being conducted to test these theories, and the potential implications for our understanding of the universe. The primary advantage of the book is its in-depth examination of theories that go beyond the standard model of particle physics, which has been the dominant framework for understanding particle interactions for several decades. The book assumes a solid understanding of quantum field theory, particle physics, and general relativity before delving into advanced topics in the field. Its extensive coverage includes various topics such as grand unification, supersymmetry, extra dimensions, and others. It also discusses the theoretical and experimental implications of these ideas, providing readers with a comprehensive overview of the current state of the field. Additionally, the book highlights the significance of experimental data in testing new theoretical ideas and refining them based on experimental results. As a result, this resource is invaluable for graduate-level researchers, physicists and students seeking to stay abreast of the most recent developments in the field.

Key Features:

- Provides extensive coverage of theories that transcend the Standard Model of particle physics,
- Connects theoretical concepts with real-world experimental data in a clear and accessible style.
- Written by an expert in the field.

Professor Shaaban Khalil is one of the most prominent High Energy Physicists in the Middle East. He is well known for his outstanding contributions to the understanding of possible new physics beyond the standard model of particle physics. He has been the founding director of the Center for Fundamental Physics at Zewail City of Science and Technology in Egypt, since 2012. Prof. Khalil has published more than 300 scientific papers in renowned scientific journals. His work enabled him to earn a D.Sc. degree in 2006. His research has been widely recognised, and he has received several national and international awards. Prof. Khalil has been a visiting professor at the University of Southampton since 2011. He is the president of the Arab Physical Society and the founding team leader of the Egyptian particle physics group at the CMS experiment at CERN, Geneva, Switzerland.

Series in Particle Physics, Cosmology, and Gravitation

Series Editors: **Brian Foster**, *Oxford University, UK*

Edward W Kolb, *Fermi National Accelerator Laboratory, USA*

This series of books covers all aspects of theoretical and experimental high energy physics, cosmology and gravitation and the interface between them. In recent years the fields of particle physics and astrophysics have become increasingly interdependent and the aim of this series is to provide a library of books to meet the needs of students and researchers in these fields.

Other recent books in the series:

An Introduction to Beam Physics
Martin Berz, Kyoko Makino, and Weishi Wan

Neutrino Physics, Second Edition
K Zuber

Group Theory for the Standard Model of Particle Physics and Beyond
Ken J Barnes

The Standard Model and Beyond
Paul Langacker

Particle and Astroparticle Physics
Utpal Sakar

Joint Evolution of Black Holes and Galaxies
M Colpi, V Gorini, F Haardt, and U Moschella (Eds)

Gravitation: From the Hubble Length to the Planck Length
I Ciufolini, E Coccia, V Gorini, R Peron, and N Vittorio (Eds)

The Galactic Black Hole: Lectures on General Relativity and Astrophysics
H Falcke, and F Hehl (Eds)

The Mathematical Theory of Cosmic Strings: Cosmic Strings in the Wire Approximation
M R Anderson

Geometry and Physics of Branes
U Bruzzo, V Gorini, and U Moschella (Eds)

Modern Cosmology
S Bonometto, V Gorini, and U Moschella (Eds)Gravitation and Gauge SymmetriesM Blagojevic

Gravitational Waves
I Ciufolini, V Gorini, U Moschella, and P Fré (Eds)

Introduction to Particle Physics and Cosmology
Grant J. Mathews and Guobao Tang

Beyond the Standard Model of Particle Physics
Shaaban Khalil

Beyond the Standard Model of Particle Physics

Advanced Theories in Particle Physics

Shaaban Khalil

CRC Press
Taylor & Francis Group
Boca Raton London New York

CRC Press is an imprint of the
Taylor & Francis Group, an **informa** business

Designed cover image: ATLAS collision event in which a microscopic-black-hole was produced in the collision of two protons. © 2008-2025 CERN

First edition published 2026
by CRC Press
2385 NW Executive Center Drive, Suite 320, Boca Raton FL 33431

and by CRC Press
4 Park Square, Milton Park, Abingdon, Oxon, OX14 4RN

CRC Press is an imprint of Taylor & Francis Group, LLC

ISBN: 978-1-032-59699-0 (hbk)
ISBN: 978-1-032-60137-3 (pbk)
ISBN: 978-1-003-45770-1 (ebk)

DOI: 10.1201/9781003457701

Typeset in Latin Modern font
by KnowledgeWorks Global Ltd.

To my late wife, with love and remembrance

Contents

Preface

The Standard Model (SM) of particle physics stands as one of the most successful scientific theories of the 20th century. It has passed numerous experimental tests with extraordinary precision and has accurately described the electromagnetic, weak, and strong interactions among elementary particles. The discovery of the Higgs boson at the Large Hadron Collider (LHC) in 2012 marked the crowning achievement of the SM, confirming its mechanism of mass generation. Yet, despite these triumphs, the SM leaves many fundamental questions unanswered.

Why do neutrinos have mass? What is the nature of dark matter? How can we resolve the hierarchy problem? What explains the baryon asymmetry of the universe? Is there a deeper symmetry behind the SM gauge structure? These puzzles, and many others, provide strong motivation for the search for physics beyond the SM (BSM).

This book offers a comprehensive introduction to BSM physics, aimed at advanced undergraduate students, graduate students, and researchers with an interest in high-energy theory and phenomenology. The topics covered span a wide range of well-motivated extensions of the SM, including:

- Two Higgs Doublet Models (2HDMs), which provide a richer scalar sector and natural explanations for dark matter and baryogenesis;
- Gauged $U(1)_{B-L}$ models, which accommodate neutrino masses and right-handed neutrinos in a theoretically consistent manner;
- Left-Right Symmetric Models (LRSMs), offering a compelling framework for restoring parity symmetry and generating neutrino masses through seesaw mechanisms;
- Grand Unified Theories (GUTs), such as $SU(5)$, which seek a unified description of the SM forces;
- Supersymmetry (SUSY), a symmetry that links fermions and bosons and addresses the hierarchy problem;

- Theories with extra spatial dimensions, which aim to resolve outstanding theoretical puzzles and may connect particle physics with quantum gravity.

Special emphasis is placed on the phenomenological implications of these models. Chapters examine how BSM physics manifests in collider signatures, flavor observables, neutrino experiments, and cosmological measurements. Particular attention is given to the testability of these models at current and future experiments, including the LHC and proposed next-generation colliders. The interplay between particle physics and cosmology, especially in addressing the nature of dark matter, dark energy, and the matter-antimatter asymmetry, remains a central focus in contemporary particle physics.

The structure of the book reflects a logical progression: from foundational motivations for BSM physics, to detailed presentations of theoretical frameworks, to their testable predictions. Throughout, both theoretical consistency and experimental viability are prioritized.

This book is the result of years of engagement with the open questions at the frontier of fundamental physics. It is intended as both a learning tool and a reference for those aiming to deepen their understanding or contribute to the development of particle physics BSM. Whether used in a classroom setting, during independent study, or as a guide to research, we hope this volume provides readers with both conceptual clarity and inspiration for further exploration of the fundamental laws of nature.

Shaaban Khalil

CHAPTER 1

Introduction to Physics BSM

The Standard Model (SM) of particle physics is a highly successful quantum field theory that describes electromagnetic, weak, and strong interactions among elementary particles. It has provided a remarkably accurate framework for understanding particle dynamics and interactions at accessible energies. This chapter summarizes both the achievements and limitations of the SM, drawing from reference [1].

1.1 SM OVERVIEW AND FEATURES

The SM is based on the gauge symmetry:

$$SU(3)_C \times SU(2)_L \times U(1)_Y,$$

where $SU(3)_C$ corresponds to quantum chromodynamics (QCD), which governs strong interactions, while $SU(2)_L \times U(1)_Y$ describes electroweak interactions.

The model respects gauge invariance and Poincaré symmetry, and it is renormalizable, ensuring well-defined predictions at various energy scales. The matter content of the SM comprises three generations of fermions, each consisting of a left-handed quark doublet and a left-handed lepton doublet transforming under $SU(2)_L$, along with right-handed singlets for the up-type quark, down-type quark, and charged lepton. The gauge sector comprises eight gluons, three weak gauge bosons, and one hypercharge gauge boson. Electroweak symmetry is spontaneously broken by the Higgs mechanism, which gives masses to

DOI: 10.1201/9781003457701-1

TABLE 1.1 The SM field content and quantum numbers under the gauge group $SU(3)_C \times SU(2)_L \times U(1)_Y$.

Fields	Components (Spin)		$SU(3)_C$	$SU(2)_L$	$U(1)_Y$
Fermions					
L	$\begin{pmatrix} \nu \\ e \end{pmatrix}_L$	(1/2)	1	2	$-\frac{1}{2}$
E	e_R	(1/2)	1	1	-1
Q	$\begin{pmatrix} u \\ d \end{pmatrix}_L$	(1/2)	3	2	$+\frac{1}{6}$
U	u_R	(1/2)	3	1	$+\frac{2}{3}$
D	d_R	(1/2)	3	1	$-\frac{1}{3}$
Gauge Bosons					
B_μ	B_μ (1)		singlet	singlet	singlet
W_μ	W^+, W^-, W^3 (1)		singlet	triplet	singlet
G_μ	G^a_μ $(a = 1 \ldots 8)$ (1)		octet	singlet	singlet
Higgs					
ϕ	$\begin{pmatrix} \phi^+ \\ \phi^0 \end{pmatrix}$	(0)	1	2	$+\frac{1}{2}$

the weak bosons and fermions. Table 1.1 summarizes the field content and quantum numbers of the SM.

A key relation in the SM is the Gell-Mann-Nishijima formula, which links the electric charge Q to the third component of the weak isospin T_3 and the hypercharge Y:

$$Q = T_3 + \frac{Y}{2}. \tag{1.1}$$

This equation emerges from electroweak unification, where the symmetry $SU(2)_L \times U(1)_Y$ is spontaneously broken to $U(1)_{\text{EM}}$ via the Higgs mechanism. The resulting massless photon and massive weak bosons reflect the decomposition of gauge degrees of freedom. The Gell-Mann-Nishijima relation provides a consistent framework for assigning electric charges to all SM particles.

1.1.1 The SM Lagrangian

Having outlined the particle content, gauge symmetries, and charge assignments in the SM, we now present its full Lagrangian formulation. The SM Lagrangian encodes the dynamics and interactions of all fundamental fields, incorporating gauge invariance under the group $SU(3)_C \times SU(2)_L \times U(1)_Y$, spontaneous symmetry breaking via the Higgs mechanism, and the generation of fermion masses through Yukawa interactions [2–7].

The total SM Lagrangian can be written as:

$$\mathcal{L}_{\text{SM}} = \mathcal{L}_{\text{YM}} + \mathcal{L}_{\text{K}} + \mathcal{L}_{\text{Higgs}} + \mathcal{L}_{\text{Y}}, \tag{1.2}$$

where:

- $\mathcal{L}_{\text{YM}}$ describes the gauge fields and their self-interactions.
- $\mathcal{L}_{\text{K}}$ contains the kinetic terms of fermions and their gauge interactions.
- $\mathcal{L}_{\text{Higgs}}$ governs the Higgs field dynamics and potential.
- $\mathcal{L}_{\text{Y}}$ accounts for the Yukawa interactions between fermions and the Higgs field.

Gauge Field Kinetic Terms

The Yang-Mills part of the SM Lagrangian describes the dynamics of the gauge bosons:

$$\mathcal{L}_{\text{YM}} = -\frac{1}{4} G^a_{\mu\nu} G^{\mu\nu a} - \frac{1}{4} W^i_{\mu\nu} W^{\mu\nu i} - \frac{1}{4} B_{\mu\nu} B^{\mu\nu}, \tag{1.3}$$

where the field strength tensors for the $SU(3)_C$, $SU(2)_L$, and $U(1)_Y$ gauge groups are:

$$G^a_{\mu\nu} = \partial_\mu G^a_\nu - \partial_\nu G^a_\mu - g_s f^{abc} G^b_\mu G^c_\nu, \tag{1.4}$$

$$W^i_{\mu\nu} = \partial_\mu W^i_\nu - \partial_\nu W^i_\mu - g\, \epsilon^{ijk} W^j_\mu W^k_\nu, \tag{1.5}$$

$$B_{\mu\nu} = \partial_\mu B_\nu - \partial_\nu B_\mu. \tag{1.6}$$

Here, g_s, g, and g_Y are the gauge couplings for $SU(3)_C$, $SU(2)_L$, and $U(1)_Y$, respectively. The gluons G^a_μ ($a = 1, \ldots, 8$), weak gauge bosons W^i_μ ($i = 1, 2, 3$), and the hypercharge field B_μ correspond to the gauge groups $SU(3)_C$, $SU(2)_L$, and $U(1)_Y$, respectively. The structure constants f^{abc} and ϵ^{ijk} satisfy the commutation relations:

$$[T^\alpha, T^\beta] = i f^{\alpha\beta\gamma} T^\gamma, \qquad [\tau^a, \tau^b] = i\varepsilon^{abc}\tau^c, \tag{1.7}$$

for $SU(3)_C$ and $SU(2)_L$, respectively.

Fermion Kinetic Terms

The kinetic terms for fermions, including their gauge interactions, are given by:

$$\begin{aligned}\mathcal{L}_{\text{K}} &= i\sum_{j=1}^{3}\Big(\overline{L}_j\gamma^\mu D_\mu L_j + \overline{E}_j\gamma^\mu D_\mu E_j + \overline{Q}_j\gamma^\mu D_\mu Q_j \\ &\quad + \overline{U}_j\gamma^\mu D_\mu U_j + \overline{D}_j\gamma^\mu D_\mu D_j\Big),\end{aligned} \tag{1.8}$$

where the covariant derivative is:

$$D_\mu = \partial_\mu - i g_s \Lambda^a G^a_\mu - i g T^i W^i_\mu - i g_Y Y B_\mu. \tag{1.9}$$

This expression depends on the specific gauge representation and hypercharge Y of each fermion, as summarized in Table 1.1. For example:

$$\overline{L}\gamma^\mu D_\mu L = \overline{L}\gamma^\mu\left(\partial_\mu - \frac{ig}{2}\sigma\cdot W_\mu + \frac{ig_Y}{2}B_\mu\right)L, \tag{1.10}$$

$$\overline{E}\gamma^\mu D_\mu E = \overline{E}\gamma^\mu\left(\partial_\mu + i g_Y B_\mu\right)E, \tag{1.11}$$

$$\begin{aligned}\overline{Q}^\alpha\gamma^\mu D_\mu Q^\beta &= \overline{Q}^\alpha\gamma^\mu\left[\left(\partial_\mu - \frac{ig}{2}\sigma\cdot W_\mu - \frac{ig_Y}{6}B_\mu\right)\delta_{\alpha\beta}\right. \\ &\quad \left. - \frac{ig_s}{2}\lambda_{\alpha\beta}\cdot G_\mu\right]Q^\beta.\end{aligned} \tag{1.12}$$

Analogous expressions hold for U and D quarks. Here, σ^i and λ^a are the Pauli and Gell-Mann matrices, and the indices α, β denote color.

Higgs Sector

The Higgs Lagrangian consists of kinetic and potential terms for the scalar doublet field ϕ:

$$\mathcal{L}_{\text{Higgs}} = (D^\mu\phi)^\dagger(D_\mu\phi) - V(\phi), \tag{1.13}$$

with the covariant derivative:

$$D_\mu\phi = \left(\partial_\mu - \frac{ig}{2}\sigma\cdot W_\mu - \frac{ig_Y}{2}B_\mu\right)\phi. \tag{1.14}$$

The scalar potential is:

$$V(\phi) = \mu^2\phi^\dagger\phi + \lambda(\phi^\dagger\phi)^2, \tag{1.15}$$

where $\lambda > 0$ ensures the potential is bounded from below. Electroweak symmetry breaking occurs when $\mu^2 < 0$, giving rise to the Higgs VEV and generating masses for the weak bosons and fermions [7–9].

Yukawa Interactions

The Yukawa Lagrangian governs the interactions between fermions and the Higgs field, which after electroweak symmetry breaking generate fermion masses:

$$\mathcal{L}_{\mathrm{Y}} = -\sum_{i,j=1}^{3}\left(y^e_{ij}\overline{L}_i\phi E_j + y^u_{ij}\overline{Q}_i\widetilde{\phi}U_j + y^d_{ij}\overline{Q}_i\phi D_j\right) + \text{h.c.} \tag{1.16}$$

Here, the dual Higgs doublet $\widetilde{\phi}$ is defined as:

$$\widetilde{\phi} \equiv i\sigma^2\phi^* = \begin{pmatrix} \phi^{0*} \\ -\phi^- \end{pmatrix}, \tag{1.17}$$

and y^e, y^u, y^d are complex 3×3 matrices of Yukawa couplings.

1.1.2 Spontaneous Symmetry Breaking and the Higgs Mechanism

In the SM Lagrangian, all fermions and gauge bosons are initially massless. The Higgs mechanism introduces mass for these particles via spontaneous symmetry breaking. The electroweak symmetry breaking pattern is given by:

$$SU(2)_L \times U(1)_Y \xrightarrow{\langle\phi\rangle} U(1)_{\mathrm{EM}}, \tag{1.18}$$

where $U(1)_{\mathrm{EM}}$ corresponds to the electromagnetic gauge symmetry.

The Higgs field in the SM is a complex scalar doublet under $SU(2)_L$, and is given by:

$$\phi = \begin{pmatrix} \phi^+ \\ \phi^0 \end{pmatrix}, \tag{1.19}$$

with quantum numbers listed in Table 1.1. For $\mu^2 < 0$ in the Higgs potential (1.15), the vacuum is degenerate and spontaneous symmetry

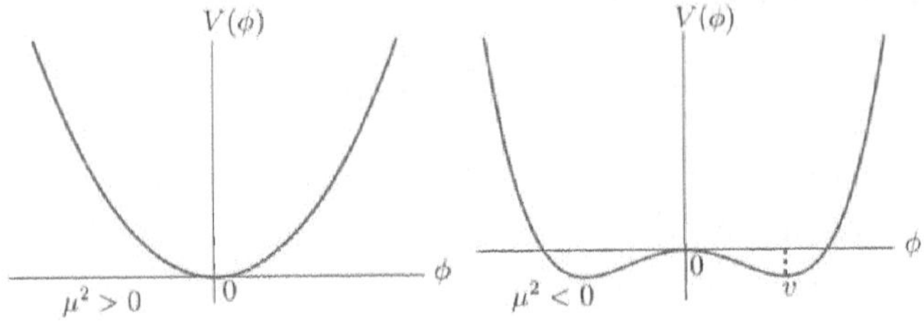

Figure 1.1 The scalar potential $V(\phi)$ for $\mu^2 > 0$ (symmetric phase) and $\mu^2 < 0$ (broken phase), respectively.

breaking occurs, as illustrated in Fig. 1.1. The VEV of the Higgs field satisfies:

$$|\phi|^2 = \frac{v^2}{2}, \quad \text{where} \quad v = \sqrt{\frac{-\mu^2}{\lambda}}. \tag{1.20}$$

Choosing a specific vacuum, we take:

$$\phi_0 = \begin{pmatrix} 0 \\ \frac{v}{\sqrt{2}} \end{pmatrix}. \tag{1.21}$$

The action of the electroweak generators on the vacuum state ϕ_0 is:

$$\frac{1}{2}\tau^1\phi_0 = \frac{1}{2}\begin{pmatrix} v/\sqrt{2} \\ 0 \end{pmatrix}, \quad \frac{1}{2}\tau^2\phi_0 = \frac{1}{2}\begin{pmatrix} -iv/\sqrt{2} \\ 0 \end{pmatrix},$$
$$\frac{1}{2}\tau^3\phi_0 = -\frac{1}{2}\phi_0, \quad Y\phi_0 = \phi_0. \tag{1.22}$$

Thus, the vacuum is not invariant under $SU(2)_L \times U(1)_Y$, indicating that these symmetries are spontaneously broken. However, the electromagnetic gauge symmetry remains unbroken since:

$$Q\phi_0 = \left(\frac{1}{2}\tau^3 + \frac{1}{2}Y\right)\phi_0 = 0. \tag{1.23}$$

The Higgs field contains four real degrees of freedom, three of which become the longitudinal components of the $W^\pm$ and Z bosons via the Higgs mechanism. These can be parameterized as:

$$\phi(x) = \frac{1}{\sqrt{2}} e^{-i\tau^i\eta^i(x)/2v} \begin{pmatrix} 0 \\ v + h(x) \end{pmatrix}, \tag{1.24}$$

where $\eta^i(x)$ (with $i = 1, 2, 3$) are Goldstone bosons and $h(x)$ is the physical Higgs field. In the unitary gauge, where $\phi \to U(\eta)\phi$, the Higgs field simplifies to:

$$\phi(x) = \frac{1}{\sqrt{2}} \begin{pmatrix} 0 \\ v + h(x) \end{pmatrix}. \tag{1.25}$$

1.1.3 Particle Mass Generation in the SM

Gauge Boson Masses

To derive gauge boson masses, consider the kinetic term of the Higgs doublet:

$$\mathcal{L}_{\text{kin}} = (D_\mu \phi)^\dagger (D^\mu \phi), \tag{1.26}$$

where the covariant derivative is:

$$D_\mu = \partial_\mu - ig\frac{T^a}{2} W_\mu^a - ig'\frac{Y}{2} B_\mu. \tag{1.27}$$

Substituting $\phi \to \phi_0$, the mass terms for gauge bosons become:

$$\mathcal{L}_{\text{mass}} \supset M_W^2 W_\mu^+ W^{-\mu} + \frac{M_Z^2}{2} Z_\mu Z^\mu, \tag{1.28}$$

with the physical eigenstates:

$$W_\mu^\pm = \frac{1}{\sqrt{2}}(W_\mu^1 \mp i W_\mu^2), \tag{1.29}$$

$$Z_\mu = \cos\theta_W W_\mu^3 - \sin\theta_W B_\mu, \tag{1.30}$$

$$A_\mu = \sin\theta_W W_\mu^3 + \cos\theta_W B_\mu. \tag{1.31}$$

The corresponding masses are:

$$M_W = \frac{1}{2} g v, \tag{1.32}$$

$$M_Z = \frac{1}{2} v \sqrt{g^2 + g'^2}, \tag{1.33}$$

$$M_A = 0. \tag{1.34}$$

The Weinberg angle θ_W is defined via:

$$\sin\theta_W = \frac{g'}{\sqrt{g^2 + g'^2}}, \tag{1.35}$$

$$\cos\theta_W = \frac{g}{\sqrt{g^2 + g'^2}}. \tag{1.36}$$

Fermion Masses

In the unitary gauge, the Yukawa Lagrangian (1.16) becomes:

$$\mathcal{L}_Y = -\frac{v + h(x)}{\sqrt{2}} \left(y_{ij}^e \overline{e_{Li}} e_{Rj} + y_{ij}^u \overline{u_{Li}} u_{Rj} + y_{ij}^d \overline{d_{Li}} d_{Rj} \right) + \text{h.c.} \tag{1.37}$$

The corresponding fermion masses are:

$$m_e = \frac{y_e v}{\sqrt{2}}, \quad m_u = \frac{y_u v}{\sqrt{2}}, \quad m_d = \frac{y_d v}{\sqrt{2}}. \tag{1.38}$$

For three families, the Yukawa matrices y_f (with $f = e, u, d$) are 3×3 complex matrices. These can be diagonalized via unitary transformations:

$$f_L \to V_f f_L, \quad f_R \to U_f f_R, \quad \nu_L \to V_e \nu_L, \tag{1.39}$$

such that:

$$y_f^{\text{diag}} = V_f^\dagger y_f U_f. \tag{1.40}$$

The physical masses are then:

$$m_{f_i} = y_{f_i}^{\text{diag}} \frac{v}{\sqrt{2}}. \tag{1.41}$$

Higgs Mass

The Higgs potential after spontaneous symmetry breaking becomes:

$$V(\phi^\dagger \phi) = -\frac{\mu^2}{2}(v + h)^2 + \frac{\lambda}{4}(v + h)^4 \tag{1.42}$$

$$= \frac{1}{2}(2\lambda v^2)h^2 + \lambda v h^3 + \frac{\lambda}{4}h^4 - \frac{\mu^2 v^2}{4}. \tag{1.43}$$

From the quadratic term, the Higgs boson mass is:

$$M_h = \sqrt{2\lambda v^2}. \tag{1.44}$$

The Higgs mass and the self-coupling λ are not predicted within the SM and must be determined experimentally.

1.1.4 SM Interactions

This section provides a systematic overview of the SM interaction structure, beginning with the electroweak and Higgs sectors, and highlighting their implications for fermion masses, gauge boson couplings, and flavor physics.

Electroweak Interactions

The electroweak interactions of leptons and quarks in the SM are derived from the fermionic kinetic terms in the Lagrangian:

$$\begin{aligned} \mathcal{L}_{\mathrm{K}} &= i\sum_{j=1}^{3}\Big(\overline{L}_j\gamma^\mu D_\mu L_j + \overline{E}_j\gamma^\mu D_\mu E_j + \overline{Q}_j\gamma^\mu D_\mu Q_j \\ &\quad + \overline{U}_j\gamma^\mu D_\mu U_j + \overline{D}_j\gamma^\mu D_\mu D_j\Big), \end{aligned} \tag{1.45}$$

where the covariant derivative D_μ encodes the electroweak gauge fields. The interactions of fermions with the gauge bosons are extracted from the covariant derivatives:

$$\begin{aligned} \overline{\psi} i\gamma^\mu D_\mu \psi &\supset \overline{\psi}\gamma^\mu\Big(\frac{g}{\sqrt{2}}(W_\mu^+ T^+ + W_\mu^- T^-)P_L + eQA_\mu \\ &\quad + \frac{g}{\cos\theta_W}(T^3 P_L - \sin^2\theta_W Q)Z_\mu\Big)\psi, \end{aligned} \tag{1.46}$$

where $P_L = \frac{1}{2}(1-\gamma_5)$, $T^\pm = T_1 \pm iT_2$, and $e = g\sin\theta_W = g_Y\cos\theta_W$.

The full electroweak interaction Lagrangian can be written in a compact form:

$$\mathcal{L}_{\mathrm{EW}} = \mathcal{L}_{\mathrm{EM}} + \mathcal{L}_{\mathrm{NC}} + \mathcal{L}_{\mathrm{CC}}, \tag{1.47}$$

where

$$\mathcal{L}_{\mathrm{EM}} = eA_\mu J_A^\mu, \tag{1.48}$$

$$\mathcal{L}_{\mathrm{NC}} = \frac{g}{\cos\theta_W} Z_\mu J_Z^\mu, \tag{1.49}$$

$$\mathcal{L}_{\mathrm{CC}} = \frac{g}{\sqrt{2}}\left(W_\mu^+ J_W^\mu + W_\mu^- J_W^{\mu\dagger}\right). \tag{1.50}$$

With the currents:

$$J_A^\mu = \overline{\psi}\gamma^\mu Q\psi, \tag{1.51}$$

$$J_Z^\mu = \overline{\psi}\gamma^\mu(T^3 P_L - \sin^2\theta_W Q)\psi, \tag{1.52}$$

$$J_W^\mu = \overline{\psi}\gamma^\mu T^+ P_L\psi, \quad J_W^{\mu\dagger} = \overline{\psi}\gamma^\mu T^- P_L\psi. \tag{1.53}$$

Electromagnetic Interactions

Electromagnetic interactions, mediated by the photon, couple only to electrically charged fermions:

$$\mathcal{L}_{\mathrm{EM}} = eA_\mu J_A^\mu, \tag{1.54}$$

with the current:

$$J^\mu_A = \sum_i Q^i \overline{f}_i \gamma^\mu f_i = -\overline{e}\gamma^\mu e + \frac{2}{3}\overline{u}\gamma^\mu u - \frac{1}{3}\overline{d}\gamma^\mu d. \tag{1.55}$$

Neutral Current Interactions

Neutral weak interactions are mediated by the Z-boson:

$$\mathcal{L}_{\text{NC}} = \frac{g}{\cos\theta_W} Z_\mu J^\mu_Z, \tag{1.56}$$

with current:

$$\begin{aligned} J^\mu_Z &= \sum_i \overline{f}_i \gamma^\mu (T^3_i P_L - Q^i \sin^2\theta_W) f_i \\ &= \frac{1}{2}\overline{\nu}_e \gamma^\mu P_L \nu_e + \overline{e}\gamma^\mu(-\frac{1}{2}P_L + \sin^2\theta_W)e \\ &+ \overline{u}\gamma^\mu \left(\frac{1}{2}P_L - \frac{2}{3}\sin^2\theta_W\right) u + \overline{d}\gamma^\mu \left(-\frac{1}{2}P_L + \frac{1}{3}\sin^2\theta_W\right) d. \end{aligned} \tag{1.57}$$

Charged Current Interactions

Charged weak interactions are mediated by the $W^\pm$ bosons:

$$\mathcal{L}_{\text{CC}} = \frac{g}{\sqrt{2}}\left(W^+_\mu J^\mu_W + W^-_\mu J^{\mu\dagger}_W\right), \tag{1.58}$$

with:

$$J^\mu_W = \sum_{i,j} \left(\overline{\nu}_{Li}\gamma^\mu e_{Lj}\delta_{ij} + \overline{u}_{Li}\gamma^\mu d_{Lj} V_{ij}\right), \tag{1.59}$$

$$J^{\mu\dagger}_W = \sum_{i,j} \left(\overline{e}_{Lj}\gamma^\mu \nu_{Li}\delta_{ij} + \overline{d}_{Lj}\gamma^\mu u_{Li} V^\dagger_{ij}\right), \tag{1.60}$$

where $V \equiv V_{\text{CKM}} = V^{u\dagger}V^d$ is the Cabibbo-Kobayashi-Maskawa (CKM) matrix [10, 11].

The CKM matrix parameterizes quark generation mixing:

$$V_{\text{CKM}} = \begin{pmatrix} c_{12}c_{13} & s_{12}c_{13} & s_{13}e^{-i\delta} \\ -s_{12}c_{23} - c_{12}s_{23}s_{13}e^{i\delta} & c_{12}c_{23} - s_{12}s_{23}s_{13}e^{i\delta} & s_{23}c_{13} \\ s_{12}s_{23} - c_{12}c_{23}s_{13}e^{i\delta} & -c_{12}s_{23} - s_{12}c_{23}s_{13}e^{i\delta} & c_{23}c_{13} \end{pmatrix}, \tag{1.61}$$

with $s_{ij} = \sin\theta_{ij}$, $c_{ij} = \cos\theta_{ij}$. Experimental values are [12]:

$$\sin\theta_{12} = 0.2229 \pm 0.0022, \tag{1.62}$$
$$\sin\theta_{23} = 0.0412 \pm 0.0002, \tag{1.63}$$
$$\sin\theta_{13} = 0.0036 \pm 0.0007, \tag{1.64}$$
$$\delta = 1.02 \pm 0.22. \tag{1.65}$$

In the SM, neutrinos are massless and right-handed neutrinos are absent. Thus, the lepton mixing matrix can be chosen trivial: $V^\nu = V^e$, leading to no observable lepton flavor mixing in the charged current sector.

Higgs Interactions in the SM

The kinetic Lagrangian (1.13) of the Higgs field contains the following interaction terms of Higgs field and gauge bosons

$$\mathcal{L}^{\rm K}_{\rm Higgs} \supset W^+_\mu W^{-\mu}(\frac{g^2}{4}h^2 + gM_W h) + Z_\mu Z^\mu(\frac{g^2}{8\cos^2\theta_W}h^2 + \frac{gM_Z}{2\cos\theta_W}h) \tag{1.66}$$

In the unitary gauge, the Yukawa Lagrangian Eq. (1.37) contains the following interaction terms of the Higgs field with fermions

$$-\mathcal{L}_Y \supset \frac{h}{v}\sum_f m_f\bar{f}f = \frac{h}{v}(m_e\bar{e}e + m_u\bar{u}u + m_d\bar{d}d) \tag{1.67}$$

As the neutrino is massless, it doesn't interact with the Higgs field. The primary Higgs interaction channels in the SM are illustrated in Figs. 1.2 and 1.3.

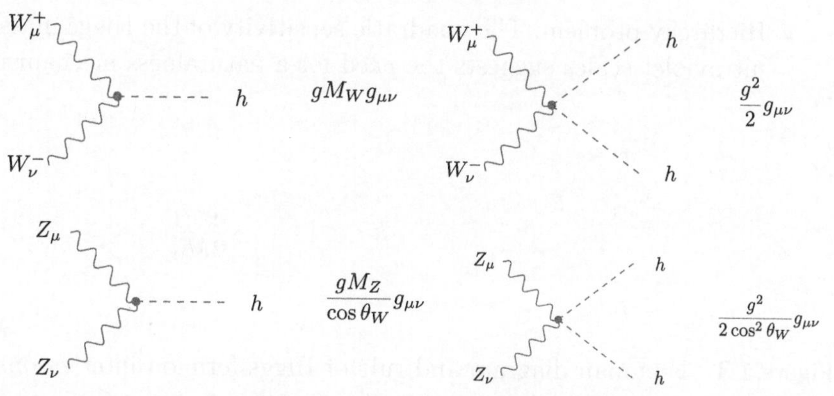

Figure 1.2 Feynman diagrams and rules of Higgs-gauge interactions in the SM.

1.2 EVIDENCE FOR PHYSICS BSM

The SM is a four-dimensional quantum field theory invariant under the Poincaré group and based on the local gauge symmetry

$$SU(3)_C \times SU(2)_L \times U(1)_Y.$$

It contains three generations of point-like fermions (quarks and leptons), gauge bosons mediating the fundamental forces, and a single complex scalar Higgs doublet responsible for electroweak symmetry breaking via the Brout-Englert-Higgs mechanism [7,8].

In its minimal form, the SM does not include right-handed neutrinos, implying that neutrinos are massless. This is in clear contradiction with the experimental observation of neutrino oscillations, which imply nonzero neutrino masses and lepton flavor violation [13,14].

Despite its tremendous success in describing particle physics phenomena up to the electroweak scale [12], there are several theoretical and experimental indications that the SM is incomplete:

- Neutrino masses: The SM cannot accommodate nonzero neutrino masses without extension.
- Dark matter: Astrophysical and cosmological observations require a stable, non-baryonic component of matter not accounted for in the SM.
- Baryon asymmetry: The observed imbalance between matter and antimatter in the universe cannot be explained by SM CP violation alone.
- Hierarchy problem: The quadratic sensitivity of the Higgs mass to ultraviolet scales suggests the need for a naturalness mechanism.

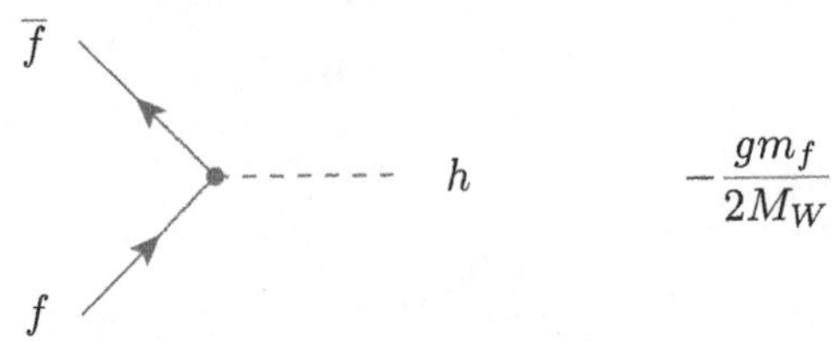

Figure 1.3 Feynman diagram and rule of Higgs-fermion interactions in the SM.

- Gauge coupling unification: The running of gauge couplings in the SM does not result in unification at a high energy scale.

In addition to these conceptual issues, several experimental anomalies provide further hints of new physics:

- Muon anomalous magnetic moment: The long-standing discrepancy between the measured value of the muon $g-2$ and the SM prediction persists at the level of about 4.2σ [15, 16].

- Flavor anomalies in B-meson decays: Deviations from lepton flavor universality in rare B decays, such as R_D and R_{D^*}, have been observed and interpreted as potential signs of new heavy gauge bosons or leptoquarks [17, 18].

In the following chapters, we will explore well-motivated extensions of the SM that aim to resolve these open questions and anomalies. These efforts represent the forefront of modern particle physics and may provide critical insights into a more fundamental underlying theory.

1.2.1 Neutrino Masses

One of the most compelling indications for physics beyond the SM is the observation of nonzero neutrino masses. In the SM, fermion masses arise through Yukawa couplings to the Higgs field, but neutrinos were originally assumed to be massless due to the absence of right-handed (RH) neutrino fields. Consequently, the SM predicts lepton flavor conservation and no neutrino mixing.

However, neutrino oscillation experiments have conclusively demonstrated that neutrinos undergo flavor transitions, which can only occur if neutrinos have non-degenerate masses and nontrivial mixing angles. These results were first established by the Super-Kamiokande experiment [13] and later confirmed by multiple independent experiments.

A minimal extension to the SM that accommodates neutrino masses is to introduce RH neutrinos ν_{R_i}, which are gauge singlets. This permits a renormalizable Yukawa interaction:

$$\mathcal{L}_\nu = y_\nu \, \overline{\ell_L} \, \phi \, \nu_R + \text{h.c.}, \tag{1.68}$$

where ℓ_L is the left-handed lepton doublet, ϕ is the Higgs doublet, and y_ν is the neutrino Yukawa coupling. After electroweak symmetry breaking, this yields a Dirac mass term $m_\nu = y_\nu v$, with $v \approx 246\,\text{GeV}$.

To reproduce the sub-eV neutrino masses inferred from oscillation data, one must assume a tiny Yukawa coupling, $y_\nu \sim 10^{-12}$, which is many orders of magnitude smaller than that of the electron. This severe hierarchy is widely regarded as unnatural and has motivated alternative explanations such as the seesaw mechanism [19,20], which naturally generates small Majorana neutrino masses through the exchange of heavy singlet fermions.

Type I Seesaw Mechanism

The Type I seesaw mechanism provides a compelling explanation for the smallness of neutrino masses within a minimal extension of the SM. It introduces heavy RH neutrinos, denoted by ν_R, which are gauge singlets and therefore do not interact via the SM gauge bosons. However, they do participate in Yukawa interactions with the SM lepton doublets and the Higgs field [19,20].

The extended Lagrangian includes the terms:

$$\mathcal{L} \supset -y_\nu \, \overline{\ell_L} \tilde{\phi} \nu_R - \frac{1}{2} \overline{(\nu_R)^c} M_R \nu_R + \text{h.c.}, \tag{1.69}$$

where y_ν is the neutrino Yukawa coupling matrix, $\tilde{\phi} = i\sigma_2\phi^*$ is the conjugate Higgs doublet, and M_R is a Majorana mass matrix for the RH neutrinos, which can be taken diagonal, real, and positive without loss of generality.

Since M_R is not constrained by SM symmetries, it can naturally lie at a very high scale. Integrating out the heavy RH neutrinos at energies below M_R generates the dimension-five Weinberg operator, yielding an effective Lagrangian:

$$\mathcal{L}^\nu_{\text{eff}} = \frac{y_\nu^2}{2M_R} \, \overline{\ell_L} \, \phi \, \phi^T \, \ell_L^c + \text{h.c.} \tag{1.70}$$

This results in a Majorana mass for the light neutrinos after electroweak symmetry breaking:

$$m_\nu \simeq -\langle \phi^0 \rangle^2 \, y_\nu \, M_R^{-1} \, y_\nu^T, \tag{1.71}$$

naturally explaining their smallness for $M_R \gg \langle \phi^0 \rangle$. This mechanism is illustrated in Fig. 1.4.

After spontaneous symmetry breaking, the neutrino mass terms can be written as:

$$\mathcal{L}^\nu_m = -\overline{\nu_L} m_D \nu_R - \frac{1}{2} \overline{(\nu_R)^c} M_R \nu_R + \text{h.c.}, \tag{1.72}$$

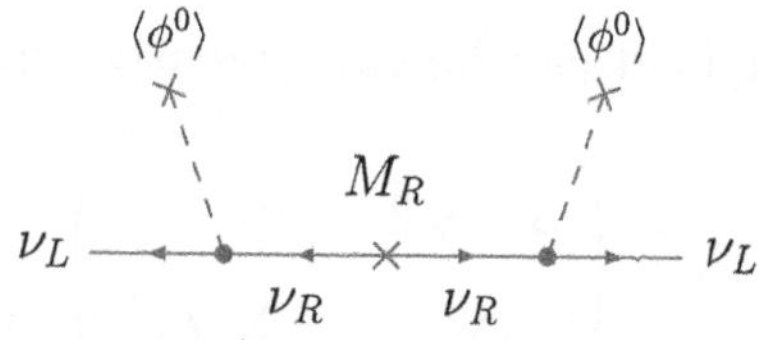

Figure 1.4 Type I seesaw realization of the small Majorana mass for the left-handed neutrino with $m_\nu \simeq -\langle\phi^0\rangle^2\, y_\nu\, M_R^{-1}\, y_\nu^T$.

where $m_D = y_\nu\langle\phi^0\rangle$ is the Dirac mass matrix. Combining terms, the full mass matrix becomes:

$$-\mathcal{L}_m^\nu = \begin{pmatrix} \overline{\nu}_L & \overline{(\nu_R)^c} \end{pmatrix} \underbrace{\begin{pmatrix} 0 & m_D \\ m_D^T & M_R \end{pmatrix}}_{M_\nu} \begin{pmatrix} (\nu_L)^c \\ \nu_R \end{pmatrix} + \text{h.c.} \tag{1.73}$$

Assuming $m_D \ll M_R$, this matrix can be block-diagonalized via a unitary transformation:

$$M_\nu^{\text{diag}} \simeq \begin{pmatrix} -m_D M_R^{-1} m_D^T & 0 \\ 0 & M_R \end{pmatrix}, \tag{1.74}$$

$$V \simeq \begin{pmatrix} -1 & m_D M_R^{-1} \\ m_D^T M_R^{-1} & 1 \end{pmatrix} \begin{pmatrix} i & 0 \\ 0 & -i \end{pmatrix}. \tag{1.75}$$

Thus, the light neutrino masses are suppressed by the large Majorana scale M_R, while the heavy states acquire masses of order M_R. This framework provides a natural mechanism for understanding the observed neutrino mass hierarchy and is a cornerstone of many Grand Unified Theories (GUTs) and leptogenesis scenarios.

Type II Seesaw Mechanism

In addition to the Type I seesaw mechanism, several extensions of the SM have been proposed to generate small Majorana neutrino masses. One such extension is the *Type II seesaw mechanism* [21, 22]. In this scenario, the SM Higgs sector is augmented by a heavy scalar triplet Δ, which carries hypercharge $Y = -2$ and transforms as a $(1, 3, -2)$ under the SM gauge group.

The triplet scalar field can be conveniently represented by a 2×2 matrix:

$$\Delta = \begin{pmatrix} \Delta^-/\sqrt{2} & \Delta^{--} \\ \Delta^0 & -\Delta^-/\sqrt{2} \end{pmatrix}. \tag{1.76}$$

In this case, the relevant terms in the Lagrangian are given by:

$$-\mathcal{L}_{\text{type-II}} = \frac{Y_\Delta}{2}\,\overline{\ell_L} i\tau_2\,\Delta\,\ell_L^c + \mu_\Delta\,\phi^T i\tau_2\Delta\,\phi + M_\Delta^2\,\text{Tr}(\Delta^\dagger\Delta) + \text{h.c.}, \tag{1.77}$$

where Y_Δ is the Yukawa coupling matrix, μ_Δ is a trilinear scalar coupling with mass dimension one, and M_Δ is the mass of the scalar triplet.

After electroweak symmetry breaking, the neutral component of the Higgs doublet acquires a non-vanishing VEV, inducing a small VEV for the neutral component of Δ through the μ_Δ term. This leads to the effective Majorana mass for light neutrinos:

$$m_{\text{eff}}^{II} \simeq \mu_\Delta\, Y_\Delta \frac{\langle\phi^0\rangle^2}{M_\Delta^2} = \lambda_\Delta\, Y_\Delta \frac{\langle\phi^0\rangle^2}{M_\Delta}, \tag{1.78}$$

where $\mu_\Delta \equiv \lambda_\Delta M_\Delta$ defines a dimensionless coupling λ_Δ. This expression mirrors the general seesaw formula, where $M_\Delta \gg \langle\phi^0\rangle$ ensures the smallness of neutrino masses. A diagram representing this process appears in the left panel of Fig. 1.5. This elegant mechanism is referred to as the Type II seesaw.

Type III Seesaw Mechanism

Another compelling approach to generate neutrino masses is the *Type III seesaw mechanism* [23]. Here, the SM is extended by introducing fermionic triplets Σ with zero hypercharge ($Y = 0$), which couple to the SM lepton and Higgs doublets via Yukawa interactions. In contrast to Type II seesaw, this scenario does not alter the scalar sector.

The $SU(2)_L$ triplet fermion field is self-conjugate and can be written in matrix form as:

$$\Sigma = \begin{pmatrix} \Sigma^0/\sqrt{2} & \Sigma^+ \\ \Sigma^- & -\Sigma^0/\sqrt{2} \end{pmatrix}. \tag{1.79}$$

The relevant Lagrangian terms are:

$$-\mathcal{L}_{\text{type-III}} = Y_\Sigma\,\overline{\ell_L}\,\Sigma\,\tilde{\phi} + \frac{1}{2} M_\Sigma\,\text{Tr}\left(\overline{\Sigma^c}\,\Sigma\right) + \text{h.c.}, \tag{1.80}$$

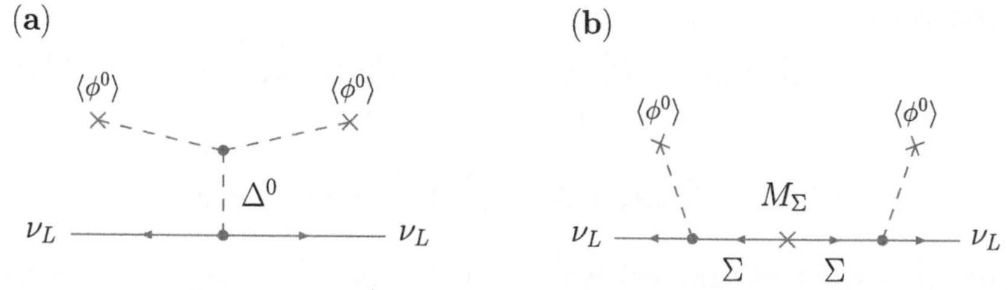

Figure 1.5 (Left) Neutrino mass generation via the Type II seesaw mechanism involving a scalar triplet Δ. (Right) Type III seesaw mechanism mediated by a heavy fermionic triplet Σ.

where Y_Σ is the Yukawa matrix and M_Σ is the mass of the fermion triplet. After electroweak symmetry breaking, integrating out the heavy Σ fields gives an effective neutrino mass matrix:

$$m_{\text{eff}}^{III} \simeq Y_\Sigma \frac{\langle \phi^0 \rangle^2}{M_\Sigma} Y_\Sigma^T. \tag{1.81}$$

This shows that, much like in the other seesaw variants, light Majorana neutrino masses naturally emerge when $M_\Sigma \gg \langle \phi^0 \rangle$. The process is depicted in the right panel of Fig. 1.5. The Type III mechanism is phenomenologically attractive due to the potential collider signatures of the charged fermion components of Σ.

Linear and Inverse Seesaw Mechanisms

In the standard type I seesaw mechanism, the smallness of neutrino masses is attributed to the exchange of heavy right-handed neutrinos with large Majorana masses $M_R \gg v$, where v is the electroweak VEV. However, this high scale is far beyond experimental reach. Alternative low-scale mechanisms such as the inverse seesaw [24, 25] and linear seesaw [26] offer phenomenologically accessible frameworks where lepton number is only softly broken, allowing $M_R \sim \mathcal{O}(\text{TeV})$.

Both mechanisms introduce two Standard Model (SM) singlet fermions: the conventional right-handed neutrino ν_R and an additional singlet S. These frameworks differ in the nature of lepton number violation and in how small neutrino masses are generated.

Inverse Seesaw

The inverse seesaw extends the neutrino sector by adding a small Majorana mass term μ_S for the singlet fermion S, leading to the Lagrangian:

$$\mathcal{L}_{\text{ISS}}^{\nu} = y_\nu \overline{\ell_L}\tilde{\phi}\nu_R + M_R\overline{(\nu_R)^c}S + \mu_S\overline{S^c}S + \text{h.c.}, \tag{1.82}$$

where y_ν is the neutrino Yukawa coupling, and $\mu_S \ll M_R$ softly breaks lepton number. After electroweak symmetry breaking, with $m_D = y_\nu v$, the light neutrino mass is approximately:

$$m_{\nu_\ell} \approx \frac{m_D^2}{M_R^2}\mu_S. \tag{1.83}$$

This allows $y_\nu \sim \mathcal{O}(1)$ and $M_R \sim$ TeV, provided $\mu_S \sim \mathcal{O}(10^{-7})$ GeV.

Linear Seesaw

In the linear seesaw, lepton number is broken via a small Dirac-type Yukawa coupling y_S between the SM lepton doublet and the new singlet S, and the Majorana mass term μ_S is absent:

$$\mathcal{L}_{\text{LSS}}^{\nu} = y_\nu \overline{\ell_L}\tilde{\phi}\nu_R + M_R\overline{(\nu_R)^c}S + y_S\overline{\ell_L}\tilde{\phi}S + \text{h.c.}, \tag{1.84}$$

with $y_S \ll 1$. The resulting light neutrino mass is:

$$m_{\nu_\ell} \approx \frac{y_\nu y_S v^2}{M_R}, \tag{1.85}$$

which is linearly suppressed by the small lepton-number-violating parameter y_S. This again allows for a low seesaw scale, potentially within reach of current collider experiments.

Both mechanisms naturally explain the smallness of neutrino masses without invoking ultra-high energy scales and open the door for rich phenomenology at the TeV scale.

1.2.2 Dark Matter

A wide consensus among astronomers, cosmologists, and particle physicists holds that approximately 95% of the matter in the Universe is composed of an invisible, non-luminous component known as Dark Matter (DM) and Dark Energy (DE). The evidence for DM, in particular, emerges from a range of astrophysical observations. One of the earliest

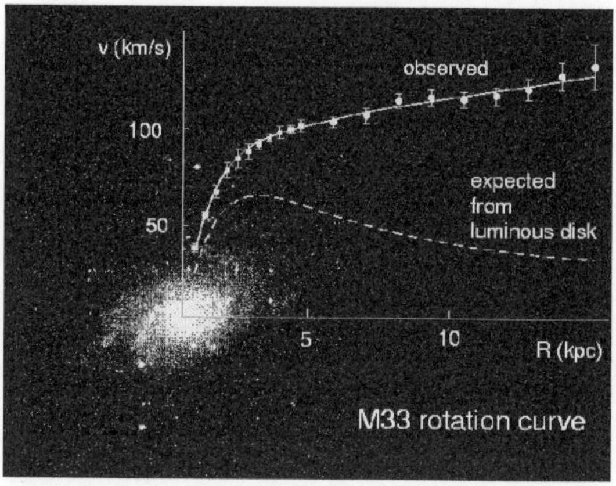

Figure 1.6 The observed rotation curve of the dwarf spiral galaxy M33 overlaid on its optical image [28].

and most compelling lines of evidence comes from the rotational dynamics of galaxies. According to Newtonian mechanics, the rotational velocity of stars in a galaxy should follow:

$$v(r) = \sqrt{\frac{GM(r)}{r}}, \tag{1.86}$$

where $v(r)$ is the velocity at radius r, G is the gravitational constant, and $M(r)$ is the enclosed mass. However, observations of numerous spiral galaxies reveal that their rotation curves remain flat or even rise with increasing distance from the galactic center, as shown in Fig. 1.6, indicating the presence of additional, unseen mass [27, 28].

Another powerful probe of DM is gravitational lensing, where the deflection of light from distant galaxies by massive foreground objects reveals much more mass than is visible. These lensing measurements confirm that the gravitational mass far exceeds the luminous matter content [29, 30].

These findings suggest that galaxies reside within massive DM halos, extending far beyond their visible components. Moreover, precise measurements of the cosmic microwave background (CMB), particularly from the Planck satellite, provide strong evidence for the abundance and non-baryonic nature of DM. The 2018 Planck results indicate [31]:

$$\Omega_{\rm m}h^2 = 0.1430 \pm 0.0011, \quad \Omega_{\rm b}h^2 = 0.02237 \pm 0.00015, \tag{1.87}$$

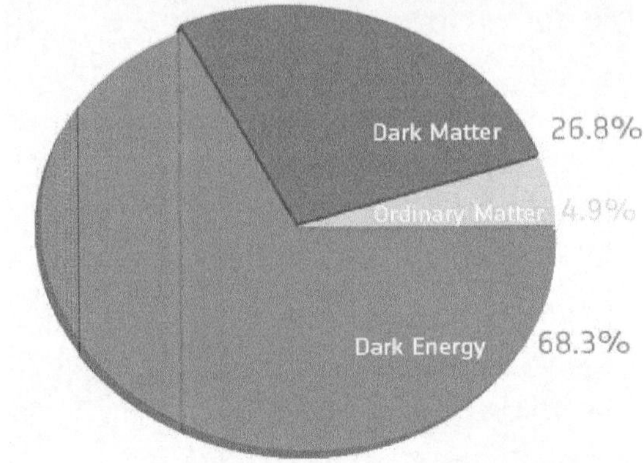

Figure 1.7 Composition of the universe based on Planck 2018 data.

where the significant difference between $\Omega_{\rm m}$ and $\Omega_{\rm b}$ implies that most matter is not composed of baryons. This is consistent with Big Bang nucleosynthesis predictions [32].

As shown in Fig. 1.7, the Universe consists of approximately 4.9% ordinary (baryonic) matter, 26.6% DM, and 68.5% dark energy. Despite the compelling evidence, the particle nature of DM remains unknown.

There are three principal methods to detect DM: direct detection, indirect detection, and collider production. These are diagrammatically represented by a generic Feynman diagram involving DM particles χ and SM particles (Fig. 1.8). Among various DM candidates, Weakly Interacting Massive Particles (WIMPs) are particularly attractive. With masses in the 10 GeV–TeV range and weak-scale interactions, WIMPs

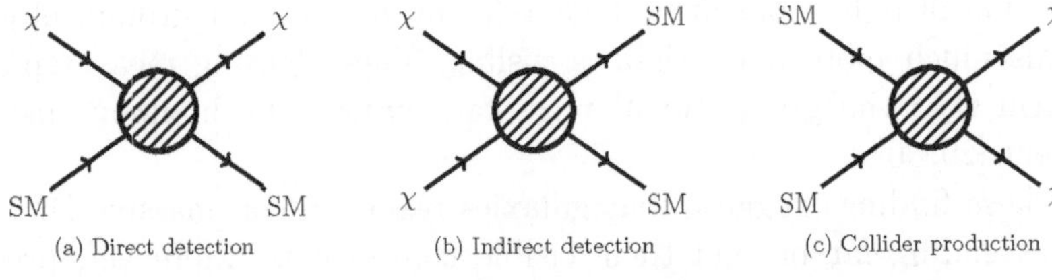

Figure 1.8 Generic Feynman diagrams for direct detection, indirect detection, and collider production of DM.

naturally lead to the correct DM abundance through the *freeze-out* mechanism [33].

In the early Universe, WIMPs were in thermal equilibrium. As the Universe cooled below $T < m_\chi$, WIMP annihilation could no longer keep up with the expansion, leading to a relic abundance. The Boltzmann equation describing the evolution of WIMP number density n_χ is:

$$\frac{dn_\chi}{dt} + 3Hn_\chi = -\langle \sigma v \rangle (n_\chi^2 - (n_\chi^{\text{eq}})^2), \tag{1.88}$$

where $\langle \sigma v \rangle$ is the thermally averaged annihilation cross-section times velocity. For non-relativistic DM, this cross-section can be expanded as [34]:

$$\langle \sigma v \rangle \simeq a + \frac{6b}{x}, \quad x \equiv \frac{m_\chi}{T}. \tag{1.89}$$

The relic density is then given by:

$$\Omega h^2 \simeq \frac{1.07 \times 10^9 \, x_f \, \text{GeV}^{-1}}{\sqrt{g_*} M_{\text{Pl}} (a + \frac{3b}{x_f})}, \tag{1.90}$$

where x_f is the freeze-out parameter, g_* is the effective number of relativistic degrees of freedom, and M_{Pl} is the Planck mass. This expression shows that a weak-scale cross-section yields the observed relic density $\Omega h^2 \approx 0.12$, a phenomenon often referred to as the WIMP miracle.

1.2.3 Higgs Mass Hierarchy

As discussed earlier, electroweak symmetry breaking in the SM is triggered by the Higgs VEV:

$$v^2 = \frac{4M_W^2}{g^2} \simeq 10^4 \text{ GeV}^2 \ll M_{\text{Pl}}^2 \simeq 10^{38} \text{ GeV}^2, \tag{1.91}$$

where M_W is the mass of the W boson, g is the SU(2) gauge coupling, and M_{Pl} is the reduced Planck mass. This enormous disparity between the electroweak scale and the Planck scale, spanning over 16 orders of magnitude, has no natural explanation within the SM and constitutes what is known as the hierarchy problem [35, 36].

The issue becomes more severe at the quantum level, as scalar masses in quantum field theory receive large radiative corrections from higher

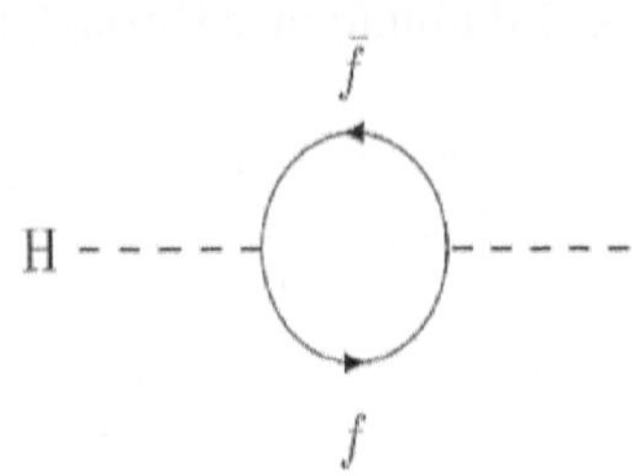

Figure 1.9 One-loop radiative correction to the Higgs boson mass from a fermion loop.

energy scales. For the Higgs boson, one-loop corrections from fermionic loops (see Fig. 1.9) lead to a quadratically divergent mass correction:

$$\delta m_H^2 = -\frac{|\lambda|^2}{8\pi^2}\left[\Lambda^2 - 2m_f^2 \ln\left(\frac{\Lambda}{m_f}\right)\right], \tag{1.92}$$

where λ is the Yukawa coupling between the Higgs field and a fermion f, m_f is the fermion mass, and Λ is the ultraviolet (UV) cutoff scale, which can be as high as the Planck scale. For $\Lambda \sim 10^{19}$ GeV, the correction becomes $\delta m_H \sim 10^{30}\,\text{GeV}^2$, vastly exceeding the physical Higgs mass $m_H \approx 125\,\text{GeV}$. Reconciling such corrections with the observed Higgs mass would require extreme fine-tuning of bare parameters, a situation considered highly unnatural [37, 38].

In contrast, fermion masses such as the electron mass receive only logarithmically divergent corrections (see Fig. 1.10):

$$\delta m_e = \frac{2\alpha_{\text{em}}}{\pi} m_e \ln\left(\frac{\Lambda}{m_e}\right), \tag{1.93}$$

where α_{em} is the electromagnetic fine-structure constant. Even for a cutoff $\Lambda \sim 10^{19}$ GeV, the correction remains modest: $\delta m_e \approx 0.24 m_e$.

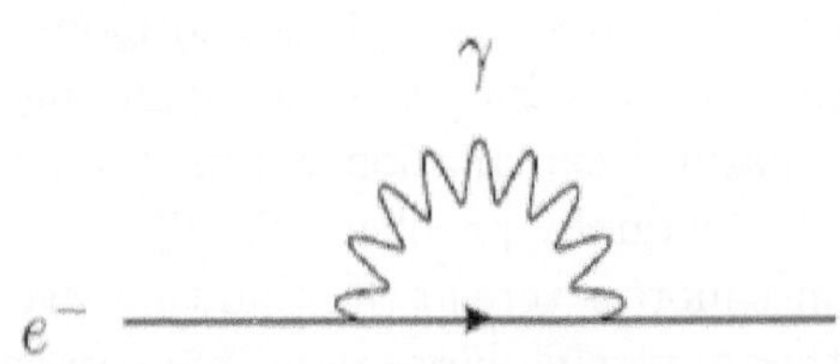

Figure 1.10 One-loop radiative correction to the electron mass.

Moreover, in the limit $m_e \to 0$, the correction vanishes due to the restoration of chiral symmetry, which protects fermion masses from large radiative corrections.

To resolve the hierarchy problem, many theoretical frameworks propose new physics at or near the TeV scale. One prominent example is supersymmetry, which introduces superpartners for all SM particles and naturally cancels the quadratic divergences in Higgs mass corrections. Other approaches include models with composite Higgs bosons, extra spatial dimensions, or invoking the anthropic principle in a multiverse framework [39].

These models aim to stabilize the Higgs mass without unnatural fine-tuning, and many of them predict new particles or interactions accessible to current or future experiments, making the hierarchy problem one of the central driving forces in the search for physics beyond the SM.

1.2.4 Baryon Asymmetry

Baryon asymmetry refers to the observed imbalance between baryons—particles composed of three quarks, such as protons and neutrons—and antibaryons, their corresponding antiparticles, in the universe. This discrepancy is one of the most profound puzzles in physics, requiring an explanation that goes beyond the SM of particle physics. The SM, in its present form, lacks a mechanism to account for this asymmetry, known as baryogenesis. According to the SM, the interactions in the early universe should have resulted in equal amounts of matter and antimatter, which would have led to their mutual annihilation, leaving behind a universe devoid of any baryons.

The enigma of baryon asymmetry, also referred to as the matter-antimatter asymmetry, raises a critical question: why does the universe consist predominantly of matter, with a clear absence of antimatter? Neither the SM of particle physics nor the theory of general relativity can provide a satisfactory explanation for this striking imbalance. In 1967, Andrei Sakharov outlined three necessary conditions for the generation of a net baryon number in the universe [40]:

- *Baryon number violation,*
- *Thermal non-equilibrium,*
- *C (charge conjugation) and CP (charge-parity) violation.*

During the early stages of the universe, all these conditions were likely satisfied. However, our understanding of CP violation in particle interactions and the ability to calculate the baryon asymmetry from first principles remain topics of active research.

The baryon-to-photon ratio, which quantifies the baryon asymmetry of the universe, is given by [41]:

$$\frac{n_B - n_{\bar{B}}}{n_\gamma} = 6.1 \times 10^{-10}, \tag{1.94}$$

indicating a small excess of baryons over antibaryons. Investigating the origin of this asymmetry, particularly the role of CP violation, and deriving it from first principles are central goals in the search for physics beyond the SM.

Several mechanisms have been proposed to explain the baryon asymmetry. One of the most well-known is leptogenesis, which involves the violation of lepton number conservation. In this scenario, the early universe may have seen the decay of heavy particles known as right-handed neutrinos. These decays could have resulted in an excess of quarks over antiquarks, which then generated the observed baryon asymmetry through sphaleron interactions. Other mechanisms beyond the SM have also been suggested, including GUTs, where the three fundamental forces of the SM (strong, weak, and electromagnetic forces) are unified at high energies [42].

1.2.5 Other Unresolved Theoretical Questions

Furthermore, when we look closely at particle physics, we find some really interesting questions that the SM can't fully answer. These questions include:

- Electroweak Symmetry Breaking: We still lack a complete understanding of how particles acquire mass, making this one of the most important questions in modern physics.

- Symmetry Group: The SM is based on a specific gauge symmetry group, $SU(3)_C \times SU(2)_L \times U(1)_Y$, but the reason for this particular choice remains unclear. Understanding its origin could provide deeper insight into the fundamental forces of nature.

- Unification: Can all the known fundamental forces be described within a single, unified framework? Achieving such unification would greatly simplify and deepen our understanding of the universe.

- Three Families of Particles: The Standard Model contains three generations of similar fermions, but the reason for this replication is unknown. Uncovering its origin could offer deeper insight into the structure and behavior of fundamental particles.

- Quantum Gravity: While we have successful quantum theories for three of the four fundamental forces, gravity remains incompatible with quantum mechanics. Developing a consistent theory of quantum gravity is one of the most profound challenges in theoretical physics.

- Cosmological Constant: The cosmological constant in Einstein's equations is vastly smaller than theoretical expectations. Understanding this discrepancy could shed light on the nature of vacuum energy and the universe's accelerated expansion.

These questions highlight the limitations of the SM and underscore the need for new theoretical frameworks to address these unresolved mysteries. The pursuit of physics BSM represents a major endeavor in the scientific community, aimed at achieving a deeper understanding of the fundamental structure and dynamics of the universe.

1.3 POSSIBLE DIRECTIONS FOR BSM

In this section, we will explore a range of avenues and approaches that physicists are actively investigating to expand our comprehension of fundamental particles and forces beyond the SM. These diverse directions encompass various theoretical frameworks and concepts, all geared toward tackling the constraints of the SM and striving for a more encompassing depiction of the universe.

- Extension of Gauge Symmetry: One direction involves extending the gauge symmetry beyond the SM gauge symmetries. This could introduce new particles and interactions that might help explain some of the unanswered questions in particle physics.

- Extension of Higgs Sector: Another avenue is to expand the Higgs sector, which is responsible for giving particles mass. Extending this sector could provide new insights into the origins of mass and the fundamental nature of the Higgs boson.

- Extension of Matter Content: Expanding the matter content involves introducing new types of particles or matter beyond what is currently described in the SM. This can lead to a richer and more diverse particle spectrum.

- Extension with Flavor Symmetry: Flavor symmetry extensions seek to understand the relationships between different types of particles and their properties, such as different flavors of quarks and leptons. This extension may reveal underlying patterns in particle physics.

- Extension of Space-time Dimensions: Some theories propose the existence of extra spatial dimensions beyond the familiar three dimensions of space. These extra dimensions could help address various questions in physics, including the unification of forces.

- Extension of Lorentz Symmetry (Supersymmetry): Supersymmetry extends the concept of spacetime symmetry, known as Lorentz symmetry. It posits a connection between particles with different intrinsic angular momenta (spin), potentially leading to a more unified theory of particles and forces.

- Incorporate Gravity (Supergravity): Incorporating gravity into the framework of particle physics is a significant challenge. Supergravity attempts to merge the principles of general relativity (gravity) with the concepts of supersymmetry from particle physics. It is a step toward a unified theory of all fundamental forces.

- One-dimensional Object (Superstring): Superstring theory proposes that the fundamental building blocks of the universe are not particles but tiny, vibrating strings. This theory aims to reconcile gravity with quantum mechanics and provide a unified framework for all fundamental forces.

These directions represent the ongoing efforts in theoretical physics to push the boundaries of our knowledge and explore new frameworks that can address the unresolved questions posed by the SM. While each direction has its own set of challenges and complexities, they collectively contribute to the pursuit of a more comprehensive and unified theory of the fundamental forces and particles in the universe.

Throughout the chapters of this book, our objective is to delve into as many potential avenues beyond the SM as we can, providing a comprehensive overview of the ongoing pursuit for a deeper comprehension of the universe.

CHAPTER 2

Two Higgs Doublet Models

The Two Higgs Doublet Model (2HDM) is a well-motivated extension of the SM of particle physics, achieved by introducing a second complex scalar doublet in addition to the one present in the SM [43]. This extension offers a promising framework to address phenomena beyond the SM, such as the baryon asymmetry of the universe and the existence of dark matter candidates. In this chapter, we examine the theoretical structure and phenomenological implications of 2HDMs, highlighting how this framework can resolve some of the SM limitations.

The two complex scalar doublets in a 2HDM are typically denoted as Φ_1 and Φ_2, and are expressed as follows:

$$\Phi_1 = \begin{pmatrix} \phi_1^+ \\ \phi_1^0 \end{pmatrix}, \quad \Phi_2 = \begin{pmatrix} \phi_2^+ \\ \phi_2^0 \end{pmatrix}.$$

Here, ϕ_1^+, ϕ_1^0, ϕ_2^+, and ϕ_2^0 represent the scalar fields associated with the two Higgs doublets. These doublets can acquire VEVs, which break the electroweak symmetry, giving mass to the W and Z bosons and to fermions through Yukawa couplings.

The 2HDM can be classified into several types based on the manner in which fermions (quarks and leptons) couple to the Higgs doublets and the specific Yukawa couplings. The most common types are:

1. Type-I 2HDM: In this model, only one Higgs doublet, Φ_2, couples to the charged fermions (both up-type and down-type quarks as

DOI: 10.1201/9781003457701-2

well as charged leptons), while Φ_1 does not couple to any fermions [44]. This is sometimes referred to as a Fermiophobic 2HDM.

2. Type-II 2HDM: In this configuration, one Higgs doublet, Φ_2, couples to the up-type quarks, while the other, Φ_1, couples to the down-type quarks and leptons [44]. This model is the one most commonly used in the Minimal Supersymmetric Standard Model (MSSM).

3. Lepton-Specific 2HDM: Here, one Higgs doublet exclusively interacts with leptons, while the other interacts with quarks. This can lead to interesting phenomena involving changes in lepton flavors. Occasionally, this is referred to as Type-X 2HDM [45].

4. Flipped 2HDM: In this variation, the roles of the two Higgs doublets are reversed compared to Type-II. The doublet that couples to up-type quarks in Type-II now couples to down-type quarks and leptons, and vice versa. This variation is sometimes referred to as Type-Y 2HDM [46].

5. Type-III 2HDM: In this model, both Φ_1 and Φ_2 couple to both up-type and down-type quarks as well as charged leptons. This type is known for its tendency to generate Flavor Changing Neutral Current (FCNC) interactions at the tree level [47]. To comply with strict experimental limits, specific assumptions on the Yukawa couplings must be made.

The choice of the 2HDM type depends on the specific physics phenomena under consideration and the experimental constraints. 2HDMs can have a much richer Higgs boson phenomenology compared to the SM, including the presence of additional Higgs bosons: a heavy neutral CP-even Higgs boson (H), a pseudo-scalar (A), and charged Higgs bosons ($H^{\pm}$). Moreover, these models allow for the potential occurrence of FCNCs and CP-violating effects.

It is important to emphasize that the 2HDM does not alter the successful predictions of the SM but rather enriches the scalar particle spectrum. This extension opens the door for the existence of new particles beyond those predicted by the SM. Additionally, 2HDMs are not limited to SM but can also be incorporated into larger frameworks such as Supersymmetry.

These models are actively studied both theoretically and experimentally to explore their predictions and assess whether they can explain observed phenomena that the SM cannot. On the experimental front, this includes searching for additional Higgs bosons and measuring their properties in particle collider experiments such as those conducted at the Large Hadron Collider (LHC).

2.1 2HDM SCALAR POTENTIAL

The most general $SU(2)_L \times U(1)_Y$-invariant scalar potential of the 2HDM involving two complex scalar doublets Φ_1 and Φ_2 is given by [48, 49]:

$$\begin{aligned} V(\Phi_1,\Phi_2) = {} & m_{11}^2\Phi_1^\dagger\Phi_1 + m_{22}^2\Phi_2^\dagger\Phi_2 - \left[m_{12}^2\Phi_1^\dagger\Phi_2 + \text{h.c.}\right] + \frac{1}{2}\lambda_1(\Phi_1^\dagger\Phi_1)^2 \\ & + \frac{1}{2}\lambda_2(\Phi_2^\dagger\Phi_2)^2 + \lambda_3(\Phi_1^\dagger\Phi_1)(\Phi_2^\dagger\Phi_2) + \lambda_4(\Phi_1^\dagger\Phi_2)(\Phi_2^\dagger\Phi_1) \\ & + \left[\frac{1}{2}\lambda_5(\Phi_1^\dagger\Phi_2)^2 + \lambda_6(\Phi_1^\dagger\Phi_1)(\Phi_1^\dagger\Phi_2) + \lambda_7(\Phi_2^\dagger\Phi_2)(\Phi_1^\dagger\Phi_2) + \text{h.c.}\right]. \end{aligned} \tag{2.1}$$

Here, $m_{11}^2, m_{22}^2, \lambda_{1,2,3,4}$ are real, while $m_{12}^2, \lambda_{5,6,7}$ are generally complex, leading to 14 real independent parameters.

Discrete Symmetry and FCNC Suppression

To suppress FCNCs at tree level, a Z_2 symmetry is imposed:

$$\Phi_1 \to -\Phi_1, \quad \Phi_2 \to \Phi_2, \tag{2.2}$$

which eliminates terms with $m_{12}^2 = \lambda_6 = \lambda_7 = 0$. Soft breaking of Z_2 is allowed via a non-zero but real m_{12}^2 to retain CP conservation.

Theoretical Constraints

To ensure physical viability, the 2HDM parameters must satisfy several theoretical constraints [48, 49]:

1. Perturbativity: Couplings must remain perturbative: $|\lambda_i| \lesssim 4\pi$.
2. Tree-Level Unitarity: Scattering amplitudes must satisfy partial wave unitarity bounds, restricting combinations of λ_i.
3. Vacuum Stability: The potential must be bounded from below. This implies:

$$\lambda_1 > 0, \quad \lambda_2 > 0, \quad \lambda_3 > -\sqrt{\lambda_1\lambda_2}, \quad \lambda_3 + \lambda_4 - |\lambda_5| > -\sqrt{\lambda_1\lambda_2}. \tag{2.3}$$

4. Electroweak Precision Constraints: The model must comply with S, T, and U parameter bounds.

5. FCNC Constraints: FCNCs are avoided by the Z_2 symmetry; further suppression can be achieved by proper Yukawa alignment.

6. Naturalness: Large hierarchies between m_H and M_Z introduce fine-tuning. Define:

$$\Delta = \max_i \left| \frac{\partial \ln M_Z^2}{\partial \ln p_i} \right| \quad \text{where } p_i \in \{m_{ij}^2, \lambda_k\}. \tag{2.4}$$

Vacuum Expectation Values and CP Conservation

To preserve electric charge and CP symmetry, only the neutral CP-even components develop VEVs:

$$\langle \Phi_1 \rangle = \frac{1}{\sqrt{2}} \begin{pmatrix} 0 \\ v_1 \end{pmatrix}, \qquad \langle \Phi_2 \rangle = \frac{1}{\sqrt{2}} \begin{pmatrix} 0 \\ v_2 \end{pmatrix}, \tag{2.5}$$

with $v^2 = v_1^2 + v_2^2 = (246 \text{ GeV})^2$. The minimization conditions yield:

$$m_{11}^2 = m_{12}^2 \frac{v_2}{v_1} - \frac{1}{2}\lambda_1 v_1^2 - \lambda_{345} v_2^2, \tag{2.6}$$

$$m_{22}^2 = m_{12}^2 \frac{v_1}{v_2} - \frac{1}{2}\lambda_2 v_2^2 - \lambda_{345} v_1^2, \tag{2.7}$$

where $\lambda_{345} = \lambda_3 + \lambda_4 + \lambda_5$. In CP-conserving 2HDMs with softly broken Z_2, the model is characterized by eight parameters:

$$\lambda_1, \lambda_2, \lambda_3, \lambda_4, \lambda_5, m_{12}^2, \tan\beta = \frac{v_2}{v_1}, \text{ and } v. \tag{2.8}$$

2.2 THE HIGGS SPECTRUM OF THE 2HDM

The scalar sector of the 2HDM consists of two $SU(2)_L$ scalar doublets, Φ_1 and Φ_2, each with hypercharge $+1$. They can be parameterized in the unitary gauge as:

$$\Phi_1 = \begin{pmatrix} \phi_1^+ \\ \frac{1}{\sqrt{2}}(v_1 + \phi_1 + i\eta_1) \end{pmatrix}, \quad \Phi_2 = \begin{pmatrix} \phi_2^+ \\ \frac{1}{\sqrt{2}}(v_2 + \phi_2 + i\eta_2) \end{pmatrix}, \tag{2.9}$$

where v_1 and v_2 are the VEVs of the doublets with $v^2 = v_1^2 + v_2^2 \simeq (246\ \text{GeV})^2$. The 8 real degrees of freedom decompose into three Goldstone bosons, eaten by the $W^\pm$ and Z bosons, and five physical Higgs states: two CP-even scalars (h, H), one CP-odd pseudoscalar (A), and two charged scalars ($H^\pm$) [48, 49].

Charged Higgs Sector

The mass matrix for the charged scalars $\phi_1^\pm, \phi_2^\pm$ is given by:

$$M^2_{H^\pm} = \left(m_{12}^2 - \frac{1}{2}(\lambda_4 + \lambda_5)v_1v_2\right)\begin{pmatrix} \frac{v_2}{v_1} & -1 \\ -1 & \frac{v_1}{v_2} \end{pmatrix}, \tag{2.10}$$

which yields one zero eigenvalue corresponding to the charged Goldstone boson $G^\pm$, and one massive eigenstate:

$$m^2_{H^\pm} = \frac{m_{12}^2}{v_1v_2}v^2 - \frac{1}{2}(\lambda_4 + \lambda_5)v^2. \tag{2.11}$$

The physical charged Higgs boson is:

$$H^\pm = -\sin\beta\ \phi_1^\pm + \cos\beta\ \phi_2^\pm, \quad \tan\beta = \frac{v_2}{v_1}. \tag{2.12}$$

CP-Odd Scalar Sector

The pseudoscalar mass matrix in the (η_1, η_2) basis reads:

$$M_A^2 = \left(\frac{m_{12}^2}{v_1v_2} - \lambda_5\right)\begin{pmatrix} v_2^2 & -v_1v_2 \\ -v_1v_2 & v_1^2 \end{pmatrix}, \tag{2.13}$$

yielding one massless Goldstone boson G^0 and one physical CP-odd scalar with mass:

$$m_A^2 = \frac{m_{12}^2}{v_1v_2}v^2 - \lambda_5 v^2, \tag{2.14}$$

and eigenstate:

$$A = -\sin\beta\ \eta_1 + \cos\beta\ \eta_2. \tag{2.15}$$

CP-Even Scalar Sector

The CP-even scalar fields ϕ_1 and ϕ_2 mix to form the mass eigenstates h and H. Their mass matrix is:

$$M_H^2 = \begin{pmatrix} m_{12}^2 \frac{v_2}{v_1} + \lambda_1 v_1^2 & -m_{12}^2 + \lambda_{345} v_1 v_2 \\ -m_{12}^2 + \lambda_{345} v_1 v_2 & m_{12}^2 \frac{v_1}{v_2} + \lambda_2 v_2^2 \end{pmatrix}, \tag{2.16}$$

where $\lambda_{345} = \lambda_3 + \lambda_4 + \lambda_5$. Diagonalizing this matrix via a mixing angle α yields the mass eigenstates:

$$h = -\sin\alpha\, \phi_1 + \cos\alpha\, \phi_2, \tag{2.17}$$

$$H = \cos\alpha\, \phi_1 + \sin\alpha\, \phi_2, \tag{2.18}$$

with $m_h^2 \leq m_H^2$ by convention. The lighter scalar h is typically identified with the observed Higgs boson at 125 GeV.

2.3 2HDM INTERACTIONS

2HDM leads to richer phenomenology, including the possibility of FCNCs and new interactions involving the Higgs and gauge bosons, as well as fermions. In this section, we explore the key interactions in the 2HDM, focusing on the gauge boson and Yukawa interactions.

Gauge Bosons in 2HDM

In the 2HDM, the gauge bosons interact with the Higgs bosons in a manner similar to the SM. The kinetic terms for the Higgs doublets are given by:

$$\mathcal{L}_\Phi = (D_\mu \Phi_1)^\dagger (D^\mu \Phi_1) + (D_\mu \Phi_2)^\dagger (D^\mu \Phi_2),$$

where D_μ represents the covariant derivative acting on the doublet fields. It accounts for the interaction of the Higgs fields with the gauge fields while maintaining gauge invariance. For a doublet field Φ_i, the covariant derivative is expressed as:

$$D_\mu \Phi_i = (\partial_\mu - ig\frac{\sigma^a}{2} W_\mu^a - ig'\frac{Y}{2} B_\mu)\Phi_i,$$

where g and g' are the gauge coupling constants for the $SU(2)_L$ and $U(1)_Y$ gauge groups, respectively, W_μ^a and B_μ are the gauge fields associated with the $SU(2)_L$ and $U(1)_Y$ groups, σ^a are the Pauli matrices, and Y represents the weak hypercharge.

After spontaneous symmetry breaking, the Higgs doublets acquire VEVs, denoted as v_1 and v_2. From the kinetic terms $\mathcal{L}_\Phi$, one derives mass terms for the gauge bosons, resulting in the following expressions for the gauge boson masses:

$$M_W = \frac{1}{4}g^2v^2, \quad M_Z = \frac{1}{4}(g^2 + g'^2)v^2, \quad M_A^2 = 0,$$

where $v = \sqrt{v_1^2 + v_2^2}$ is the total vacuum expectation value. Additionally, the neutral Higgs bosons couple to the W and Z gauge bosons. These couplings are given by:

$$\begin{aligned}\mathcal{L}_\Phi \supset \frac{g^2 + g'^2}{4} v \left[\sin(\beta - \alpha)ZZh + \cos(\beta - \alpha)ZZH\right] \\ + \frac{g^2}{2} v \left[\sin(\beta - \alpha)WWh + \cos(\beta - \alpha)WWH\right],\end{aligned} \tag{2.19}$$

where h and H denote the light and heavy CP-even Higgs mass eigenstates, respectively, and α and β are the mixing angles in the Higgs sector.

Yukawa Interactions in 2HDM

In the 2HDM, the most general gauge-invariant Lagrangian for the Yukawa interactions between the Higgs doublets and SM fermions is given by:

$$\mathcal{L}_{\text{Yukawa}} = -\sum_{i=1,2} Y_i^d \overline{Q}_L \Phi_i d_R + Y_i^u \overline{Q}_L \tilde{\Phi}_i u_R + Y_i^\ell \overline{L}_L \Phi_i \ell_R + \text{h.c.},$$

where i indexes the two Higgs doublets (Φ_1 and Φ_2), and $\tilde{\Phi}_i = i\sigma_2 \Phi_i^*$ denotes the conjugate of the Higgs doublet Φ_i. The Yukawa coupling matrices for the up-type quarks, down-type quarks, and leptons are represented by Y_i^u, Y_i^d, and Y_i^ℓ, respectively.

After electroweak symmetry breaking, the Higgs fields acquire VEVs, resulting in fermion masses. The Yukawa couplings can be related to the mass matrices. For example, the up-type quark mass matrix M_u is given by:

$$M_u = \frac{v}{\sqrt{2}} \left(Y_1^u + \tan\beta Y_2^u\right),$$

where $\tan\beta = v_2/v_1$ is the ratio of the Higgs VEVs. This relation shows that diagonalizing the fermion mass matrix does not necessarily diagonalize the Yukawa coupling matrices, thus allowing for FCNCs at tree level. These FCNCs can have observable consequences, such as in $H \to ds$ decays, which contribute to $K-\bar{K}$ oscillations [48]. To suppress these effects, it is often required that the neutral flavor-changing mediator be heavy, with a mass scale around 10 TeV. One common method to reduce FCNCs is to impose a Z_2 symmetry, which can lead to Type I or Type II 2HDMs.

As previously noted, in the Type I 2HDM, all quarks couple to a single Higgs doublet. In contrast, the Type II 2HDM assigns fermions to different doublets based on their electric charge: up-type quarks ($Q = +2/3$) couple to Φ_1, while down-type quarks ($Q = -1/3$) couple to Φ_2. This structure plays a crucial role in suppressing FCNCs. The corresponding Z_2 symmetry charge assignments for Type I and Type II models are summarized in Table 2.1.

In Type I and Type II 2HDMs, the Yukawa couplings to the neutral Higgs bosons h, H, and A Higgs bosons are given by:

$$\begin{aligned}\mathcal{L}_{\text{Yukawa}} = &- \sum_{f=u,d,\ell} \frac{m_f}{v} \left(\xi_h^f \overline{f} f h + \xi_H^f \overline{f} f H - i \xi_A^f \overline{f} \gamma_5 f A \right) \\ &- \Big\{ \frac{\sqrt{2} V_{ud}}{v} \, \overline{u} \left(m_u \xi_A^u P_L + m_d \xi_A^d P_R \right) d H^+ \\ &+ \frac{\sqrt{2} m_\ell \xi_A^\ell}{v} \overline{\nu_L} \ell_R H^+ + h.c. \Big\}, \end{aligned} \tag{2.20}$$

where $P_{L/R}$ denotes the projection operators for left-/right-handed fermions, and the corresponding factors ξ are outlined in Table 2.2.

TABLE 2.1 Z_2 charge assignments for the Higgs doublets, up-type quarks, down-type quarks, and leptons in both Type I and Type II 2HDMs.

Field	Type I Z_2 Charge	Type II Z_2 Charge
Φ_1	+1	+1
Φ_2	-1	-1
Up-type quarks	+1	+1
Down-type quarks	+1	-1
Leptons	+1	-1

TABLE 2.2 Yukawa couplings of u, d, ℓ to the neutral Higgs bosons h, H, A in the four different models. The couplings to the charged Higgs bosons follow Eq. 2.20.

	Type I	Type II
ξ_h^u	$\cos\alpha/\sin\beta$	$\cos\alpha/\sin\beta$
ξ_h^d	$\cos\alpha/\sin\beta$	$-\sin\alpha/\cos\beta$
ξ_h^ℓ	$\cos\alpha/\sin\beta$	$-\sin\alpha/\cos\beta$
ξ_H^u	$\sin\alpha/\sin\beta$	$\sin\alpha/\sin\beta$
ξ_H^d	$\sin\alpha/\sin\beta$	$\cos\alpha/\cos\beta$
ξ_H^ℓ	$\sin\alpha/\sin\beta$	$\cos\alpha/\cos\beta$
ξ_A^u	$\cot\beta$	$\cot\beta$
ξ_A^d	$-\cot\beta$	$\tan\beta$
ξ_A^ℓ	$-\cot\beta$	$\tan\beta$

2.4 PHENOMENOLOGICAL CONSEQUENCES

The 2HDM gives rise to distinctive signatures that can be tested through collider experiments and precision measurements. This section highlights key phenomenological implications, focusing on the decay modes of the Higgs bosons, their production channels at the LHC, and potential contributions to the anomalous magnetic moment of the muon.

2.4.1 Higgs Decays

We examine the dominant decay channels of the heavy neutral scalars H and A, as well as the charged Higgs boson $H^\pm$, in Type I and Type II 2HDMs. Our analysis is performed in the alignment limit, $\sin(\beta - \alpha) \simeq 1$, in which the light CP-even scalar h exhibits SM-like couplings and reproduces the observed Higgs boson properties [48]. In this limit, the tree-level couplings of H to WW and ZZ vanish, while A and $H^\pm$ do not couple to gauge boson pairs at tree level for any value of $\sin(\beta - \alpha)$.

The partial decay widths of the heavy scalars into fermion pairs are given by [50]:

$$\Gamma(\varphi \to f\bar{f}) = N_C \frac{G_F m_\varphi m_f^2}{4\sqrt{2}\pi} (\xi_\varphi^f)^2 \times \begin{cases} \beta_f^3, & \varphi = h, H, \\ \beta_f, & \varphi = A, \end{cases} \tag{2.21}$$

$$\Gamma(H^+ \to f\bar{f}') = N_C \frac{G_F m_{H^\pm}}{4\sqrt{2}\pi} |V_{ff'}|^2 \beta_{ff'} \Big[\Big(m_f^2 (\xi_A^f)^2 + m_{f'}^2 (\xi_A^{f'})^2 \Big) \left(1 - \frac{m_f^2 + m_{f'}^2}{m_{H^\pm}^2}\right) - \frac{4 m_f^2 m_{f'}^2 \xi_A^f \xi_A^{f'}}{m_{H^\pm}^2} \Big], \tag{2.22}$$

where:

- $N_C = 3$ for quarks and $N_C = 1$ for leptons,
- G_F is the Fermi constant,
- m_φ denotes the mass of the decaying scalar $\varphi = h, H, A$,
- $m_{H^\pm}$ is the mass of the charged Higgs,
- $m_f, m_{f'}$ are the masses of the final-state fermions,
- $V_{ff'}$ is the CKM matrix element (or 1 for leptonic decays),
- $\beta_f = \sqrt{1 - 4m_f^2/m_\varphi^2}$, and
- $\beta_{ff'} = \lambda^{1/2}\left(\frac{m_f^2}{m_{H^\pm}^2}, \frac{m_{f'}^2}{m_{H^\pm}^2}\right)$,
- ξ_φ^f are the model-dependent Yukawa coupling modifiers [48].

We also consider the loop-induced decays of the scalar particles, including the decay modes $\varphi \to \gamma\gamma$, $\varphi \to Z\gamma$, and $\varphi \to gg$, where φ represents any of the heavy scalars H, A, or $H^\pm$. The corresponding decay widths are given by [51]:

$$\Gamma(\varphi \to \gamma\gamma) = \frac{G_F \alpha_{\mathrm{EM}}^2 m_\varphi^3}{128\sqrt{2}\pi^3} \left| \sum_f Q_f^2 I_f^\varphi(m_f, N_C) + I_W^\varphi + I_{H^\pm}^\varphi \right|^2, \tag{2.23}$$

$$\Gamma(\varphi \to Z\gamma) = \frac{G_F \alpha_{\mathrm{EM}}^2 m_\varphi^3}{64\sqrt{2}\pi^3} \left(1 - \frac{m_Z^2}{m_\varphi^2}\right)^3 \left| \sum_f Q_f J_f^\varphi(m_f, N_C) + J_W^\varphi + J_{H^\pm}^\varphi \right|^2, \tag{2.24}$$

$$\Gamma(\varphi \to gg) = \frac{G_F \alpha_S^2 m_\varphi^3}{64\sqrt{2}\pi^3} \left| \sum_{f=q} I_f^\varphi(m_f, 1) \right|^2, \tag{2.25}$$

where the fermionic loop functions I_f^{φ} and J_f^{φ} depend on the particle masses and Yukawa couplings. The fermionic loop functions are as follows [51]:

$$I_f^{\varphi}(m_f, N_C) = \xi_{\varphi}^f \times \begin{cases} -N_C \frac{4m_f^2}{m_\varphi^2}\left[2-\beta_f^2 m_\varphi^2 C_0(0,0,m_\varphi^2,m_f^2,m_f^2,m_f^2)\right] \\ \text{for } \varphi = h, H, \\ +4N_C m_f^2 C_0(0,0,m_\varphi^2,m_f^2,m_f^2,m_f^2) \\ \text{for } \varphi = A, \end{cases} \tag{2.26}$$

$$J_f^{\varphi}(m_f, N_C) = \xi_{\varphi}^f \times \begin{cases} -4N_C c_V^f \left[J_1(m_f) - J_2(m_f)\right] & \text{for } \varphi = h, H, \\ -4N_C c_V^f \left[-J_2(m_f)\right] & \text{for } \varphi = A, \end{cases} \tag{2.27}$$

where c_V^f is the vector coupling of the fermion f to the Z-boson, given by $c_V^f = \frac{1}{2s_W c_W}\left(T_{3L}^f - 2Q_f s_W^2\right)$, with T_{3L}^f denoting the third component of the weak isospin for fermion f. The functions $J_1(m_f)$ and $J_2(m_f)$ are given by:

$$J_1(m_f) = \frac{2m_f^2}{m_\varphi^2 - m_Z^2}\left\{1 + 2m_f^2 C_0(0,m_Z^2,m_\varphi^2,m_f^2,m_f^2,m_f^2)\right.$$
$$\left. + \frac{m_Z^2}{m_\varphi^2 - m_Z^2}\left[B_0(m_\varphi^2,m_f^2,m_f^2) - B_0(m_Z^2,m_f^2,m_f^2)\right]\right\}, \tag{2.28}$$

$$J_2(m_f) = m_f^2 C_0(0,m_Z^2,m_\varphi^2,m_f^2,m_f^2,m_f^2), \tag{2.29}$$

where B_0 and C_0 are the Passarino-Veltman functions [52]. These functions are defined as:

$$B_0(m_\varphi^2, m^2, m^2) = \Delta - 2g\left(\frac{4m^2}{m_\varphi^2}\right), \tag{2.30}$$

$$C_0(0,0,m_\varphi^2,m^2,m^2,m^2) = \frac{-2}{m_\varphi^2} f\left(\frac{4m^2}{m_\varphi^2}\right), \tag{2.31}$$

with Δ denoting the regularization constant and $g(x)$ and $f(x)$ defined as:

$$g(x) = \begin{cases} (x-1)\arcsin\left(\sqrt{\frac{1}{x}}\right) & \text{for } x \geq 1, \\ \frac{1}{2}(1-x)\left[\ln\left(\frac{1+\sqrt{1-x}}{1-\sqrt{1-x}}\right) - i\pi\right] & \text{for } x < 1, \end{cases} \tag{2.32}$$

$$f(x) = \begin{cases} \left[\arcsin\left(\sqrt{\frac{1}{x}}\right)\right]^2 & \text{for } x \geq 1, \\ -\frac{1}{4}\left[\ln\left(\frac{1+\sqrt{1-x}}{1-\sqrt{1-x}}\right) - i\pi\right]^2 & \text{for } x < 1. \end{cases} \tag{2.33}$$

Finally, the loop functions I_W^φ, $I_{H^\pm}^\varphi$, J_W^φ, and $J_{H^\pm}^\varphi$ are common to all types of Yukawa interactions and can be found in Ref. [51].

In the type-I 2HDM, the decay of the neutral Higgs boson H into a gauge boson pair, such as $\gamma\gamma$ or $Z\gamma$, can be enhanced for large values of $\tan\beta$. This enhancement occurs because, while the fermionic decays (including the gg mode) are suppressed at large $\tan\beta$, the charged scalar loop contribution to the $\gamma\gamma$ and $Z\gamma$ decay modes is not always suppressed. On the other hand, such an enhancement does not occur in the decay of A, as there is no AH^+H^- coupling.

For charged Higgs boson decays with $m_{H^\pm} = 150\,\text{GeV}$, the dominant decay mode in both the type-I and type-II 2HDM is into $\tau\nu$ for $\tan\beta \gtrsim 1$, while hadronic decay modes can also dominate in the type-Y 2HDM [53].

2.4.2 Higgs Production at the LHC

At the LHC, neutral and charged Higgs bosons can be produced through several key mechanisms, depending on the Higgs sector structure. In the type-II 2HDM, the production of the additional neutral Higgs bosons H and A occurs predominantly via:

- Gluon fusion: $gg \to H, A$, a loop-induced process mediated primarily by top and bottom quarks. The effective coupling is enhanced at low $\tan\beta$ by top loops and at high $\tan\beta$ by bottom loops.
- *Associated production with bottom quarks:* $pp \to b\bar{b}H, b\bar{b}A$, relevant at large $\tan\beta$ where the Yukawa couplings to down-type quarks are enhanced.
- Associated production with top quarks: $pp \to t\bar{t}H, t\bar{t}A$, most sensitive at $\tan\beta \sim 1$.

On the other hand, vector boson fusion $VV \to H$ and associated production with gauge bosons are suppressed in the alignment limit ($\sin(\beta - \alpha) \simeq 1$), and forbidden for the CP-odd Higgs A due to the absence of VVA couplings.

The dominant decay channels of H and A depend on mass and $\tan\beta$. At high $\tan\beta$, decays to down-type fermions dominate: $H/A \to b\bar{b}, \tau^+\tau^-$, though these suffer from significant QCD and Drell-Yan backgrounds. Decays to $\mu^+\mu^-$ offer better mass resolution despite lower branching ratios, and can be competitive at low Higgs masses [54]. Simulations show that the $b\bar{b}H/A \to b\bar{b}\tau^+\tau^-$ channel provides better sensitivity than gluon fusion in the high $\tan\beta$ regime [55, 56].

The charged Higgs bosons $H^{\pm}$ provide distinctive signatures that can be probed at colliders:

- For $m_{H^{\pm}} < m_t$, they can be produced through top quark decays, $t \to H^+ b$. In the type-II Two-Higgs-Doublet Model (2HDM), such decays offer discovery potential for $\tan\beta \lesssim 2$ and $\tan\beta \gtrsim 20$, given sufficient integrated luminosity [57].

- For $m_{H^{\pm}} > m_t$, direct production channels such as $gb \to tH^-$ and pair production via $q\bar{q}$, $gg \to H^+H^-$ become significant. These processes are governed by the Yukawa interaction:

$$\mathcal{L}_{H^{\pm}tb} = \frac{g}{\sqrt{2}M_W} H^+ \bar{t}\,(m_b \tan\beta\, P_R + m_t \cot\beta\, P_L)\, b + \text{h.c.} \tag{2.34}$$

The cross-sections are enhanced for both low and high values of $\tan\beta$, making these channels complementary in probing the parameter space of the model.

2.4.3 Muon Anomalous Magnetic Moment

The anomalous magnetic moment of the muon, defined as $a_\mu \equiv \frac{1}{2}(g_\mu - 2)$, is a precisely measured observable that provides a sensitive probe of physics beyond the SM. The latest combined measurement by the Fermilab Muon $g-2$ experiment and the earlier E821 experiment at Brookhaven reports a discrepancy with the SM prediction at the level of $4.2\,\sigma$ [15, 16]:

$$\Delta a_\mu \equiv a_\mu^{\text{exp}} - a_\mu^{\text{SM}} = (251 \pm 59) \times 10^{-11}. \tag{2.35}$$

This suggests that potential new physics contributions should lie in the range:

$$215 \times 10^{-11} \leq \Delta a_\mu^{\text{NP}} \leq 637 \times 10^{-11}. \tag{2.36}$$

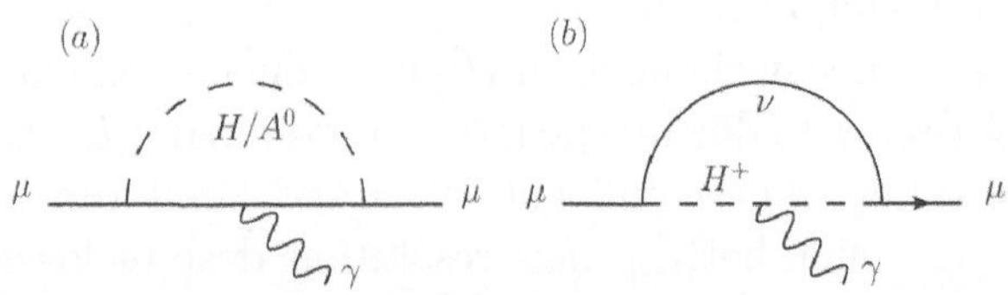

Figure 2.1 One-loop Feynman diagrams for the muon $g-2$.

In the 2HDM, additional contributions to a_μ arise from both one-loop and two-loop diagrams involving the extra scalar states $H, A, H^\pm$. At one loop, the electromagnetic vertex correction for a lepton ℓ has the general form:

$$\bar{\ell}(p')\Gamma^\mu \ell(p) = \bar{\ell}(p')\left[\gamma^\mu F_1(k^2) + \frac{i\sigma^{\mu\nu}k_\nu}{2m_\ell}F_2(k^2)\right]\ell(p), \tag{2.37}$$

where the anomalous magnetic moment is given by $a_\ell = F_2(0)$.

The 2HDM one-loop contribution is [58]:

$$\Delta a_\mu^{\text{1-loop}} = \frac{G_F}{4\pi^2}\frac{m_\mu^2}{\sqrt{2}}\sum_j (\xi_\mu^j)^2\, r_\mu^j\, f_j(r_\mu^j), \tag{2.38}$$

where $j = h, H, A, H^\pm$, $r_\mu^j = m_\mu^2/M_j^2$, and the loop functions $f_j(r)$ are:

$$f_{h,H}(r) = \int_0^1 dx \frac{x^2(2-x)}{1-x+rx^2}, \tag{2.39}$$

$$f_A(r) = \int_0^1 dx \frac{-x^3}{1-x+rx^2}, \tag{2.40}$$

$$f_{H^\pm}(r) = \int_0^1 dx \frac{-x(1-x)}{1-(1-x)r}. \tag{2.41}$$

However, one-loop effects are typically too small to explain the observed anomaly unless the scalar masses are of order a few tens of GeV. A more significant contribution arises from two-loop Barr-Zee type diagrams [59, 60], especially when involving heavy fermion loops.

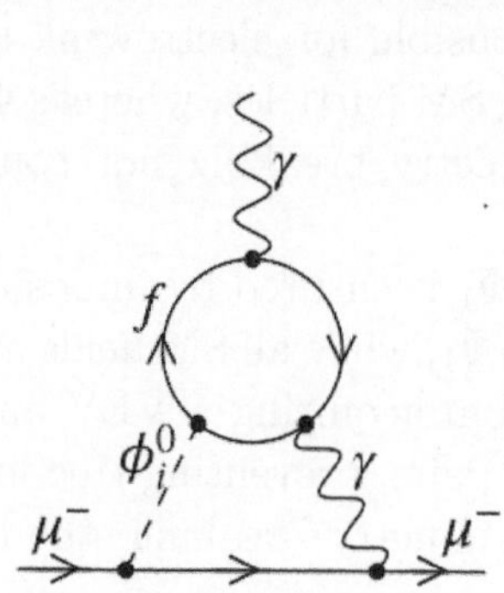

Figure 2.2 Barr-Zee two-loop Feynman diagrams for the muon $g-2$.

The dominant two-loop Barr-Zee contribution is:

$$\Delta a_\mu^{\mathrm{BZ}} = \frac{G_F\,\alpha_{\mathrm{em}}}{4\sqrt{2}\pi^3} m_\mu^2 \sum_{i,f} N_c^f Q_f^2\, \xi_\mu^i \xi_f^i\, \frac{m_f^2}{M_i^2}\, g_i\left(\frac{m_f^2}{M_i^2}\right), \tag{2.42}$$

where $i = h, H, A$, and Q_f, m_f, N_c^f are the electric charge, mass, and color factor of the loop fermion, respectively. The loop functions are:

$$g_{h,H}(x) = \int_0^1 dx \frac{2x(1-x)-1}{x(1-x)-x} \ln\left[\frac{x(1-x)}{x}\right], \tag{2.43}$$

$$g_A(x) = \int_0^1 dx \frac{1}{x(1-x)-x} \ln\left[\frac{x(1-x)}{x}\right]. \tag{2.44}$$

The pseudoscalar A yields a positive contribution to Δa_μ, while the CP-even scalars h, H give negative ones. Thus, a light pseudoscalar with enhanced couplings to leptons, particularly in Type-II 2HDM with large $\tan\beta$, can provide a viable explanation for the observed anomaly.

2.5 INERT HIGGS DOUBLET

Within the framework of the 2HDM, the Inert Higgs Doublet Model (IHDM) arises when one of the Higgs doublets, typically denoted as Φ_1, does not acquire a VEV, while the second doublet Φ_2 does. This leads to the vacuum structure:

$$\langle\Phi_1\rangle = \begin{pmatrix} 0 \\ 0 \end{pmatrix}, \quad \langle\Phi_2\rangle = \begin{pmatrix} 0 \\ v \end{pmatrix}. \tag{2.45}$$

As a result, Φ_2 is responsible for electroweak symmetry breaking and generates masses for the SM particles, whereas Φ_1 remains inert: it does not participate in symmetry breaking nor couple to fermions at tree level.

The inert nature of Φ_1 is ensured by imposing a discrete $\mathbb{Z}_2$ symmetry under which $\Phi_1 \to -\Phi_1$, while all SM fields and Φ_2 remain even. This symmetry forbids Φ_1 from acquiring a VEV and ensures that its components interact only in pairs, preventing tree-level mixing with Φ_2 [61].

After electroweak symmetry breaking, the doublets are parameterized as:

$$\Phi_1 = \begin{pmatrix} H^+ \\ \frac{1}{\sqrt{2}}(S + iA) \end{pmatrix}, \quad \Phi_2 = \begin{pmatrix} G^+ \\ \frac{1}{\sqrt{2}}(v + h + iG^0) \end{pmatrix},$$

where G^+ and G^0 are the Goldstone bosons absorbed by the W^+ and Z bosons, respectively. The fields S, A, and $H^\pm$ originating from Φ_1 are physical scalars, which are stabilized by the unbroken $\mathbb{Z}_2$ symmetry and thus become potential DM candidates.

The scalar potential, constrained by the exact $\mathbb{Z}_2$ symmetry (which forbids the $m_{12}^2\Phi_1^\dagger\Phi_2$ term), takes the form:

$$\begin{aligned} V &= \mu_1^2|\Phi_1|^2 + \mu_2^2|\Phi_2|^2 + \lambda_1|\Phi_1|^4 + \lambda_2|\Phi_2|^4 + \lambda_3|\Phi_1|^2|\Phi_2|^2 \\ &\quad + \lambda_4|\Phi_1^\dagger\Phi_2|^2 + \frac{\lambda_5}{2}\left[(\Phi_1^\dagger\Phi_2)^2 + \text{h.c.}\right]. \end{aligned} \tag{2.46}$$

After minimizing the potential, the physical scalar masses are:

$$m_h^2 = 2\lambda_2 v^2, \tag{2.47}$$

$$m_S^2 = \mu_1^2 + \frac{1}{2}(\lambda_3 + \lambda_4 + \lambda_5)v^2 = \mu_1^2 + \lambda_L v^2, \tag{2.48}$$

$$m_A^2 = \mu_1^2 + \frac{1}{2}(\lambda_3 + \lambda_4 - \lambda_5)v^2 = \mu_1^2 + \lambda_A v^2, \tag{2.49}$$

$$m_{H^\pm}^2 = \mu_1^2 + \frac{1}{2}\lambda_3 v^2, \tag{2.50}$$

where $\lambda_{L,A} = \frac{1}{2}(\lambda_3 + \lambda_4 \pm \lambda_5)$. The model's independent parameters can be expressed as $\{m_h, m_S, m_A, m_{H^\pm}, \lambda_2, \lambda_L\}$, with the DM candidate identified as the lighter neutral scalar between S and A. Here, we assume the CP-odd scalar A is the lightest and thus serves as the DM candidate.

Assuming a standard thermal history of the Universe, the relic abundance of inert scalar DM is determined by its annihilation cross-sections into SM final states, including:

$$\begin{aligned} &AA \to f\bar{f} \quad (f = t, b, c, \tau, \mu), \\ &AA \to W^+W^-,\ ZZ,\ \gamma\gamma,\ \gamma Z, \\ &AA \to H^+H^-. \end{aligned}$$

For $m_A < m_{W,Z}$, three- and four-body final states from off-shell vector bosons (VV^* and V^*V^*, with $V = W, Z$) become relevant. The relic density Ωh^2 is obtained by solving the Boltzmann equation, which at freeze-out yields the approximate expression:

$$\Omega h^2 \approx \frac{1.07 \times 10^9\ \text{GeV}^{-1}}{M_{\text{Pl}}\sqrt{g_*}\,\langle\sigma v\rangle}, \tag{2.51}$$

where M_{Pl} is the Planck mass, g_* is the effective number of relativistic degrees of freedom, and $\langle\sigma v\rangle$ is the thermally averaged annihilation cross-section. Detailed relic density computations often require numerical tools such as `MicrOMEGAs` [62], ensuring consistency with the latest observational data from Planck:

$$\Omega_{\text{CDM}} h^2 = 0.120 \pm 0.001 \tag{2.52}$$

The IHDM thus presents a minimal, well-motivated framework where a stable scalar, protected by an exact symmetry, can account for the observed DM in the Universe, while remaining consistent with current collider and cosmological constraints.

CHAPTER 3

Minimal *B-L* extension of the Standard Mode

As discussed in Chapter 1, the SM of electroweak and strong interactions has been remarkably successful in accounting for a wide range of experimental observations. Nevertheless, he discovery of non-zero neutrino masses provides direct and compelling evidence for physics BSM. The phenomenon of neutrino oscillations, confirmed by solar and atmospheric neutrino experiments, implies that at least two neutrino species must possess non-zero mass [13,14].

A particularly well-motivated and minimal extension of the SM that accommodates neutrino masses is based on the enlarged gauge group:

$$SU(3)_C \times SU(2)_L \times U(1)_Y \times U(1)_{B-L},$$

where $B-L$ denotes the difference between baryon (B) and lepton (L) numbers. This extension, commonly referred to as the B–L extended SM (BLSM), is particularly attractive from a theoretical standpoint because it becomes anomaly-free upon the inclusion of three SM singlet fermions, one per generation. These additional fermions are naturally identified as RH neutrinos [63,64].

In the minimal B-L model, the breaking of the extra $U(1)_{B-L}$ gauge symmetry is achieved via a complex scalar SM-singlet field that develops a VEV at the TeV scale [64]. This gives rise to of an additional neutral gauge boson Z', and a new scalar particle. These new states can lead to observable signals at the LHC, such as resonant production of Z' bosons or extended Higgs-sector signatures [65,66].

DOI: 10.1201/9781003457701-3

The minimal $B-L$ model is considered "minimal" in both its scalar and fermion sectors:

- A single additional complex scalar singlet is introduced to spontaneously break the $U(1)_{B-L}$ symmetry.
- Exactly three RH neutrinos, transforming as $\nu_R \sim (1,1,0,-1)$, are included to cancel all gauge and gravitational anomalies associated with the extended gauge symmetry.

As we will discuss, this model offers a simple, testable, and elegant framework to explain the origin of neutrino masses via the type-I seesaw mechanism, while predicting new particles accessible at current or near-future collider energies.

3.1 CONSTRUCTING THE BLSM

The particle content of the BLSM includes, in addition to the SM fields, the following new states:

(i) A scalar field χ, which is a SM singlet carrying a $B-L$ charge of $+2$. Its VEV spontaneously breaks the $U(1)_{B-L}$ gauge symmetry.

(ii) Three SM-singlet fermions ν_i $(i = 1,2,3)$, each with a $B-L$ charge of -1. These fermions serve a dual purpose: they cancel the $B-L$ gauge anomalies and enable the implementation of the type-I seesaw mechanism for generating light neutrino mass generation.

Table 3.1 summarizes the particle content of the BLSM, along with their charge assignments under the extended gauge group.

The minimal gauge-invariant Lagrangian of the $B-L$ extension of the SM can be written as [64]:

$$\mathcal{L}_{B-L} = \mathcal{L}_{\rm YM} + \mathcal{L}_f + \mathcal{L}_s + \mathcal{L}_Y, \tag{3.1}$$

TABLE 3.1 The $SU(2)_L \times U(1)_Y$ and $U(1)_{B-L}$ quantum numbers of the particles in the model.

	l	ν_R	e_R	Q	u_R	d_R	ϕ	χ
$SU(2)_L \times U(1)_Y$	$(\mathbf{2},-1/2)$	$(\mathbf{1},0)$	$(\mathbf{1},-1)$	$(\mathbf{2},1/6)$	$(\mathbf{1},2/3)$	$(\mathbf{1},-1/3)$	$(\mathbf{2},1/2)$	$(\mathbf{1},0)$
$U(1)_{B-L}$	-1	-1	-1	$1/3$	$1/3$	$1/3$	0	2

where the Yang-Mills (gauge kinetic) part includes the kinetic terms for all gauge bosons:

$$\begin{aligned}\mathcal{L}_{\rm YM} = &-\frac{1}{4}G^{\alpha}_{\mu\nu}G^{\mu\nu\alpha} - \frac{1}{4}W^{a}_{\mu\nu}W^{\mu\nu a} - \frac{1}{4}B_{\mu\nu}B^{\mu\nu} \\ &-\frac{1}{4}B'_{\mu\nu}B'^{\mu\nu} - \frac{k}{2}B_{\mu\nu}B'^{\mu\nu},\end{aligned} \tag{3.2}$$

where k is a real parameter representing kinetic mixing between the two Abelian gauge fields. The field strengths are defined as

$$B_{\mu\nu} = \partial_\mu B_\nu - \partial_\nu B_\mu, \quad B'_{\mu\nu} = \partial_\mu B'_\nu - \partial_\nu B'_\mu. \tag{3.3}$$

The kinetic mixing term can be removed via a suitable field redefinition, leading to canonically normalized field strengths [66]. After diagonalization of the kinetic terms, the most general covariant derivative takes the form:

$$D_\mu = \partial_\mu - ig_S T^\alpha G^\alpha_\mu - ig\tau^a W^a_\mu - ig'YB_\mu - i(\tilde{g}Y + g''Y_{B-L})B'_\mu, \tag{3.4}$$

where g' and g'' are the gauge couplings of $U(1)_Y$ and $U(1)_{B-L}$, respectively, $\tilde{g}$ is a mixing parameter arising from the diagonalization procedure, and Y, Y_{B-L} are the respective charges.

The fermionic part of the Lagrangian is

$$\mathcal{L}_f = \sum_\psi \bar{\psi} i\gamma^\mu D_\mu \psi, \tag{3.5}$$

where the sum runs over all SM fermions, including the RH neutrinos ν_R, which are necessary for anomaly cancellation and for implementing the seesaw mechanism. The $B-L$ charges are:

$$\begin{aligned}Y_{B-L}(q_L, u_R, d_R) &= \frac{1}{3}, \\ Y_{B-L}(\ell_L, e_R, \nu_R) &= -1.\end{aligned}$$

The scalar content consists of the usual SM Higgs doublet H and an additional SM-singlet scalar χ, which is charged under $U(1)_{B-L}$ and is responsible for spontaneously breaking the $B-L$ symmetry. The scalar Lagrangian is

$$\mathcal{L}_s = |D_\mu \phi|^2 + |D_\mu \chi|^2 - V(\phi, \chi), \tag{3.6}$$

with the scalar potential

$$V(\phi,\chi) = m_1^2\phi^\dagger\phi + m_2^2\chi^\dagger\chi + \lambda_1(\phi^\dagger\phi)^2 + \lambda_2(\chi^\dagger\chi)^2 + \lambda_3(\chi^\dagger\chi)(\phi^\dagger\phi). \tag{3.7}$$

Recall that the scalar field ϕ is a Higgs doublet, and χ is a singlet, with $B-L$ charges of 0 and +2, respectively.

Finally, the Yukawa interactions for quarks and charged leptons are as in the SM:

$$\mathcal{L}_Y \supset -y_u \bar{q}_L \tilde{H} u_R - y_d \bar{q}_L \phi d_R - y_e \bar{\ell}_L \phi e_R + \text{h.c.} \tag{3.8}$$

And for neutrinos, the presence of ν_R allows both Dirac and Majorana mass terms:

$$\mathcal{L}_Y \supset -y_\nu \bar{\ell}_L \tilde{H} \nu_R - \frac{1}{2} y_M \bar{\nu}_R^c \chi \nu_R + \text{h.c.} \tag{3.9}$$

The first term generates a Dirac mass after EW symmetry breaking, while the second provides a Majorana mass for ν_R when χ acquires a VEV, realizing the type-I seesaw mechanism.

3.2 $B-L$ BREAKING, HIGGS AND GAUGE MASSES

The spontaneous breaking of the $B-L$ symmetry proceeds analogously to electroweak symmetry breaking. It is triggered by a non-vanishing VEV of a scalar field that is a singlet under the SM gauge group but charged under $U(1)_{B-L}$. In this setup, the Higgs sector consists of:

- A complex $SU(2)_L$ singlet scalar field, $\chi \sim (1, 1, 0, 2)$, responsible for breaking the $U(1)_{B-L}$ symmetry, and
- The usual SM Higgs doublet, $\phi \sim (1, 2, 1, 0)$, which breaks the $SU(2)_L \times U(1)_Y$ symmetry down to $U(1)_{\text{em}}$.

The most general renormalizable scalar potential involving both fields can be written as:

$$V(\phi, \chi) = m_1^2 \phi^\dagger \phi + m_2^2 \chi^\dagger \chi + \begin{pmatrix} \phi^\dagger \phi & \chi^\dagger \chi \end{pmatrix} \begin{pmatrix} \lambda_1 & \frac{\lambda_3}{2} \\ \frac{\lambda_3}{2} & \lambda_2 \end{pmatrix} \begin{pmatrix} \phi^\dagger \phi \\ \chi^\dagger \chi \end{pmatrix}. \tag{3.10}$$

As in the SM, for the potential to be bounded from below, the quartic terms must satisfy positivity conditions. This requirement translates into the following stability conditions:

$$\lambda_1, \lambda_2 > 0, \tag{3.11}$$

$$4\lambda_1 \lambda_2 - \lambda_3^2 > 0. \tag{3.12}$$

To achieve spontaneous symmetry breaking with non-zero VEVs v and v', the mass-squared parameters must satisfy $m_1^2 < 0$ and $m_2^2 < 0$. Minimizing the potential with respect to the fields yields:

$$\frac{\partial V}{\partial \phi} = 0 \Rightarrow m_1^2 + 2\lambda_1 \phi^\dagger \phi + \lambda_3 \chi^\dagger \chi = 0, \tag{3.13}$$

$$\frac{\partial V}{\partial \chi} = 0 \Rightarrow m_2^2 + 2\lambda_2 \chi^\dagger \chi + \lambda_3 \phi^\dagger \phi = 0. \tag{3.14}$$

Solving these equations leads to the VEVs [64]:

$$v^2 = \frac{-4\lambda_2 m_1^2 + 2\lambda_3 m_2^2}{4\lambda_1\lambda_2 - \lambda_3^2}, \tag{3.15}$$

$$v'^2 = \frac{-4\lambda_1 m_2^2 + 2\lambda_3 m_1^2}{4\lambda_1\lambda_2 - \lambda_3^2}. \tag{3.16}$$

The vacuum configuration is taken to be:

$$\langle \phi \rangle = \frac{1}{\sqrt{2}} \begin{pmatrix} 0 \\ v \end{pmatrix}, \quad \langle \chi \rangle = \frac{v'}{\sqrt{2}}. \tag{3.17}$$

Since the denominators in Eqs. (3.15) and (3.16) are positive by Eq. (3.12), the numerators must also be positive to ensure real VEVs:

$$-2\lambda_2 m_1^2 + \lambda_3 m_2^2 > 0, \qquad -2\lambda_1 m_2^2 + \lambda_3 m_1^2 > 0. \tag{3.18}$$

These inequalities imply that both $m_1^2, m_2^2 > 0$ are not allowed, regardless of the sign of λ_3. Scenarios with mixed signs, such as $m_1^2 > 0, m_2^2 < 0$, require $\lambda_3 < 0$. For $m_1^2, m_2^2 < 0$, both signs of λ_3 are permitted. In the decoupling limit $\lambda_3 \to 0$, the two sectors become independent, and v^2 reduces to the familiar SM expression. Combining Eqs. (3.15) and (3.16), one can express v'^2 in terms of v:

$$v'^2 = \frac{-2(m_1^2 + \lambda_1 v^2)}{\lambda_3}. \tag{3.19}$$

This relation shows that if $v' \gg v$, then a fine-tuned cancellation between $\lambda_3 v'^2$ and $2m_1^2$ is needed to obtain the correct electroweak scale $v \approx 246\,\text{GeV}$. To avoid such fine-tuning, the $B-L$ symmetry should be broken at a relatively low scale, typically $v' \sim \mathcal{O}(1)\,\text{TeV}$. We assume that the singlet VEV v' is larger than the doublet VEV v, i.e.,

$$|\langle \chi \rangle| = \frac{v'}{\sqrt{2}} \gg |\langle \phi^0 \rangle| = \frac{v}{\sqrt{2}}.$$

It is important to note that the scale v' is not fixed a priori. However, the mass of the $U(1)_{B-L}$ gauge boson B'_μ, as well as the masses of the right-handed neutrinos, depend directly on v'. Thus, experimental bounds on extra neutral gauge bosons can be used to set lower limits on the $B-L$ breaking scale.

3.2.1 Higgs Masses

To determine the scalar mass spectrum, we expand the potential (3.7) around the VEVs, given in (3.15) and (3.16). Working in the unitary gauge, the scalar fields can be written as:

$$\phi(x) = \begin{pmatrix} 0 \\ \frac{v+h_1(x)}{\sqrt{2}} \end{pmatrix}, \quad \chi(x) = \frac{v' + h_2(x)}{\sqrt{2}}. \tag{3.20}$$

The scalar potential then becomes:

$$V(h_1, h_2) = \lambda_1 v^2 h_1^2 + \lambda_2 v'^2 h_2' 2 + \lambda_3 v v' h_1 h_2' + \frac{\lambda_3}{2}(v' h_1^2 h_2 + v h_1 h_2^2) \\ + \lambda_1 v h_1^3 + \lambda_2 v' h_2^3 + \frac{\lambda_1}{4} h_1^4 + \frac{\lambda_2}{4} h_2^4. \tag{3.21}$$

The mass terms are given by the quadratic part of the potential:

$$\frac{1}{2} M^2(h_1, h_2) = \begin{pmatrix} h_1 & h_2 \end{pmatrix} \begin{pmatrix} \lambda_1 v^2 & \frac{\lambda_3}{2} v v' \\ \frac{\lambda_3}{2} v v' & \lambda_2 v'^2 \end{pmatrix} \begin{pmatrix} h_1 \\ h_2 \end{pmatrix}. \tag{3.22}$$

This matrix is diagonalized by an orthogonal transformation:

$$\begin{pmatrix} h \\ H \end{pmatrix} = \begin{pmatrix} \cos\theta & -\sin\theta \\ \sin\theta & \cos\theta \end{pmatrix} \begin{pmatrix} h_1 \\ h_2 \end{pmatrix}, \tag{3.23}$$

where the mixing angle θ is defined as:

$$\tan 2\theta = \frac{|\lambda_3| v v'}{\lambda_1 v^2 - \lambda_2 v'^2}. \tag{3.24}$$

The masses of the physical Higgs bosons h and H are:

$$m_{h,H}^2 = \lambda_1 v^2 + \lambda_2 v'^2 \mp \sqrt{(\lambda_1 v^2 - \lambda_2 v'^2)^2 + \lambda_3^2 v^2 v'^2}. \tag{3.25}$$

Here, h corresponds to the SM-like Higgs and H is the heavy Higgs associated with the $B-L$ breaking sector. The parameter λ_3 controls the mixing between the two Higgs states [65]. In the limit $\lambda_3 \to 0$ or $v' \gg v$, the mixing vanishes, and we recover the SM Higgs mass $m_h = \sqrt{2\lambda_1}v$ and the $B-L$ Higgs mass $m_H = \sqrt{2\lambda_2}v'$. However, for non-zero λ_3 and moderate v', the SM-like Higgs mass receives corrections and can be lighter than in the pure SM case, which has important implications for Higgs phenomenology [67].

3.2.2 Gauge Sector and Mass Mixing

The gauge boson masses arise from the scalar kinetic terms after spontaneous symmetry breaking. In this $U(1)_{B-L}$ extension, the charged gauge bosons $W^{\pm} = (W^1_\mu \mp iW^2_\mu)/\sqrt{2}$ acquire their standard masses $M_{W^\pm} = gv/2$, unaffected by the Abelian extension.

The neutral gauge bosons, however, mix due to the extra $U(1)_{B-L}$ symmetry. In the unitary gauge, the scalar kinetic terms yield:

$$\begin{aligned}\mathcal{L}^{\text{kin}}_{\text{scalar}} &= \frac{1}{2}\partial^\mu h_1\,\partial_\mu h_1 + \frac{1}{8}(h_1+v)^2\Big[g^2|W_1^\mu - iW_2^\mu|^2 + (gW_3^\mu - g'B^\mu \\ &\quad - \tilde{g}B'^\mu)^2\Big] + \frac{1}{2}\partial^\mu h_2\,\partial_\mu h_2 + \frac{1}{2}(h_2+v')^2(2g''B'^\mu)^2, \qquad (3.26)\end{aligned}$$

In the basis (B^μ, W_3^μ, B'^μ), the neutral gauge boson mass matrix is:

$$\mathcal{M}^2 = \frac{v^2}{2}\begin{pmatrix} g'^2 & -gg' & g'\tilde{g} \\ -gg' & g^2 & -g\tilde{g} \\ g'\tilde{g} & -g\tilde{g} & \tilde{g}^2 + \left(\frac{4v'g''}{v}\right)^2 \end{pmatrix}. \qquad (3.27)$$

Applying the electroweak rotation:

$$R_{\text{EW}}(\theta_w) = \begin{pmatrix} \cos\theta_w & \sin\theta_w & 0 \\ -\sin\theta_w & \cos\theta_w & 0 \\ 0 & 0 & 1 \end{pmatrix}, \qquad (3.28)$$

we isolate the massless photon and obtain the 2×2 submatrix for Z and Z'. The physical eigenstates are:

$$\begin{pmatrix} A^\mu \\ Z^\mu \\ Z'^\mu \end{pmatrix} = \begin{pmatrix} \cos\theta_w & \sin\theta_w & 0 \\ -\sin\theta_w\cos\theta' & \cos\theta_w\cos\theta' & \sin\theta' \\ \sin\theta_w\sin\theta' & -\cos\theta_w\sin\theta' & \cos\theta' \end{pmatrix}\begin{pmatrix} B^\mu \\ W_3^\mu \\ B'^\mu \end{pmatrix}, \qquad (3.29)$$

where the mixing angle θ' satisfies:

$$\tan 2\theta' = \frac{2\tilde{g}\sqrt{g^2+g'^2}}{\tilde{g}^2 + \left(\frac{4v'g''}{v}\right)^2 - (g^2+g'^2)}. \tag{3.30}$$

The physical masses are:

$$M_A = 0, \tag{3.31}$$

$$M_Z = \frac{v\sqrt{g^2+g'^2}}{2}\left[\frac{1}{2}\left(1+\frac{\tilde{g}^2+\left(\frac{4v'g''}{v}\right)^2}{g^2+g'^2}\right) - \frac{\tilde{g}}{s_{2\theta'}\sqrt{g^2+g'^2}}\right], \tag{3.32}$$

$$M_{Z'} = \frac{v\sqrt{g^2+g'^2}}{2}\left[\frac{1}{2}\left(1+\frac{\tilde{g}^2+\left(\frac{4v'g''}{v}\right)^2}{g^2+g'^2}\right) + \frac{\tilde{g}}{s_{2\theta'}\sqrt{g^2+g'^2}}\right], \tag{3.33}$$

where

$$s_{2\theta'} = \frac{2\tilde{g}\sqrt{g^2+g'^2}}{\left(\tilde{g}^2+\left(\frac{4v'g''}{v}\right)^2 - g^2 - g'^2\right)^2 + (2\tilde{g})^2}. \tag{3.34}$$

In the limit $\tilde{g} \to 0$, corresponding to the pure $B-L$ model, the mixing vanishes ($\theta' = 0$), and one recovers the SM result $M_Z = 1/2v\sqrt{g^2+g'^2}2$. We adopt $\tilde{g} = 0$ in what follows, consistent with experimental limits on Z–Z' mixing, which require it to be less than 10^{-3} [68].

3.3 NEUTRINO MASSES IN THE BLSM

In this section, we analyze neutrino masses and mixing in the $B-L$ extension of the SM, where the neutrino masses arise through a TeV-scale seesaw mechanism [64]. After $U(1)_{B-L}$ symmetry breaking, the field χ can be shifted around its VEV:

$$\chi = \frac{1}{\sqrt{2}}(h_2 + v').$$

The term $\frac{1}{2}y_M\overline{(\nu_R)^c}\chi\nu_R$ leads to a right-handed neutrino mass:

$$M_R = \frac{1}{2\sqrt{2}}y_M v'.$$

Similarly, after electroweak symmetry breaking, the field ϕ can be shifted as:

$$\phi = \begin{pmatrix} 0 \\ \frac{h_1+v}{\sqrt{2}} \end{pmatrix},$$

giving rise to a Dirac neutrino mass term:

$$m_D = \frac{1}{\sqrt{2}} y_\nu v.$$

Thus, the mass matrix for the left- and right-handed neutrinos is:

$$\mathcal{M} = \begin{pmatrix} 0 & m_D \\ m_D^T & M_R \end{pmatrix}.$$

The masses of the light and heavy neutrinos are given by diagonalizing this mass matrix:

$$m_{\nu_\ell} \simeq m_D M_R^{-1} m_D^T, \quad m_{\nu_H} \simeq M_R.$$

This is the seesaw mechanism, where large M_R results in very light neutrino masses, m_{ν_ℓ}, of the order 10^{-2} eV.

In the usual seesaw mechanism, M_R is typically assumed to be of the order of the Planck scale, $\sim 10^{14}$ GeV, to accommodate a Dirac neutrino mass m_D of order 100 GeV. However, there is no direct evidence that such a large scale is required, and a "low scale seesaw" with $M_R \sim$ TeV is an acceptable alternative, especially when Dirac neutrino masses are small, of order 10^{-4} GeV.

A commonly used parametrization for the Dirac neutrino mass matrix is [69]:

$$m_D = U_{PMNS} \sqrt{m_{\nu_\ell}^{\rm diag}} R \sqrt{M_R},$$

where $m_{\nu_\ell}^{\rm diag}$ is the diagonal light neutrino mass matrix, U_{PMNS} is the lepton mixing matrix, and R is an orthogonal matrix parametrized by three complex angles. A possible parametrization for R is:

$$R = \begin{pmatrix} c_2 c_1 & s_1 c_2 & s_2 \\ -s_1 c_3 - c_1 s_2 s_3 & c_1 c_3 - s_1 s_2 s_3 & s_1 c_2 \\ s_1 s_3 - c_3 s_2 c_1 & -c_1 s_3 - s_1 s_2 c_3 & c_3 c_2 \end{pmatrix},$$

where $c_i = \cos(\theta_i)$ and $s_i = \sin(\theta_i)$, for $i = 1, 2, 3$.

3.4 Z' GAUGE BOSON IN THE BLSM

The Lagrangian describing the dynamics and interactions of the additional $U(1)_{B-L}$ gauge boson, denoted by Z', is given by:

$$\mathcal{L}_{Z'} = -\frac{1}{4}Z'_{\mu\nu}Z'^{\mu\nu} + \frac{1}{2}M^2_{Z'}Z'_\mu Z'^\mu + 2g''^2 Z'_\mu Z'^\mu \chi^2 + 4g''^2 v'\chi Z'_\mu Z'^\mu \\ - g'' J^{B-L}_\mu Z'^\mu, \tag{3.35}$$

where the $B-L$ current is

$$J^{B-L}_\mu = \frac{1}{3}\bar{u}\gamma_\mu u + \frac{1}{3}\bar{d}\gamma_\mu d - \bar{e}\gamma_\mu e - \bar{\nu}\gamma_\mu \nu. \tag{3.36}$$

Experimental constraints from LEP II and LHC searches place a lower bound on the Z' mass in $B-L$ models [70]. These constraints typically yield

$$\frac{M_{Z'}}{g''} \gtrsim 6 \text{ TeV}, \tag{3.37}$$

implying a scale for breaking symmetry $v' \sim \mathcal{O}(\text{TeV})$, since $M_{Z'} = 2g''v'$. Here, the electroweak scale remains $v \simeq 246$ GeV.

The interactions of the Z' boson with SM fermions are governed by their $B-L$ charges:

$$\mathcal{L}^{Z'}_{\text{int}} = \sum_f g'' Y^f_{B-L} Z'_\mu \bar{f}\gamma^\mu f. \tag{3.38}$$

In this minimal $B-L$ model, tree-level Z–Z' mixing is absent, and Z' couples directly to SM fermions via their $B-L$ quantum numbers [64]. The partial decay widths of the Z' boson are approximately given by:

$$\Gamma(Z' \to \ell^+\ell^-) \simeq \frac{(g''Y^\ell_{B-L})^2}{24\pi} M_{Z'}, \tag{3.39}$$

$$\Gamma(Z' \to q\bar{q}) \simeq \frac{(g''Y^q_{B-L})^2}{8\pi} M_{Z'}\left(1 + \frac{\alpha_s}{\pi}\right), \tag{3.40}$$

$$\Gamma(Z' \to t\bar{t}) \simeq \frac{(g''Y^t_{B-L})^2}{8\pi} M_{Z'}\left(1 - \frac{m_t^2}{M^2_{Z'}}\right)\sqrt{1 - \frac{4m_t^2}{M^2_{Z'}}}\left(1 + \frac{\alpha_s}{\pi}\right). \tag{3.41}$$

As shown in Fig. 3.1, the leptonic branching ratio of the Z' boson is improved relative to its hadronic decays. This is due to $|Y^\ell_{B-L}| = 3|Y^q_{B-L}|$, leading to:

$$\text{BR}(Z' \to \ell^+\ell^-) \simeq 20\%, \quad \text{versus} \quad \text{BR}(Z \to \ell^+\ell^-) \simeq 3\%.$$

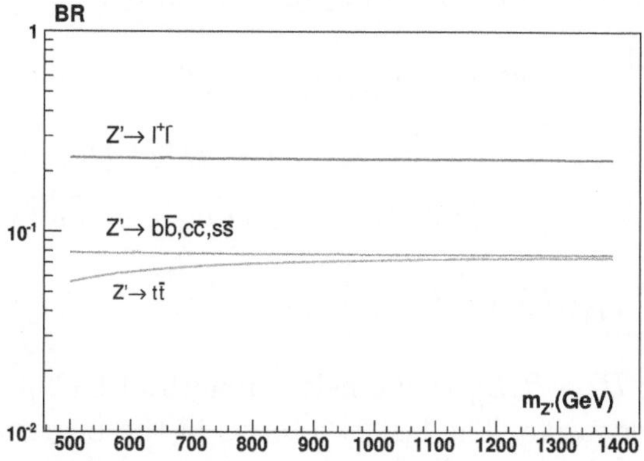

Figure 3.1 Branching ratios of the Z' boson as functions of $M_{Z'}$. Figure adapted from [65].

Hence, the clean di-lepton signal makes Z' a promising early discovery channel at the LHC [64, 65, 71].

3.5 BLSM WITH INVERSE SEESAW MECHANISM

In this section, we review the TeV-scale $B{-}L$ extension of the SM incorporating the Inverse Seesaw (ISS) Mechanism, as proposed in Ref. [72]. The model is based on the extended gauge group $SU(3)_C \times SU(2)_L \times U(1)_Y \times U(1)_{B-L}$, where the $U(1)_{B-L}$ symmetry is spontaneously broken by a SM singlet scalar field χ carrying a $B-L$ charge of -1. This breaking gives rise to a new neutral gauge boson Z'. To ensure anomaly cancellation, the model introduces three RH neutrinos ν_{R_i}, each with a $B{-}L$ charge of -1. In order to implement the ISS mechanism, the model also includes three additional SM-singlet fermions S_1 with $B-L$ charge -2, and three fermions S_2 with charge $+2$. The complete set of $B-L$ charge assignments for the fermions and scalar fields is summarized in Table 3.2.

TABLE 3.2 $B-L$ quantum numbers of fermions and Higgs particles.

Particle	Q	u_R	d_R	L	e_R	ν_R	ϕ	χ	S_1	S_2
Y_{B-L}	1/3	1/3	1/3	-1	-1	-1	0	-1	-2	$+2$

The relevant part of the Lagrangian is given by:

$$\begin{aligned}\mathcal{L}_{B-L} &= -\frac{1}{4}B'_{\mu\nu}B'^{\mu\nu} + i\overline{\ell_L}D_\mu\gamma^\mu\ell_L + i\overline{e_R}D_\mu\gamma^\mu e_R + i\overline{\nu_R}D_\mu\gamma^\mu\nu_R \\ &+ i\overline{S_1}D_\mu\gamma^\mu S_1 + i\overline{S_2}D_\mu\gamma^\mu S_2 + (D^\mu\phi)^\dagger D_\mu\phi + (D^\mu\chi)^\dagger D_\mu\chi \\ &- V(\phi,\chi) - \left(y_e\overline{\ell_L}\phi e_R + y_\nu\overline{\ell_L}\tilde{\phi}\nu_R + y_S\overline{(\nu_R)^c}\chi S_2 + \text{h.c.}\right) \\ &- \frac{1}{M^3}\overline{(S_1)^c}\chi^{\dagger 4}S_1 - \frac{1}{M^3}\overline{(S_2)^c}\chi^4 S_2, \end{aligned} \tag{3.42}$$

where $B'_{\mu\nu} = \partial_\mu B'_\nu - \partial_\nu B'_\mu$ is the field strength of $U(1)_{B-L}$, and D_μ is the covariant derivative. The last two terms in the Lagrangian are non-renormalizable and are required for generating small masses for S_1 and S_2 at the TeV scale, essential for the ISS.

After $B-L$ symmetry breaking, the Z' gauge boson acquires the mass:

$$M_{Z'} = g''v' \quad \text{with} \quad v' \sim \mathcal{O}(1 \text{ TeV}), \tag{3.43}$$

where $v' = \langle\chi\rangle/\sqrt{2}$ is the vacuum expectation value (VEV) of the χ field. The constraints from LEP II on $Z-Z'$ mixing lead to $M_{Z'}/g'' > 6$ TeV [73].

The neutrino mass terms arising from the Yukawa interactions are:

$$\mathcal{L}_m^\nu = m_D\overline{\nu_L}\nu_R + M_R\overline{(\nu_R)^c}S_2 + \text{h.c.}, \tag{3.44}$$

where $m_D = \frac{1}{\sqrt{2}}y_\nu v$ and $M_R = \frac{1}{\sqrt{2}}y_S v'$. These terms give rise to the neutrino mass matrix in the flavor basis:

$$\mathcal{M}_\nu = \begin{pmatrix} 0 & m_D & 0 \\ m_D^T & 0 & M_R \\ 0 & M_R^T & \mu_s \end{pmatrix}, \tag{3.45}$$

where $\mu_s \sim 10^{-10}$ GeV is the Majorana mass term for S_2, and M_R is of order the TeV scale.

The diagonalization of the neutrino mass matrix leads to the light neutrino masses:

$$m_{\nu_\ell} = m_D M_R^{-1}\mu_s(M_R^T)^{-1}m_D^T. \tag{3.46}$$

Thus, the light neutrino masses can be of order eV, with a TeV-scale M_R if $\mu_s \ll M_R$. The heavy neutrino physics in this model can be probed at the LHC [64].

CHAPTER 4

Left-Right Symmetric Model

One of the most well-motivated extensions of the SM is the Left-Right Symmetric Model (LRSM) . Originally introduced by Pati and Salam in 1974 [74], the LRSM was further developed and analyzed in detail by Mohapatra and Senjanović [22, 75]. These models provide a natural framework for addressing parity violation in weak interactions, as well as the origin of neutrino masses via the seesaw mechanism.

The LRSM is based on the extended gauge group:

$$SU(3)_C \times SU(2)_L \times SU(2)_R \times U(1)_{B-L}. \tag{4.1}$$

At low energies, this symmetry is spontaneously broken down to the SM gauge group, and eventually to electromagnetism. The symmetry breaking proceeds in two stages:

$$SU(2)_L \times SU(2)_R \times U(1)_{B-L} \longrightarrow SU(2)_L \times U(1)_Y \longrightarrow U(1)_{\text{EM}},$$

where the electromagnetic charge operator is given by a generalized Gell-Mann–Nishijima relation:

$$Q = T_L^3 + T_R^3 + \frac{1}{2}(B - L). \tag{4.2}$$

This expression emerges naturally from the symmetry-breaking sequence in which the gauge group $SU(2)_R \times U(1)_{B-L}$ reduces to the SM hypercharge group $U(1)_Y$. The hypercharge itself is defined as a specific linear combination of the right-handed isospin and the $B-L$ quantum number. When substituted into the conventional expression for electric charge,

DOI: 10.1201/9781003457701-4

this leads directly to the above generalized Gell-Mann–Nishijima relation. This formulation not only reproduces the correct electric charges for all SM fermions but also illustrates the consistent embedding of hypercharge within the extended gauge symmetry of the LRSM.

Like the SM, the LRSM includes strong interactions via the $SU(3)_C$ gauge group. However, its weak interaction sector is extended to $SU(2)_L \times SU(2)_R \times U(1)_{B-L}$, which implies the existence of additional heavy gauge bosons. In particular, the LRSM predicts two charged right-handed gauge bosons $W_R^{\pm}$ and a neutral gauge boson Z_R, in addition to the SM gauge bosons.

The phenomenology of the LRSM strongly depends on the scalar (Higgs) sector responsible for symmetry breaking. Various choices for the scalar content, such as bidoublets, triplets, or doublets under $SU(2)_{L,R}$, lead to different variants of the model with distinct features. These choices impact not only the mass spectrum of gauge and scalar bosons but also the nature of neutrino masses (type-I, type-II, or hybrid seesaw mechanisms), flavor structure, and potential collider signatures.

4.1 CONVENTIONAL LEFT-RIGHT MODEL

In the LRSM, RH fermions are grouped into doublets under $SU(2)_R$, and the $B-L$ quantum number plays a crucial role in anomaly cancellation and neutrino mass generation. The spontaneous symmetry breaking of $SU(2)_R \times U(1)_{B-L}$ to $U(1)_Y$ is achieved via a triplet scalar field Δ_R. The field content of this model and the associated quantum numbers are given in Table 4.1. Contrary to the SM, all RH components of fermion fields transform as doublets under $SU(2)_R$, reflecting the left-right symmetric nature of the model.

The full Lagrangian of the LRSM includes gauge, kinetic, scalar, and Yukawa parts:

$$\mathcal{L}_{\text{LRSM}} = \mathcal{L}_{\text{Gauge}} + \mathcal{L}_{\text{Kinetic}} + \mathcal{L}_{\text{Scalar}} + \mathcal{L}_{\text{Yukawa}}. \tag{4.3}$$

The gauge kinetic terms for the gauge bosons are

$$\mathcal{L}_{\text{Gauge}} = -\frac{1}{4} G^a_{\mu\nu} G^{a\mu\nu} - \frac{1}{4} W^i_{L\mu\nu} W^{i\mu\nu}_L - \frac{1}{4} W^i_{R\mu\nu} W^{i\mu\nu}_R - \frac{1}{4} V_{\mu\nu} V^{\mu\nu}, \tag{4.4}$$

TABLE 4.1 Field content of the LRSM and respective quantum numbers.

Fields	Components	$SU(2)_L$	$\times SU(2)_R$	$\times U(1)_{B-L}$
Fermions				
L_L	$\begin{pmatrix} \nu \\ e \end{pmatrix}_L$	2	1	$-\frac{1}{2}$
L_R	$\begin{pmatrix} \nu \\ e \end{pmatrix}_R$	1	2	$-\frac{1}{2}$
Q_L	$\begin{pmatrix} u \\ d \end{pmatrix}_L$	2	1	$+\frac{1}{6}$
Q_R	$\begin{pmatrix} u \\ d \end{pmatrix}_R$	1	2	$+\frac{1}{6}$
Gauge bosons				
W_L	$W_L^\pm, W_L^3$	triplet	singlet	singlet
W_R	$W_R^\pm, W_R^3$	singlet	triplet	singlet
B	B_μ	singlet	singlet	singlet
Scalars				
Φ	$\begin{pmatrix} \Phi_1^0 & \Phi_1^+ \\ \Phi_2^- & \Phi_2^0 \end{pmatrix}$	2	2	0
Δ_L	$\begin{pmatrix} \frac{\Delta^+}{\sqrt{2}} & \Delta^{++} \\ \Delta^0 & -\frac{\Delta^+}{\sqrt{2}} \end{pmatrix}_L$	3	1	+1
Δ_R	$\begin{pmatrix} \frac{\Delta^+}{\sqrt{2}} & \Delta^{++} \\ \Delta^0 & -\frac{\Delta^+}{\sqrt{2}} \end{pmatrix}_R$	1	3	+1

with field strength tensors:

$$G^a_{\mu\nu} = \partial_\mu G^a_\nu - \partial_\nu G^a_\mu - g_s f^{abc} G^b_\mu G^c_\nu, \tag{4.5}$$

$$W^i_{L\mu\nu} = \partial_\mu W^i_{L\nu} - \partial_\nu W^i_{L\mu} - g_L \epsilon^{ijk} W^j_{L\mu} W^k_{L\nu}, \tag{4.6}$$

$$W^i_{R\mu\nu} = \partial_\mu W^i_{R\nu} - \partial_\nu W^i_{R\mu} - g_R \epsilon^{ijk} W^j_{R\mu} W^k_{R\nu}, \tag{4.7}$$

$$V_{\mu\nu} = \partial_\mu V_\nu - \partial_\nu V_\mu. \tag{4.8}$$

The fermionic kinetic terms and their gauge interactions are encoded as:

$$\mathcal{L}_{\text{Kinetic}} = \sum_{j=1}^{3} \left[\overline{L}_{Lj} i\gamma^{\mu} D_{\mu} L_{Lj} + \overline{L}_{Rj} i\gamma^{\mu} D_{\mu} L_{Rj} \right. \\ \left. + \overline{Q}_{Lj} i\gamma^{\mu} D_{\mu} Q_{Lj} + \overline{Q}_{Rj} i\gamma^{\mu} D_{\mu} Q_{Rj} \right], \tag{4.9}$$

where the covariant derivatives for the leptons and quarks take the form:

$$D_{\mu} L_L = \left(\partial_{\mu} - \frac{ig_L}{2} \sigma \cdot W_{L\mu} + \frac{ig_{BL}}{2} V_{\mu} \right) L_L, \tag{4.10}$$

$$D_{\mu} L_R = \left(\partial_{\mu} - \frac{ig_R}{2} \sigma \cdot W_{R\mu} + \frac{ig_{BL}}{2} V_{\mu} \right) L_R, \tag{4.11}$$

$$D_{\mu} Q_L = \left(\partial_{\mu} - \frac{ig_L}{2} \sigma \cdot W_{L\mu} - \frac{ig_{BL}}{6} V_{\mu} - ig_s \lambda \cdot G_{\mu} \right) Q_L, \tag{4.12}$$

$$D_{\mu} Q_R = \left(\partial_{\mu} - \frac{ig_R}{2} \sigma \cdot W_{R\mu} - \frac{ig_{BL}}{6} V_{\mu} - ig_s \lambda \cdot G_{\mu} \right) Q_R. \tag{4.13}$$

The scalar sector in the LRSM plays a crucial role in the sequential symmetry breaking:

$$SU(2)_L \times SU(2)_R \times U(1)_{B-L} \xrightarrow{\langle \Delta_R \rangle} SU(2)_L \times U(1)_Y \xrightarrow{\langle \Phi \rangle} U(1)_{\text{EM}}.$$

In addition to breaking the gauge symmetry, the scalar VEVs are responsible for generating particle masses. The electroweak symmetry is broken by the VEV of the bidoublet scalar Φ, giving rise to the masses of SM fermions and gauge bosons, while the heavy masses of the RH gauge bosons W_R, Z_R and RH neutrinos are generated through the VEV of the RH triplet Δ_R.

The gauge-invariant Yukawa interaction $\overline{\psi}_L \Phi \psi_R$ requires that fermions transform as doublets under both $SU(2)_L$ and $SU(2)_R$, and the scalar field Φ transforms as a bidoublet under the extended gauge group. Specifically, Φ is a 2×2 complex matrix with $B - L = 0$, whose columns form doublets under $SU(2)_L$ and rows under $SU(2)_R$. As a result, the VEV of Φ does not break the $U(1)_{B-L}$ symmetry.

To spontaneously break $SU(2)_R \times U(1)_{B-L}$ and reduce the symmetry to that of the SM, an additional scalar field is introduced. This is typically chosen to be a right-handed triplet $\Delta_R \sim (1, 3, +2)$, which acquires a large VEV. For the restoration of discrete parity symmetry at high energies, a left-handed counterpart $\Delta_L \sim (3, 1, +2)$ is also included in the scalar sector [74].

The Lagrangian of scalar fields is given by

$$\mathcal{L}_{\text{Scalar}} = \mathcal{L}^{\text{Kin}}_{\text{Scalar}} - V(\Phi, \Delta_L, \Delta_R), \tag{4.14}$$

where the kinetic Lagrangian is given by

$$\mathcal{L}^{\text{Kin}}_{\text{Scalar}} = \text{Tr}[|D_\mu \Phi|^2] + \text{Tr}[|D_\mu \Delta_L|^2] + \text{Tr}[|D_\mu \Delta_R|^2]. \tag{4.15}$$

The most general gauge invariant and renormalizable Higgs potential given as follows [76]:

$$\begin{aligned}
V(\Phi, \Delta_L, \Delta_R) &= -\mu_1^2 \text{Tr}(\Phi^\dagger \Phi) - \mu_2^2 \text{Tr}(\widetilde{\Phi}^\dagger \Phi + \Phi^\dagger \widetilde{\Phi}) - \mu_3^2 \text{Tr}(\Delta_L^\dagger \Delta_L + \Delta_R^\dagger \Delta_R) \\
&+ \lambda_1 [\text{Tr}(\Phi^\dagger \Phi)]^2 + \lambda_2 \{[\text{Tr}(\widetilde{\Phi}^\dagger \Phi)]^2 + [\text{Tr}(\Phi^\dagger \widetilde{\Phi})]^2\} + \lambda_3 \text{Tr}(\widetilde{\Phi}^\dagger \Phi)\text{Tr}(\Phi^\dagger \widetilde{\Phi}) \\
&+ \lambda_4 \text{Tr}(\Phi^\dagger \Phi)\text{Tr}(\widetilde{\Phi}^\dagger \Phi + \Phi^\dagger \widetilde{\Phi}) + \rho_1 \{[\text{Tr}(\Delta_L^\dagger \Delta_L)]^2 + [\text{Tr}(\Delta_R^\dagger \Delta_R)]^2\} \\
&+ \rho_2 \{\text{Tr}(\Delta_L^\dagger \Delta_L^\dagger)\text{Tr}(\Delta_L \Delta_L) + \text{Tr}(\Delta_R^\dagger \Delta_R^\dagger)\text{Tr}(\Delta_R \Delta_R)\} + \rho_3 \{\text{Tr}(\Delta_L^\dagger \Delta_L) \\
&\times \text{Tr}(\Delta_R^\dagger \Delta_R)\} + \rho_4 \{\text{Tr}(\Delta_L \Delta_L)\text{Tr}(\Delta_R^\dagger \Delta_R^\dagger) + \text{Tr}(\Delta_R \Delta_R)\text{Tr}(\Delta_L^\dagger \Delta_L^\dagger)\} \\
&+ \alpha_1 \text{Tr}(\Phi \Phi^\dagger)[\text{Tr}(\Delta_L^\dagger \Delta_L + \Delta_R^\dagger \Delta_R)] + \alpha_2 \{\text{Tr}(\Phi^\dagger \widetilde{\Phi})\text{Tr}(\Delta_L^\dagger \Delta_L) \\
&+ \text{Tr}(\Phi \widetilde{\Phi}^\dagger)\text{Tr}(\Delta_R^\dagger \Delta_R)\} + \alpha_2^* \{\text{Tr}(\Phi^\dagger \widetilde{\Phi})\text{Tr}(\Delta_R^\dagger \Delta_R) + \text{Tr}(\Phi \widetilde{\Phi}^\dagger)\text{Tr}(\Delta_L^\dagger \Delta_L)\} \\
&+ \alpha_3 \text{Tr}(\Phi \Phi^\dagger \Delta_L \Delta_L^\dagger + \Phi^\dagger \Phi \Delta_R \Delta_R^\dagger) + \beta_1 \text{Tr}(\Delta_L^\dagger \Phi \Delta_R \Phi^\dagger + \Delta_R^\dagger \Phi^\dagger \Delta_L \Phi) \\
&+ \beta_2 \text{Tr}(\Delta_L^\dagger \widetilde{\Phi} \Delta_R \Phi^\dagger + \Delta_R^\dagger \widetilde{\Phi}^\dagger \Delta_L \Phi)\} + \beta_3 \text{Tr}(\Delta_L^\dagger \Phi \Delta_R \widetilde{\Phi}^\dagger + \Delta_R^\dagger \Phi^\dagger \Delta_L \widetilde{\Phi})\},
\end{aligned} \tag{4.16}$$

where the conjugate Higgs bidoublet is defined as

$$\widetilde{\Phi} = \sigma_2 \Phi^* \sigma_2 = \begin{pmatrix} \Phi_2^{0*} & -\Phi_2^+ \\ -\Phi_1^- & \Phi_1^{0*} \end{pmatrix}, \tag{4.17}$$

which transforms in the same way as Φ.

Finally, the Yukawa interactions responsible for fermion masses are given by:

$$\begin{aligned}
\mathcal{L}_Y = \overline{Q}_L \left(Y_q \Phi + \tilde{Y}_q \tilde{\Phi} \right) Q_R + \overline{L}_L \left(Y_\ell \Phi + \tilde{Y}_\ell \tilde{\Phi} \right) L_R \\
+ f_L L_L^T C i \sigma_2 \Delta_L L_L + f_R L_R^T C i \sigma_2 \Delta_R L_R + \text{h.c.}
\end{aligned} \tag{4.18}$$

Here, $Y_q, \tilde{Y}_q, Y_\ell, \tilde{Y}_\ell$ are Dirac-type Yukawa matrices for quarks and leptons, and f_L, f_R are symmetric Majorana-type couplings for neutrinos. After symmetry breaking, the Lagrangian generates:

- Dirac masses for charged fermions and neutrinos via $\langle \Phi \rangle$,
- Majorana masses for neutrinos via $\langle \Delta_{L,R} \rangle$,

naturally realizing both the type-I and type-II seesaw mechanisms [20]. The LRSM thus provides a compelling framework for addressing open questions in the SM, including the origin of neutrino masses, the nature of parity violation, and new physics at the TeV scale [77].

4.2 SPONTANEOUS SYMMETRY BREAKING IN THE LRSM

The gauge symmetries imply that all gauge bosons and fermions must be massless, thus the symmetry with massive gauge bosons (short-range forces like the weak interaction) must be spontaneously broken. The spontaneous symmetry breaking (SSB) in the LRSM occurs in two stages, and the mechanism is similar to that of the SM. Specifically, it is achieved by the VEV of scalar Higgs multiplets. The symmetry breaking scheme is as follows:

$$SU(2)_L \times SU(2)_R \times U(1)_{B-L} \xrightarrow{\langle \Delta_R \rangle} SU(2)_L \times U(1)_Y, \tag{4.19}$$

at this stage, only the RH scalar Higgs triplet gets a VEV and breaks the LR symmetry down to the electroweak (EW) symmetry. After this stage of symmetry breaking, the physical $W_R^{\pm}$ and Z_R gauge bosons acquire masses through interactions with the right-handed triplet Higgs. Additionally, the hypercharge operator Y appears as a linear combination of the difference between baryon and lepton numbers and the third component of the right-handed isospin:

$$Y = T_R^3 + Q_{B-L}, \tag{4.20}$$

where Q_{B-L} is the generator of the $U(1)_{B-L}$ group, as defined in Eq. (4.2). The values of the new gauge couplings g_R and g_{B-L} are related to the SM hypercharge coupling g_Y by:

$$g_Y = \frac{g_R g_{B-L}}{\sqrt{g_R^2 + g_{B-L}^2}}. \tag{4.21}$$

From the definition of the hypercharge, it follows that $Y(\Delta_R^0) = 0$, ensuring that hypercharge remains an exact symmetry. Note that Δ_R is a singlet under the $SU(2)_L$ gauge group, so it remains an exact symmetry at this stage as well.

The second stage of symmetry breaking further reduces the symmetry, breaking the electroweak group down to electromagnetism, just like in the SM but with an extended Higgs sector. The breaking is achieved by the neutral components of the bidoublet Higgs Φ and the left-handed triplet Higgs Δ_L, which acquire VEVs, as follows:

$$SU(2)_L \times U(1)_Y \xrightarrow{\langle \Phi \rangle, \langle \Delta_L \rangle} U(1)_{\text{EM}}. \tag{4.22}$$

Here the neutral components of the bidoublet Higgs and LH triplet get VEVs and break the EW symmetry down to EM. After this stage of symmetry breaking, physical $W^{\pm}_{L\mu}$ and $Z_{L\mu}$ gauge bosons acquire masses. Without loss of generality, the VEVs of the neutral components of the Higgs fields can be written as

$$\langle\Phi\rangle = \frac{1}{\sqrt{2}}\begin{pmatrix} k_1 & 0 \\ 0 & k_2 \end{pmatrix}, \quad \langle\Delta_R\rangle = \frac{1}{\sqrt{2}}\begin{pmatrix} 0 & 0 \\ v_R & 0 \end{pmatrix}, \quad \langle\Delta_L\rangle = \frac{1}{\sqrt{2}}\begin{pmatrix} 0 & 0 \\ v_L & 0 \end{pmatrix}. \tag{4.23}$$

The minimization conditions of the scalar potential $V(k_1, k_2, v_L, v_R)$ with respect to the neutral VEVs yield four equations from the derivatives with respect to k_1, k_2, v_L, and v_R. Solving these, one obtains the following relations among the potential parameters and VEVs:

$$\beta_3 = -\frac{\beta_2 k_1^2 + \beta_1 k_1 k_2 - \rho_{\bar{13}} v_L v_R}{k_2^2}, \tag{4.24}$$

$$\mu_3^2 = \frac{1}{2}\left(\alpha_1 k_1^2 + 4\alpha_2 k_1 k_2 + \alpha_{13} k_2^2 + 2\rho_1 (v_L^2 + v_R^2)\right), \tag{4.25}$$

$$v_L v_R = \frac{\beta_2 k_1^2 + \beta_1 k_1 k_2 + \beta_3 k_2^2}{\rho_{\bar{13}}}, \tag{4.26}$$

$$v_L^2 + v_R^2 = \frac{1}{2\rho_1}\left(2\mu_3^2 - \alpha_1 k_1^2 - 4\alpha_2 k_1 k_2 - \alpha_{13} k_2^2\right), \tag{4.27}$$

where the shorthand notations are defined as:

$$\alpha_{13} = \alpha_1 + \alpha_3, \quad \rho_{\bar{13}} = 2\rho_1 - \rho_3, \quad v_+^2 = v_L^2 + v_R^2, \quad v_\times^2 = v_L v_R.$$

The expressions for the mass parameters μ_1^2 and μ_2^2 are more involved and are omitted for brevity. In the phenomenologically relevant limit where $v_L \ll k_{1,2} \ll v_R$, the SM VEV is given by

$$\sqrt{k_1^2 + k_2^2} \equiv v \simeq 246 \text{ GeV}.$$

4.2.1 Gauge Boson Masses in the LRSM

After SSB, the gauge bosons masses are generated as follows, from Higgs kinetic term $\text{Tr}|D_\mu \langle\Delta_R\rangle|^2$ one obtains the mass terms:

$$\frac{g_R^2 v_R^2}{2} W_R^{\mu-} W_{R\mu}^+ + \frac{v_R^2}{2}(g_R W_R^{\mu 3} - g_{BL} V^\mu)(g_R W_{R\mu}^3 - g_{BL} V_\mu) \tag{4.28}$$

Defining the mass eigenstates via:

$$W_R^{\pm} = \frac{W_R^1 \mp i W_R^2}{\sqrt{2}}, \quad \begin{pmatrix} Z_R \\ B \end{pmatrix} = \begin{pmatrix} \cos\varphi & -\sin\varphi \\ \sin\varphi & \cos\varphi \end{pmatrix}\begin{pmatrix} W_R^3 \\ V \end{pmatrix}, \tag{4.29}$$

where the mixing angle φ is defined as:

$$\sin\varphi = \frac{g_{BL}}{\sqrt{g_R^2 + g_{BL}^2}}, \qquad \cos\varphi = \frac{g_R}{\sqrt{g_R^2 + g_{BL}^2}}, \tag{4.30}$$

the massless field B is identified with the hypercharge gauge boson, while Z_R is a heavy neutral boson that decouples at low energies.

In the second stage, the electroweak symmetry is broken by the bidoublet Higgs Φ acquiring a VEV $\langle\Phi\rangle = \text{diag}(k_1, k_2)$ and the left-handed triplet Δ_L developing a small VEV $\langle\Delta_L\rangle = v_L$. The kinetic terms yield masses for the SM gauge bosons:

$$\begin{aligned}\text{Tr}|D_\mu\langle\Phi\rangle|^2 &\supset \frac{v^2}{4}(g_L^2 W_L^+ W_L^- + g_R^2 W_R^+ W_R^-) - \frac{g_L g_R k_1 k_2}{2}(W_L^+ W_R^- + \text{h.c.}) \\ &\quad + \frac{v^2}{8}(g_L W_L^3 - g_Y B)^2, &(4.31)\\ \text{Tr}|D_\mu\langle\Delta_L\rangle|^2 &\supset \frac{g_L^2 v_L^2}{2} W_L^+ W_L^- + \frac{v_L^2}{2}(g_L W_L^3 - g_Y B)^2, &(4.32)\end{aligned}$$

where g_Y is the hypercharge coupling:

$$g_Y = \frac{g_R g_{BL}}{\sqrt{g_R^2 + g_{BL}^2}}. \tag{4.33}$$

Neutral gauge bosons in the LRSM

After diagonalization, the neutral gauge bosons mix via the Weinberg angle θ_W as in the Standard Model:

$$\begin{pmatrix} Z_L \\ A \end{pmatrix} = \begin{pmatrix} \cos\theta_W & -\sin\theta_W \\ \sin\theta_W & \cos\theta_W \end{pmatrix} \begin{pmatrix} W_L^3 \\ B \end{pmatrix}, \tag{4.34}$$

with

$$\sin\theta_W = \frac{g_Y}{\sqrt{g_L^2 + g_Y^2}}, \qquad \cos\theta_W = \frac{g_L}{\sqrt{g_L^2 + g_Y^2}}. \tag{4.35}$$

The full neutral gauge boson mass matrix in the basis (W_L^3, W_R^3, V) is given by [20]:

$$M_{V^0}^2 = \frac{1}{4}\begin{pmatrix} g_L^2(4v_L^2 + v^2) & -g_L g_R v^2 & -4 g_L g_{BL} v_L^2 \\ -g_L g_R v^2 & g_R^2(4v_R^2 + v^2) & -4 g_R g_{BL} v_R^2 \\ -4 g_L g_{BL} v_L^2 & -4 g_R g_{BL} v_R^2 & g_{BL}^2(4v_L^2 + 4v_R^2) \end{pmatrix}. \tag{4.36}$$

This matrix can be diagonalized to extract the masses of the physical bosons (Z_L, A, Z_R) by

$$\begin{pmatrix} W^3_{L\mu} \\ W^3_{R\mu} \\ V_\mu \end{pmatrix} = \begin{pmatrix} \frac{e}{g_Y} & \frac{e}{g_L} & 0 \\ \frac{-eg_Y}{g_L\, g_R} & \frac{e}{g_R} & \frac{g_Y}{g_{BL}} \\ \frac{-eg_Y}{g_L\, g_{BL}} & \frac{e}{g_{BL}} & \frac{-g_Y}{g_R} \end{pmatrix} \begin{pmatrix} Z_{L\mu} \\ A_\mu \\ Z_{R\mu} \end{pmatrix}. \tag{4.37}$$

The bidoublet Φ induces mixing between the neutral gauge bosons Z_L and Z_R, resulting in the mass-squared matrix in the basis $\{Z_{L\mu}, Z_{R\mu}\}$:

$$M^2_{Z_L Z_R} = \begin{pmatrix} M^0_{LL} & M^0_{LR} \\ M^0_{LR} & M^0_{RR} \end{pmatrix}, \tag{4.38}$$

where the matrix elements depend on the gauge couplings (g_L, g_R, g_{BL}) and the VEVs $(k_{1,2}, v_{L,R})$. The physical states Z and Z' are obtained via a mixing angle ϑ,

$$\tan 2\vartheta = \frac{2M^0_{LR}}{M^0_{LL} - M^0_{RR}}, \qquad \begin{pmatrix} Z \\ Z' \end{pmatrix} = \begin{pmatrix} \cos\vartheta & \sin\vartheta \\ -\sin\vartheta & \cos\vartheta \end{pmatrix} \begin{pmatrix} Z_L \\ Z_R \end{pmatrix}, \tag{4.39}$$

with corresponding eigenvalues:

$$M^2_{Z,Z'} = \frac{1}{2}\left(M^0_{LL} + M^0_{RR} \mp \sqrt{(M^0_{LL} - M^0_{RR})^2 + 4(M^0_{LR})^2} \right). \tag{4.40}$$

In the limit $v_R \gg v$, one finds $Z \simeq Z_L$ and $Z' \simeq Z_R$. Experimental constraints require $\vartheta \lesssim 10^{-3}$ and $M_{Z'} \gtrsim 2$ TeV [78].

Charged Gauge Bosons in the LRSM

The bidoublet Φ, transforming under both $SU(2)_L \times SU(2)_R$, induces mixing between $W_L^\pm$ and $W_R^\pm$ through the mass matrix:

$$M_{V^\pm} = \frac{1}{4} \begin{pmatrix} g_L^2(k_1^2 + k_2^2 + 2v_L^2) & -2g_L g_R k_1 k_2 \\ -2g_L g_R k_1 k_2 & g_R^2(k_1^2 + k_2^2 + 2v_R^2) \end{pmatrix}. \tag{4.41}$$

The physical states $W_1^\pm$ and $W_2^\pm$ arise via an orthogonal rotation:

$$\begin{pmatrix} W_1^\pm \\ W_2^\pm \end{pmatrix} = \begin{pmatrix} \cos\xi & e^{-i\omega}\sin\xi \\ -\sin\xi & e^{-i\omega}\cos\xi \end{pmatrix} \begin{pmatrix} W_L^\pm \\ W_R^\pm \end{pmatrix}, \tag{4.42}$$

where ξ is the mixing angle, constrained to $\xi < 10^{-3}$ [79].

Defining $k_1 = v\cos\beta$, $k_2 = v\sin\beta$ with $v = 246$ GeV, the trace and determinant of the matrix are:

$$T^{g\pm} = \frac{1}{4}\left((g_L^2 + g_R^2)v^2 + 2(g_L^2 v_L^2 + g_R^2 v_R^2)\right), \tag{4.43}$$

$$D^{g\pm} = \frac{1}{16} g_L^2 g_R^2 \left[(v^2 + 2v_L^2)(v^2 + 2v_R^2) - 4v^4 \sin^2\beta\cos^2\beta\right]. \tag{4.44}$$

The masses are given by:

$$M_{W_1,W_2}^2 = \frac{1}{2}\left(T^{g\pm} \mp \sqrt{(T^{g\pm})^2 - 4D^{g\pm}}\right), \tag{4.45}$$

with

$$\tan 2\xi = \frac{2M_{LR}}{M_{LL} - M_{RR}} \approx \frac{4g_{RL}k_1k_2}{2g_{RL}^2 v_R^2 + (g_{RL}^2 - 1)v^2}$$

and

$$g_{RL} \equiv g_R/g_L.$$

In the limit $\xi \to 0$, the mass eigenstates reduce to:

$$M_{W_L} = \frac{vg_L}{2}, \quad M_{W_R} = \frac{v_R g_R}{\sqrt{2}}. \tag{4.46}$$

4.2.2 Fermion Masses in the LRSM

Fermion masses arise from Yukawa couplings to the bidoublets Φ, $\widetilde{\Phi}$, and the triplets $\Delta_{L,R}$:

$$-\mathcal{L}_{\text{Yukawa}} \supset \overline{L}_L(y^L\langle\Phi\rangle + \widetilde{y}^L\langle\widetilde{\Phi}\rangle)L_R + \overline{Q}_L(y^Q\langle\Phi\rangle + \widetilde{y}^Q\langle\widetilde{\Phi}\rangle)Q_R \\ + iy^M\left(L_L^T C\sigma^2\langle\Delta_L\rangle L_L + L_R^T C\sigma^2\langle\Delta_R\rangle L_R\right) + \text{h.c.} \tag{4.47}$$

The Dirac mass matrices for quarks and leptons are given by:

$$M_u = \frac{1}{\sqrt{2}}(y^Q k_1 + \widetilde{y}^Q k_2), \qquad M_d = \frac{1}{\sqrt{2}}(y^Q k_2 + \widetilde{y}^Q k_1), \tag{4.48}$$

$$M_\nu = \frac{1}{\sqrt{2}}(y^L k_1 + \widetilde{y}^L k_2), \qquad M_e = \frac{1}{\sqrt{2}}(y^L k_2 + \widetilde{y}^L k_1), \tag{4.49}$$

while the Majorana mass matrices for the left- and right-handed neutrinos are:

$$M_{\nu_L} = \frac{1}{\sqrt{2}} y^M v_L, \quad M_{\nu_R} = \frac{1}{\sqrt{2}} y^M v_R. \tag{4.50}$$

The charged fermion mass matrices can be diagonalized using biunitary transformations:

$$V_L^{u\dagger} M_u V_R^u = M_u^{\text{diag}}, \quad V_L^{d\dagger} M_d V_R^d = M_d^{\text{diag}}, \quad V_L^{e\dagger} M_e V_R^e = M_e^{\text{diag}}. \tag{4.51}$$

Neutrino mass terms receive contributions from both Dirac (y^L, $\widetilde{y}^L$) and Majorana (y^M) couplings. It is convenient to express them using self-conjugate Majorana spinors [20, 22, 76]:

$$\nu = \frac{\nu_L + \nu_L^c}{\sqrt{2}}, \quad N = \frac{\nu_R^c + \nu_R}{\sqrt{2}}. \tag{4.52}$$

We also define the Dirac Yukawa coupling as:

$$y_D \equiv \frac{1}{\sqrt{2}} \frac{y^L k_1 + \widetilde{y}^L k_2}{v} = \frac{M_\nu}{v}, \tag{4.53}$$

which governs the magnitude of the Dirac-type neutrino mass.

In terms of y_D and y^M, the full neutrino mass matrix in the (ν, N) basis becomes:

$$\begin{pmatrix} \overline{\nu} & \overline{N} \end{pmatrix} \begin{pmatrix} \sqrt{2} y^M v_L & y_D v \\ y_D v & \sqrt{2} y^M v_R \end{pmatrix} \begin{pmatrix} \nu \\ N \end{pmatrix}. \tag{4.54}$$

For natural values of Yukawa couplings $y^M \sim y_D \sim M_e/v$, and the hierarchy $v_R \gg v \gg v_L$ (required for $M_{W_R} \gg M_{W_L}$), the light and heavy neutrino states become approximate flavor eigenstates with masses given by the seesaw mechanism [76]:

$$m_N \simeq \sqrt{2} y^M v_R, \quad m_\nu \simeq \sqrt{2} \left(y^M v_L - \frac{v^2}{2 v_R} y_D (y^M)^{-1} y_D^T \right). \tag{4.55}$$

Finally, the light and heavy neutrino mass matrices m_ν and m_N are diagonalized by unitary matrices V_L^ν and V_R^N:

$$V_L^{\nu\dagger} m_\nu V_L^\nu = m_\nu^{\text{diag}}, \quad V_R^{N\dagger} m_N V_R^N = m_N^{\text{diag}}. \tag{4.56}$$

4.3 ELECTROWEAK INTERACTIONS IN THE LRSM

In the LRSM, the electroweak interactions arise from the covariant derivatives acting on fermion fields in the kinetic Lagrangian:

$$\begin{aligned}\mathcal{L}_{\text{kin}} &= i\sum_{j=1}^{3}\Big(\bar{L}_{Lj}\gamma^{\mu}D_{\mu}L_{Lj}+\bar{L}_{Rj}\gamma^{\mu}D_{\mu}L_{Rj}+\bar{Q}_{Lj}\gamma^{\mu}D_{\mu}Q_{Lj}\\ &\quad+\bar{Q}_{Rj}\gamma^{\mu}D_{\mu}Q_{Rj}\Big), \end{aligned}\tag{4.57}$$

where $L_{L,R}$ and $Q_{L,R}$ are the left- and right-handed lepton and quark doublets, respectively. The electric charge of any fermion in LRSM is given by:

$$Q = T_L^3 + T_R^3 + \frac{B-L}{2}, \tag{4.58}$$

with $T_{L,R}^3$ being the third generators of $SU(2)_{L,R}$ and $B-L$ the baryon-minus-lepton number. This leads to the correct embedding of electromagnetic and weak interactions [75].

After symmetry breaking, the gauge bosons mix to yield the photon, the SM-like Z, a heavier Z', and the charged bosons $W_L^{\pm}$ and $W_R^{\pm}$. The gauge interactions of fermions then include:

- Charged currents: mediated by $W_L^{\pm}$ and $W_R^{\pm}$, involving left- and right-handed fermions respectively, with mixing encoded in CKM and RH-CKM matrices.
- Neutral currents: mediated by Z, Z', and the photon A, coupling to combinations of T_L^3, T_R^3, and $B-L$.

Summarizing the weak interaction terms:

$$\begin{aligned}\mathcal{L}_{\text{weak}} \supset \bar{\psi}\gamma^{\mu}\left[\frac{g_L}{\sqrt{2}}\left(W_L^+T_L^+ + W_L^-T_L^-\right)P_L + \frac{g_R}{\sqrt{2}}\left(W_R^+T_R^+ + W_R^-T_R^-\right)P_R\right]\psi \\ +\bar{\psi}\gamma^{\mu}\left[g_L T_L^3 W_L^3 + g_R T_R^3 W_R^3 + g_{BL}\frac{B-L}{2}V_{\mu}\right]\psi, \end{aligned}\tag{4.59}$$

where ψ represents any SM fermion field. After diagonalization, these interactions yield the familiar electromagnetic coupling and the neutral/charged weak currents, now extended by right-handed counterparts [22].

4.4 HIGGS SECTOR IN THE LRSM

The scalar content of the LRSM consists of a complex bidoublet Φ and two triplets $\Delta_{L,R}$, all complex. As emphasized above, the neutral scalar components are expanded around their VEVs as:

$$\begin{aligned}\Phi_1^0 &= \frac{1}{\sqrt{2}}(k_1 + h_1^0 + i\varphi_1^0),\\ \Phi_2^0 &= \frac{1}{\sqrt{2}}(k_2 + h_2^0 + i\varphi_2^0),\\ \Delta_{L,R}^0 &= \frac{1}{\sqrt{2}}(v_{L,R} + h_{L,R}^0 + i\varphi_{L,R}^0).\end{aligned} \tag{4.60}$$

The mass spectrum of the physical Higgs scalar states is summarized as follows [76]:

Doubly-Charged Scalars

In the basis $(\Delta_L^{\pm\pm}, \Delta_R^{\pm\pm})$, the doubly-charged mass matrix is a 2×2 symmetric matrix $M^{\pm\pm}$, whose eigenvalues are:

$$m_{1,2\pm\pm}^2 = \frac{1}{2}\left(\text{Tr}(M^{\pm\pm}) \pm \sqrt{[\text{Tr}(M^{\pm\pm})]^2 - 4\det(M^{\pm\pm})}\right). \tag{4.61}$$

The physical states are obtained via a unitary transformation Z^D:

$$Z^{D\dagger} M^{\pm\pm} Z^D = \text{diag}(m_{1\pm\pm}^2, m_{2\pm\pm}^2). \tag{4.62}$$

Singly-Charged Scalars

In the basis $(\phi_1^\pm, \phi_2^\pm, \Delta_L^\pm, \Delta_R^\pm)$, the singly-charged mass matrix $M^\pm$ is 4×4 and has two physical eigenstates and two Goldstone modes. The non-zero eigenvalues are:

$$m_{1,2\pm}^2 = \frac{1}{2}\left(T^\pm \pm \sqrt{(T^\pm)^2 - 4D_2^\pm}\right), \tag{4.63}$$

where $T^\pm = \text{Tr}(M^\pm)$ and $D_2^\pm$ is the second principal invariant: $D_2 = \frac{1}{2}\left[(\text{Tr}(M^\pm))^2 - \text{Tr}((M^\pm)^2)\right]$. The matrix is diagonalized by $Z^\pm$:

$$Z^{\pm\dagger} M^\pm Z^\pm = \text{diag}(m_{1\pm}^2, m_{2\pm}^2, 0, 0). \tag{4.64}$$

with two physical charged states and two Goldstone modes. The eigenvalues follow:

Neutral Pseudoscalars

The pseudoscalar sector consists of four fields $(\varphi_1^0, \varphi_2^0, \varphi_L^0, \varphi_R^0)$ forming a 4×4 symmetric mass matrix. Two eigenstates correspond to Goldstone bosons; the other two are physical pseudoscalars with masses determined analogously.

Neutral Scalars

The CP-even neutral scalars form a 4×4 real symmetric mass matrix in the basis

$$S_i^{0^t} = (\phi_1^{\text{re}}, \phi_2^{\text{re}}, \Delta_L^{\text{re}}, \Delta_R^{\text{re}}),$$

with entries determined by the scalar potential parameters and VEVs $k_{1,2}, v_{L,R}$. The mass eigenstates are obtained by diagonalizing the matrix via a unitary transformation:

$$Z^{S\dagger} M^H Z^S = \text{diag}(m_1^2, m_2^2, m_3^2, m_4^2),$$

where one eigenstate corresponds to the SM-like Higgs boson with $m_h \approx$ 125 GeV. The characteristic equation is quartic and solvable analytically [80], allowing constraints on model parameters.

4.5 HIGGS INTERACTIONS IN THE LRSM

The kinetic terms of the Higgs fields in the LRSM encode their interactions with the gauge bosons via the covariant derivatives. For the bidoublet field Φ, the covariant derivative is given by

$$D_\mu \Phi = \partial_\mu \Phi - i g_{_L} (W_{L\mu} \cdot T_L) \Phi + i g_{_R} \Phi (W_{R\mu} \cdot T_R), \tag{4.65}$$

where $W_{L\mu}^a$ and $W_{R\mu}^a$ are the gauge fields of $SU(2)_L$ and $SU(2)_R$, respectively. In terms of the physical gauge bosons $W_{L,R}^\pm, Z_L, Z_R$, and A, the derivative becomes

$$\begin{aligned} D_\mu \Phi = \partial_\mu \Phi &- i \frac{g_L}{\sqrt{2}} \left(W_{L\mu}^+ T_L^+ + W_{L\mu}^- T_L^- \right) \Phi \\ &+ i \frac{g_R}{\sqrt{2}} \Phi \left(W_{R\mu}^+ T_R^+ + W_{R\mu}^- T_R^- \right) - ie \left(T_L^3 \Phi - \Phi T_R^3 \right) A_\mu \\ &- i Z_{L\mu} \left(\frac{g_L}{\cos\theta_W} T_L^3 \Phi + \frac{g_Y}{\tan\varphi} \Phi T_R^3 \right) + i \frac{g_Y}{\tan\varphi} \Phi T_R^3 Z_{R\mu}. \end{aligned} \tag{4.66}$$

Expanding the Higgs fields in components allows extraction of the interaction terms between the physical gauge bosons and scalar fields, which are essential for phenomenological analyses.

In the LRSM, the bidoublet Φ couples to quarks through Yukawa interactions:

$$\mathcal{L}_Y \supset \overline{Q}_L(y^Q\Phi + \widetilde{y}^Q\widetilde{\Phi})Q_R + \text{h.c.} \tag{4.67}$$

After symmetry breaking, the Yukawa couplings are expressed in terms of the physical quark masses:

$$y^Q = \frac{\sqrt{2}}{k_1^2 - k_2^2}(k_1 V_L^u M_u^{\text{diag}} V_R^{u\dagger} - k_2 V_L^d M_d^{\text{diag}} V_R^{d\dagger}), \tag{4.68}$$

$$\widetilde{y}^Q = \frac{\sqrt{2}}{k_1^2 - k_2^2}(-k_2 V_L^u M_u^{\text{diag}} V_R^{u\dagger} + k_1 V_L^d M_d^{\text{diag}} V_R^{d\dagger}). \tag{4.69}$$

This leads to flavor-changing couplings between the neutral components of Φ and the mass eigenstate quarks:

$$\mathcal{L}_Y \supset \overline{u}_L(y^Q\Phi_1^0 + \widetilde{y}^Q\Phi_2^{0*})u_R + \overline{d}_L(y^Q\Phi_2^0 + \widetilde{y}^Q\Phi_1^{0*})d_R + \text{h.c.} \tag{4.70}$$

With the scalar and pseudoscalar fields rotated to their mass eigenstates, the quark couplings are given by:

$$\mathcal{L}_Y \supset \overline{u}_i\left(\lambda_{ijk}^{H\overline{U}U} H_k + i\lambda_{ijk}^{A\overline{U}U} A_k\gamma^5\right)u_j + \overline{d}_i\left(\lambda_{ijk}^{H\overline{D}D} H_k - i\lambda_{ijk}^{A\overline{D}D} A_k\gamma^5\right)d_j, \tag{4.71}$$

with couplings

$$\lambda_{ijk}^{H\overline{U}U} = \frac{k_1 Z_{S,1k} - k_2 Z_{S,2k}}{k_1^2 - k_2^2} M_{u_i}\delta_{ij} + \frac{-k_2 Z_{S,1k} + k_1 Z_{S,2k}}{k_1^2 - k_2^2}\sum_l V_{il}^L M_{d_l} V_{jl}^{R*}, \tag{4.72}$$

$$\lambda_{ijk}^{H\overline{D}D} = \frac{k_1 Z_{S,1k} - k_2 Z_{S,2k}}{k_1^2 - k_2^2} M_{d_i}\delta_{ij} + \frac{-k_2 Z_{S,1k} + k_1 Z_{S,2k}}{k_1^2 - k_2^2}\sum_l V_{li}^{L*} M_{u_l} V_{lj}^{R}, \tag{4.73}$$

$$\lambda_{ijk}^{A\overline{U}U} = \frac{k_1 Z_{P,1k} + k_2 Z_{P,2k}}{k_1^2 - k_2^2} M_{u_i}\delta_{ij} - \frac{k_2 Z_{P,1k} + k_1 Z_{P,2k}}{k_1^2 - k_2^2}\sum_l V_{il}^L M_{d_l} V_{jl}^{R*}, \tag{4.74}$$

$$\lambda_{ijk}^{A\overline{D}D} = \frac{k_1 Z_{P,1k} + k_2 Z_{P,2k}}{k_1^2 - k_2^2} M_{d_i}\delta_{ij} - \frac{k_2 Z_{P,1k} + k_1 Z_{P,2k}}{k_1^2 - k_2^2}\sum_l V_{li}^{L*} M_{u_l} V_{lj}^{R}. \tag{4.75}$$

The off-diagonal terms in these couplings are induced by the nontrivial structure of the mixing matrices V^L and V^R, leading to tree-level flavor-changing neutral currents (FCNC). These interactions are tightly constrained by experimental limits on FCNC processes [76]. A similar mechanism applies in the leptonic sector due to the non-diagonality of the PMNS matrix.

4.6 PROBLEMS IN THE LRSM

While the LRSM elegantly addresses several shortcomings of the SM, such as restoring parity symmetry at high energies and explaining non-zero neutrino masses through a natural seesaw mechanism, it also faces significant theoretical and phenomenological challenges:

- Doubly Charged Higgs Bosons: The scalar triplet fields $\Delta_{L,R}$ introduced to break the LR symmetry and implement the seesaw mechanism predict the existence of doubly charged Higgs bosons, $H^{\pm\pm}$. These exotic states have yet to be observed at collider experiments, such as the LHC, and current bounds place stringent constraints on their masses and couplings. The absence of detection challenges the minimal LRSM or pushes the relevant scale beyond current experimental reach.

- Fine-Tuning of the Left-Handed Triplet VEV: To generate light neutrino masses via the type-I or type-II seesaw mechanism, the VEV of the LH triplet, v_L, must be highly suppressed ($v_L \lesssim \mathcal{O}(1 \text{ eV})$). However, the scalar potential naturally induces a non-zero v_L through terms proportional to the RH triplet VEV, v_R. Achieving a sufficiently small v_L requires delicate fine-tuning of parameters in the scalar potential, raising concerns about naturalness and stability.

- Extended Higgs Sector: The minimal LRSM contains a large number of scalar fields: a bidoublet Φ and two triplets Δ_L and Δ_R. This rich scalar content leads to a complicated scalar potential with many parameters and mass eigenstates. Diagonalizing the full scalar mass matrices and analyzing the physical spectrum and couplings becomes a technically challenging task. Moreover, the mixing among scalars complicates the phenomenology and makes it harder to identify SM-like Higgs signatures.

- Tree-Level FCNC: In the minimal LRSM, both the up-type and down-type quark mass matrices receive contributions from two Yukawa couplings due to the bidoublet Φ, leading to a non-diagonal structure in the physical basis. As a result, the Higgs-mediated neutral current interactions are not flavor-diagonal at tree level. This gives rise to FCNC processes such as $K^0 - \bar{K}^0$ and $B^0 - \bar{B}^0$ mixing, which are tightly constrained by experiment. Suppressing these FCNCs either requires fine-tuning the Yukawa matrices or introducing additional symmetries or mechanisms.

Due to these challenges, various extensions and modifications of the LRSM have been proposed. In the following chapter, we explore one such alternative: the LRSM with an inverse seesaw mechanism, which alleviates some of the aforementioned problems while preserving the appealing features of LR symmetry.

CHAPTER 5

Doublet Left-Right Symmetric Model

In the previous chapter, it was emphasized that the conventional LRSM suffers from a non-minimal Higgs sector, typically involving a bidoublet and two triplet scalar fields. This setup complicates the model and often leads to tensions with low-scale phenomenology. An alternative approach is to consider a simplified Higgs sector consisting of a single bidoublet and one $SU(2)_R$ doublet Higgs field, with the possible inclusion of an $SU(2)_L$ doublet in certain extensions. This minimal configuration defines the *Doublet LRSM*, which offers a more economical structure and improved compatibility with experimental constraints [81, 82].

In this framework, neutrino masses are generated via the Inverse Seesaw (IS) or Linear Seesaw (LS) mechanisms, as discussed in Chapter 1.

5.1 MODEL STRUCTURE

The DLRSM is based on the extended gauge group:

$$SU(3)_C \times SU(2)_L \times SU(2)_R \times U(1)_{B-L}.$$

DOI: 10.1201/9781003457701-5

The fermion content mirrors that of the conventional LRSM, with left- and RH doublets for both quarks and leptons:

$$Q_L = \begin{pmatrix} u_L \\ d_L \end{pmatrix} \sim (3,2,1,1/3), \qquad Q_R = \begin{pmatrix} u_R \\ d_R \end{pmatrix} \sim (3,1,2,1/3),$$

$$L_L = \begin{pmatrix} \nu_L \\ e_L \end{pmatrix} \sim (1,2,1,-1), \qquad L_R = \begin{pmatrix} \nu_R \\ e_R \end{pmatrix} \sim (1,1,2,-1). \quad (5.1)$$

Parity symmetry is manifest at high energies, implying invariance of the Lagrangian under the exchange $Q_L \leftrightarrow Q_R$, $L_L \leftrightarrow L_R$. Notably, the presence of the RH neutrino ν_R is required for the completeness of the lepton doublet.

To implement the IS mechanism, the model introduces three SM singlet fermions S_1 with $B-L=-2$ and three singlets S_2 with $B-L=+2$. This pairwise inclusion ensures cancellation of $U(1)_{B-L}$ anomalies [72]. To avoid a direct Majorana mass term $M\bar{S}_1^c S_2$, which would spoil the structure of the IS, a discrete Z_2 symmetry is imposed, under which all fields are even except S_1, which is odd. Here, we consider the following Higgs sector consists of:

- A scalar bidoublet ϕ, responsible for EW symmetry breaking,
- An $SU(2)_R$ scalar doublet χ_R, which breaks $SU(2)_R \times U(1)_{B-L}$ down to $U(1)_Y$.

The hypercharge is defined via the relation:

$$Y = 2T_R^3 + (B-L),$$

where T_R^3 is the third component of the $SU(2)_R$ isospin. The full field content and their quantum numbers under the gauge group are summarized in Table 5.1.

This LR extension with IS, henceforth referred to as LRIS, leads to rich phenomenological implications, including TeV-scale RH gauge bosons and observable lepton flavor violation, which have been studied in Refs. [72,81–84]. The symmetry breaking in the LRIS model proceeds

TABLE 5.1 Field content of the DLRSM and their gauge quantum numbers.

Fields	$SU(3)_C \times SU(2)_L \times SU(2)_R \times U(1)_{B-L}$
$Q_L = \begin{pmatrix} u_L \\ d_L \end{pmatrix}$	$(\mathbf{3}, \mathbf{2}, \mathbf{1}, \frac{1}{3})$
$Q_R = \begin{pmatrix} u_R \\ d_R \end{pmatrix}$	$(\mathbf{3}, \mathbf{1}, \mathbf{2}, \frac{1}{3})$
$L_L = \begin{pmatrix} \nu_L \\ e_L \end{pmatrix}$	$(\mathbf{1}, \mathbf{2}, \mathbf{1}, -1)$
$L_R = \begin{pmatrix} \nu_R \\ e_R \end{pmatrix}$	$(\mathbf{1}, \mathbf{1}, \mathbf{2}, -1)$
S_1	$(\mathbf{1}, \mathbf{1}, \mathbf{1}, -2)$
S_2	$(\mathbf{1}, \mathbf{1}, \mathbf{1}, 2)$
$\phi = \begin{pmatrix} \phi_1^0 & \phi_1^+ \\ \phi_2^- & \phi_2^0 \end{pmatrix}$	$(\mathbf{1}, \mathbf{2}, \mathbf{2}, 0)$
$\chi_R = \begin{pmatrix} \chi_R^+ \\ \chi_R^0 \end{pmatrix}$	$(\mathbf{1}, \mathbf{1}, \mathbf{2}, 1)$

in two stages. The first stage reduces the gauge symmetry $SU(2)_L \times SU(2)_R \times U(1)_{B-L}$ to the SM gauge group via the VEV of a RH scalar doublet χ_R:

$$SU(2)_L \times SU(2)_R \times U(1)_{B-L} \xrightarrow{\langle \chi_R \rangle} SU(2)_L \times U(1)_Y. \tag{5.2}$$

This breaking occurs when χ_R develops a VEV in its neutral component:

$$\langle \chi_R \rangle = \frac{1}{\sqrt{2}} \begin{pmatrix} 0 \\ v_R \end{pmatrix}, \tag{5.3}$$

where $v_R \gg 1$ TeV ensures that the associated gauge bosons $W_R^{\pm}$ and Z_R acquire large masses [85, 86].

The SM hypercharge is identified as a combination of the $SU(2)_R$ and $U(1)_{B-L}$ generators:

$$Y = T_R^3 + Q_{B-L}, \tag{5.4}$$

which is unbroken by $\langle \chi_R \rangle$, since:

$$Y(\langle \chi_R \rangle) = T_R^3(\langle \chi_R \rangle) + Q_{B-L}(\langle \chi_R \rangle) = 0. \tag{5.5}$$

The corresponding gauge couplings obey:

$$\frac{1}{g_Y^2} = \frac{1}{g_R^2} + \frac{1}{g_{B-L}^2}. \tag{5.6}$$

The second stage of symmetry breaking, from the SM to electromagnetism, is driven by the VEVs of the neutral components of a Higgs bidoublet Φ:

$$SU(2)_L \times U(1)_Y \xrightarrow{\langle \Phi \rangle} U(1)_{\text{EM}}, \tag{5.7}$$

with

$$\langle \Phi \rangle = \frac{1}{\sqrt{2}} \begin{pmatrix} k_1 & 0 \\ 0 & k_2 \end{pmatrix}, \tag{5.8}$$

where $\sqrt{k_1^2 + k_2^2} = v = 246$ GeV is the electroweak scale. At this stage, the gauge bosons $W_L^{\pm}$ and Z_L acquire masses, and the electromagnetic coupling constant satisfies:

$$\frac{1}{e^2} = \frac{1}{g_L^2} + \frac{1}{g_R^2} + \frac{1}{g_{B-L}^2} = \frac{1}{g_L^2} + \frac{1}{g_Y^2}. \tag{5.9}$$

5.2 NEUTRINO AND FERMION MASSES IN THE LRIS

The Yukawa Lagrangian gives Dirac masses for fermions and light neutrino masses via the IS mechanism for all three generations. The relevant terms are:

$$\begin{aligned} \mathcal{L}_Y = \sum_{i,j=1}^{3} \Big[& y_{ij}^L \bar{L}_{Li} \phi L_{Rj} + \tilde{y}_{ij}^L \bar{L}_{Li} \tilde{\phi} L_{Rj} + y_{ij}^Q \bar{Q}_{Li} \phi Q_{Rj} \\ & + \tilde{y}_{ij}^Q \bar{Q}_{Li} \tilde{\phi} Q_{Rj} + y_{ij}^s \bar{L}_{Ri} \tilde{\chi}_R S_{2j}^c \Big] + \text{h.c.} \end{aligned} \tag{5.10}$$

The bidoublet ϕ and its conjugate $\tilde{\phi}$ couple to both quarks and leptons, generating Dirac masses. The RH doublet χ_R couples only to leptons and sterile singlet fermions S_2, generating the IS mechanism for neutrino masses. After $B-L$ breaking, a small Majorana mass term $\mu_s \bar{S}_2^c S_2$ arises from a dimension-seven non-renormalizable operator, yielding:

$$\mu_s = \lambda_s \frac{v_R^4}{M^3} \lesssim \mathcal{O}(1) \text{ keV}, \tag{5.11}$$

where λ_s is a dimensionless coupling. For $v_R \sim \mathcal{O}(10^3)$ GeV, $\mu_s \sim 10^{-7}$ GeV for $M \sim 10^3$ TeV and $\lambda_s \sim \mathcal{O}(1)$ [87]. Tus the neutrino mass terms are:

$$\mathcal{L}_m^\nu = M_D \bar{\nu}_L \nu_R + M_R \bar{\nu}_R^c S_2 + \mu_s \bar{S}_2^c S_2 + \text{h.c.}, \tag{5.12}$$

where $M_D = \frac{v}{\sqrt{2}}(y^L s_\beta + \tilde{y}^L c_\beta)$ is the Dirac mass matrix, and $M_R = \frac{v_R}{\sqrt{2}} y^s$. The complete 9×9 neutrino mass matrix in the basis $\psi = (\nu_L^c, \nu_R, S_2)$ is:

$$\mathcal{M}_\nu = \begin{pmatrix} 0 & M_D & 0 \\ M_D^T & 0 & M_R \\ 0 & M_R^T & \mu_s \end{pmatrix}. \tag{5.13}$$

This matrix yields three light neutrino masses:

$$m_{\nu_\ell} = M_D M_R^{-1} \mu_s (M_R^T)^{-1} M_D^T, \tag{5.14}$$

and six heavy states with masses:

$$m_{\nu_h}^2 = M_R^2 + M_D^2. \tag{5.15}$$

The matrix $\mathcal{M}_\nu$ is diagonalized by a unitary matrix V such that:

$$V^T \mathcal{M}_\nu V = \mathcal{M}_\nu^{\text{diag}}. \tag{5.16}$$

The light neutrino mass matrix in Eq. (5.14) is diagonalized by the physical lepton mixing matrix, known as the PMNS matrix U_{MNS} [88], according to the relation:

$$U_{MNS}^T \, m_{\nu_l} \, U_{MNS} = m_{\nu_l}^{\text{diag}} \equiv \text{diag}(m_{\nu_e}, m_{\nu_\mu}, m_{\nu_\tau}). \tag{5.17}$$

This allows us to express the Dirac neutrino mass matrix as:

$$m_D = U_{MNS} \sqrt{m_{\nu_l}^{\text{diag}}} \, R \sqrt{\mu_s^{-1}} \, M_N, \tag{5.18}$$

where R is a general orthogonal matrix. This expression extends the well-known Casas-Ibarra parameterization for type-I seesaw [69], which corresponds to

$$m_D = U_{MNS}\sqrt{m_{\nu_l}^{\text{diag}}}\,R\,\sqrt{M_N}.$$

The full 9×9 neutrino mass matrix $\mathcal{M}_\nu$ is diagonalized by a unitary matrix V such that

$$V^T\mathcal{M}_\nu V = \mathcal{M}_\nu^{\text{diag}}, \tag{5.19}$$

where V has the block structure:

$$V = \begin{pmatrix} V_{3\times 3} & V_{3\times 6} \\ V_{6\times 3} & V_{6\times 6} \end{pmatrix}. \tag{5.20}$$

The submatrix $V_{3\times 3}$, which governs the mixing among the light neutrinos, is approximately given by

$$V_{3\times 3} \simeq \left(1 - \frac{1}{2}FF^T\right)U_{MNS}, \tag{5.21}$$

where the correction term $\frac{1}{2}FF^T$ characterizes the deviation from unitarity due to mixing with the heavy neutrino states. Here, F is defined as

$$F = m_D M_N^{-1}. \tag{5.22}$$

Hence, in general, $V_{3\times 3}$ is not exactly unitary, and the extent of non-unitarity is governed by the size of FF^T. The matrix $V_{3\times 6}$, which encodes the mixing between light and heavy neutrinos, takes the form:

$$V_{3\times 6} = (\mathbf{0}_{3\times 3},\ F)\,V_{6\times 6}, \tag{5.23}$$

while $V_{6\times 6}$ is the matrix that diagonalizes the mass matrix in the heavy neutrino sector involving ν_R and S_2.

Furthermore, after electroweak symmetry breaking, quarks and charged leptons acquire masses via the Higgs mechanism:

$$M_u = \frac{v}{\sqrt{2}}(y^Q s_\beta + \tilde{y}^Q c_\beta), \tag{5.24}$$

$$M_d = \frac{v}{\sqrt{2}}(y^Q c_\beta + \tilde{y}^Q s_\beta), \tag{5.25}$$

$$M_\ell = \frac{v}{\sqrt{2}}(y^L c_\beta + \tilde{y}^L s_\beta), \tag{5.26}$$

with $k_1 = vs_\beta$, $k_2 = vc_\beta$, and $v = 246$ GeV. Diagonalization of these matrices gives the physical masses:

$$M_f^{\text{diag}} = V_L^{f\dagger} M_f' V_R^f, \quad f = u, d, \ell. \tag{5.27}$$

The original Yukawa matrices are expressed as:

$$y^Q = -\frac{\sqrt{2}}{vc_{2\beta}}(s_\beta V_L^u M_u V_R^{u\dagger} - c_\beta V_L^d M_d V_R^{d\dagger}), \tag{5.28}$$

$$\tilde{y}^Q = -\frac{\sqrt{2}}{vc_{2\beta}}(c_\beta V_L^u M_u V_R^{u\dagger} - s_\beta V_L^d M_d V_R^{d\dagger}). \tag{5.29}$$

Unlike the SM, fermion masses arise from two distinct VEVs, leading to a non-trivial flavor structure and the potential for tree-level FCNCs via neutral Higgs exchange.

5.3 HIGGS SECTOR IN LRIS

The Higgs sector of the LRIS model consists of a bidoublet Φ and a RH doublet χ_R. The symmetry breaking occurs in two stages:

$$SU(2)_L \times SU(2)_R \times U(1)_{B-L} \to SU(2)_L \times U(1)_Y, \tag{5.30}$$

$$SU(2)_L \times U(1)_Y \to U(1)_{\text{EM}}. \tag{5.31}$$

The neutral components of these Higgs fields expand around their VEVs as:

$$\Phi_1^0 = \frac{k_1 + h_1^0 + i\varphi_1^0}{\sqrt{2}}, \quad \Phi_2^0 = \frac{k_2 + h_2^0 + i\varphi_2^0}{\sqrt{2}}, \quad \chi_R^0 = \frac{v_R + h_R^0 + i\varphi_R^0}{\sqrt{2}}.$$

Before symmetry breaking, the model contains 12 scalar degrees of freedom (8 from Φ and 4 from χ_R). After breaking, 6 scalars remain as physical Higgs bosons: two charged Higgs bosons, one pseudoscalar, and three neutral CP-even Higgs bosons.

Singly Charged Scalars

The mass matrix for the charged Higgs bosons in the basis $(\phi_1^\pm, \phi_2^\pm, \chi_R^\pm)$ is given by:

$$M_{H^\pm}^2 = \frac{\alpha_{32}}{2}\begin{pmatrix} \frac{v_R^2 s_\beta^2}{c_{2\beta}} & \frac{v_R^2 s_{2\beta}}{2c_{2\beta}} & -vv_R s_\beta \\ . & \frac{v_R^2 c_\beta^2}{c_{2\beta}} & -vv_R c_\beta \\ . & . & v^2 c_{2\beta} \end{pmatrix}.$$

Diagonalizing this matrix leads to a physical charged Higgs mass:

$$m_{H^\pm}^2 = \frac{\alpha_{32}}{2}\left(\frac{v_R^2}{c_{2\beta}} + v^2 c_{2\beta}\right),$$

where $\alpha_{32} = \alpha_3 - \alpha_2$. The mass is typically in the range of hundreds of GeV for $v_R \sim \mathcal{O}(\text{TeV})$.

The physical charged Higgs is a combination of $\phi_1^\pm$, $\phi_2^\pm$, and $\chi_R^\pm$, with the following mixing:

$$H^\pm = Z_{13}^{H^\pm}\phi_1^\pm + Z_{23}^{H^\pm}\phi_2^\pm + Z_{33}^{H^\pm}\chi_R^\pm.$$

Neutral Pseudoscalars

For the neutral CP-odd Higgs bosons, the mass matrix is:

$$M_A^2 = \frac{1}{2}\left(\frac{v_R^2\alpha_{32}}{c_{2\beta}} - 4v^2(2\lambda_2 - \lambda_3)\right)\begin{pmatrix} c_\beta^2 & s_\beta c_\beta & 0 \\ . & s_\beta^2 & 0 \\ . & . & 0 \end{pmatrix}.$$

The eigenvalue for the pseudoscalar mass is:

$$m_A^2 = \frac{1}{2}\left(\frac{v_R^2}{c_{2\beta}}\alpha_{32} - 4v^2(2\lambda_2 - \lambda_3)\right),$$

which is typically in the range of a few hundred GeV for $v_R \sim \mathcal{O}(\text{TeV})$.

The physical pseudoscalar is given by:

$$A = c_\beta\phi_1^{0I} + s_\beta\phi_2^{0I},$$

where $s_\beta \ll 1$, so $A \approx \phi_1^{0I}$.

Neutral Scalar

The mass matrix for the neutral CP-even Higgs bosons is:

$$M_H^2 = \frac{\partial^2 V(\phi, \chi_{L,R})}{\partial\phi_i^{0R}\partial\phi_j^{0R}}\Big|_{\langle\phi_{i,j}^{0R}\rangle=0}.$$

This matrix can be diagonalized to yield the masses of the physical neutral Higgs bosons. In particular, one of the eigenvalues corresponds to the SM-like Higgs boson, while the other three eigenvalues correspond to the heavy Higgs bosons. However, there is the possibility that one of the new scalar states could be lighter than the SM Higgs, with a mass of order 95 GeV, potentially explaining some observed anomalies (such as deviations in Higgs-like particle searches at the LHC). The mass spectrum can therefore exhibit a light new Higgs state below the standard 125 GeV threshold, which could also affect phenomenology at colliders.

5.4 PHENOMENOLOGICAL IMPLICATIONS OF THE LRIS MODEL

The LRIS model offers a rich and testable phenomenology at both low-energy precision experiments and high-energy collider searches. Its extended gauge structure, additional Higgs fields, and the presence of quasi-Dirac heavy neutrinos lead to distinctive experimental BSM signatures. In this section, we explore three key phenomenological aspects of the LRIS framework. First, we examine its potential to address the long-standing discrepancy in the muon's anomalous magnetic moment, $(g-2)_\mu$, through loop contributions involving new gauge and scalar degrees of freedom. Second, we investigate the prospects for detecting heavy Higgs bosons predicted by the model at the LHC, focusing on their production channels and decay patterns. Finally, we analyze the lepton flavor violating process $\mu \to e\gamma$, which can receive significant enhancements in the LRIS scenario due to the interplay between the IS mechanism and new flavor-violating interactions. These complementary probes collectively constrain the parameter space of the LRIS model and offer promising avenues for its experimental verification.

5.4.1 Muon $g-2$ in the LRIS Framework

In this section, we investigate the contributions of the LRIS model to the muon anomalous magnetic moment, a_μ, arising from the exchange of both light and heavy gauge bosons (W, W', Z, Z'), neutral scalars (h_i, A), and charged Higgs bosons $(H^\pm)$, as illustrated in Fig. 5.1 [89].

The total new physics contribution to a_μ in this framework is given by

$$\delta a_\mu \equiv a_\mu^{\rm LRIS} = a_\mu^W + a_\mu^{W'} + a_\mu^Z + a_\mu^{Z'} + a_\mu^h + a_\mu^A + a_\mu^{H^\pm}, \tag{5.32}$$

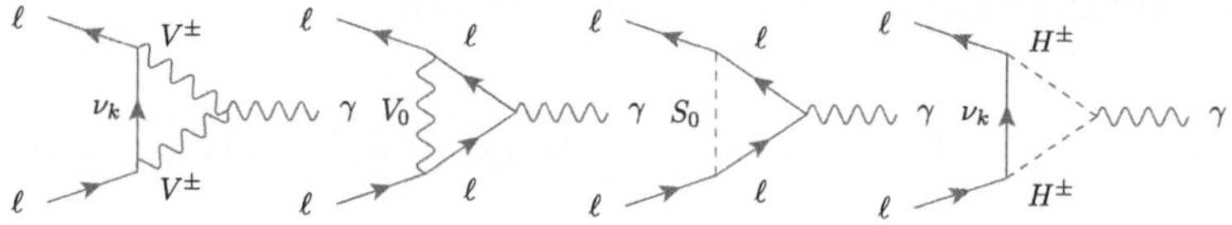

Figure 5.1 One-loop LRIS contributions to lepton $g_\ell - 2$ via massive neutrino loops. The exchanged particles include $V^\pm = W, W'$, $V^0 = Z, Z'$, $S^0 = h_i, A$, and the charged Higgs boson $H^\pm$.

where the individual terms, neglecting $W - W'$ and $Z - Z'$ mixing, are given by

$$a_\ell^W = G_{\rm F}^\ell \sum_{k=1}^{9} |U_{k,|\ell|}|^2 \left(\frac{10}{3} + \mathcal{F}_2(x_W^{\nu_k})\right), \tag{5.33}$$

$$a_\ell^{W'} = G_{\rm F}^\ell \sum_{k=1}^{9} |U_{k,3+|\ell|}|^2 \left(\frac{10}{3} + \mathcal{F}_2(x_{W'}^{\nu_k})\right) \left[\frac{1}{c_w} x_{W'}^W\right], \tag{5.34}$$

$$a_\ell^Z = G_{\rm F}^\ell \left(c_{4w} - 5\right) \left(\frac{1}{3}\right), \tag{5.35}$$

$$a_\ell^{Z'} = G_{\rm F}^\ell \left(c_{4w'} - 12 c_{2w'} - 5\right) \left[\frac{t_w^2}{48 s_{2w'}^2} x_{Z'}^W\right], \tag{5.36}$$

$$a_\ell^h = G_{\rm F}^\ell \sum_{i=1}^{3} x_{h_i}^\ell (\Gamma_{\ell\ell}^{h_i})^2 \left(\frac{7}{6} + \log x_{h_i}^\ell\right), \tag{5.37}$$

$$a_\ell^A = -\frac{1}{2} G_{\rm F}^\ell x_A^\ell (\Gamma_{\ell\ell}^A)^2 \left(\frac{7}{6} + \log x_A^\ell\right), \tag{5.38}$$

with $|\ell| = 1, 2$ for $\ell = e, \mu$, and $G_{\rm F}^\ell = \frac{G_{\rm F} m_\ell^2}{8\sqrt{2}\pi^2}$. The mass ratios are defined as $x_b^a = m_a^2/m_b^2$. The mixing angles are $s_{w'} = g_Y/g_R$ and $s_w = e/g_L$, with $\tan\theta_{w'} \lesssim 10^{-4}$ [90], and the charged gauge boson mass satisfies $m_{W'} \gtrsim \mathcal{O}(4\text{ TeV})$.

Among these, the Z', W', h_i, and A contributions are highly suppressed due to small mass ratios and can be neglected. Additionally, the second term in Eq. (5.33), which involves heavy RH neutrino loops, contributes only $\mathcal{O}(10^{-2})$, less than 0.4% of the dominant term, and is further suppressed by the GIM mechanism [91] due to the unitarity of the full 9×9 neutrino mixing matrix [87, 92, 93].

The charged Higgs contribution $a_\ell^{H^\pm}$ takes the form

$$a_\ell^{H^\pm} = G_{\rm F}^\ell\, \Gamma_\gamma^{H^\pm} \sum_{k=1}^{9} \left(|\zeta'_{k\ell}|^2\, \mathcal{F}_2(x_{H^\pm}^{\nu_k}) + 2\,\zeta'_{k\ell}\xi'_{k\ell}\, \mathcal{F}_1(x_{H^\pm}^{\nu_k}) \right), \qquad (5.39)$$

where $\zeta'_{k\ell} = \frac{v}{m_{\nu_k}}\zeta_{k\ell}$, $\xi'_{k\ell} = \frac{v}{m_\ell}\xi_{k\ell}$, and the loop functions are

$$\mathcal{F}_k(x) = \frac{x\mathcal{P}_k(x)}{(x-1)^{k+1}} - \frac{6x^{k+1}\log x}{(x-1)^{k+2}}, \qquad (5.40)$$

$$\mathcal{P}_1(x) = 3x+3, \quad \mathcal{P}_2(x) = 2x^2+5x-1. \qquad (5.41)$$

In the limit $x \to 1$, $\mathcal{F}_1(1) = 1$ and $\mathcal{F}_2(1) = \frac{1}{2}$. Over all x, the ratio $\mathcal{F}_2/\mathcal{F}_1$ remains $\mathcal{O}(1)$, ranging between 1/3 and 2/3.

The effective photon coupling is approximated by

$$\Gamma_\gamma^{H^\pm} \simeq \frac{1}{6e}(g_L U_{21}^0 + g_R U_{31}^0), \qquad (5.42)$$

valid in the limit $v_R \gg v$, with a typical value $\Gamma_\gamma^{H^\pm} \sim 0.076$. The first term in the sum contributes only about 0.02% of the second term, which dominates. Thus, the leading contribution is

$$a_\ell^{H^\pm} \simeq 2G_{\rm F}^\ell\, \Gamma_\gamma^{H^\pm} \sum_{k=4}^{9} \zeta'_{k\ell}\xi'_{k\ell}\, \mathcal{F}_1(x_{H^\pm}^{\nu_k}) \lesssim \frac{3\Gamma_\gamma^{H^\pm}}{8\pi^2}\, m_\ell \sum_{k=4}^{9} \frac{\zeta_{k\ell}\xi_{k\ell}}{m_{\nu_k}}. \qquad (5.43)$$

This contribution is enhanced by the inverse dependence on the charged lepton mass in $\xi'_{k\ell}$, making it potentially significant in explaining the a_μ anomaly, as shown in Fig. 5.2.

5.4.2 Heavy Higgs Search in the LRIS Model at the LHC

As discussed above, following Left-Right and electroweak symmetry breaking, the CP-even neutral Higgs mass matrix M_H^2 is diagonalized by the unitary matrix Z^H. The lightest eigenstate, $H_1 \equiv h$, is identified with the SM-like Higgs boson. The remaining two eigenvalues are given by

$$m_{H_{2,3}}^2 = \frac{1}{2}\left(T^h - m_h^2 \mp \sqrt{(T^h - m_h^2)^2 - \frac{4D^h}{m_h^2}} \right), \qquad (5.44)$$

where $T^h = {\rm Tr}(M_H^2)$ and $D^h = \det(M_H^2)$ denote the trace and determinant of the mass matrix, respectively.

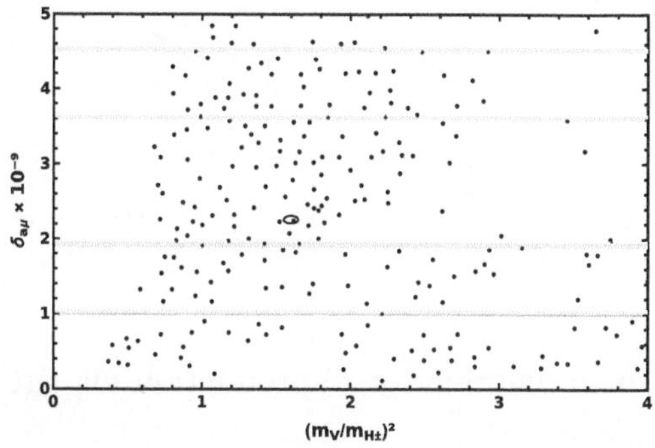

Figure 5.2 Muon magnetic moment anomaly δa_μ as a function of the mass ratio parameter $x^{\nu_5}_{H^\pm} = m^2_{\nu_5}/m^2_{H^\pm}$, where m_{ν_5} and $m_{H^\pm}$ denote the second-heaviest neutrino and charged Higgs masses, respectively. The 1σ and 2σ confidence regions from experimental measurements are shown with bands between horizontal lines. The benchmark point, circulated in *dark–shaded region*, was considered in Ref. [89].

From Eq. (5.44), the next-to-lightest CP-even neutral Higgs boson, $H_2 \equiv h'$, can acquire a mass in the few hundred GeV range. In our analysis, we scan the scalar potential parameters within the ranges:

$$\begin{aligned} &\lambda_1 \in [0.18, 0.30], \quad \lambda_4 \in [0.70, 0.99], \quad \alpha_1 \in [0.06, 0.16], \\ &\alpha_4 \in [0.60, 0.99], \quad \rho_1 \in [0.08, 0.14], \quad \lambda_{23} \in [-0.1, 3]. \end{aligned} \tag{5.45}$$

The physical eigenstate h' is composed of real parts of neutral scalar fields:

$$h' = Z^H_{21}\,\phi^{0R}_1 + Z^H_{22}\,\phi^{0R}_2 + Z^H_{23}\,\chi^{0R}_R, \tag{5.46}$$

where Z^H is the CP-even Higgs mixing matrix. Benchmark points with $m_{h'} = 250$, 400, and 600 GeV are adopted as in Ref. [94]. The h' state is dominantly composed of the ϕ_1 component, with subdominant contributions from ϕ_2 and χ_R.

The trilinear coupling of h' with two SM-like Higgs bosons is given by:

$$g_{h'hh} \approx -2iZ^H_{21}Z^H_{12}\left\{v\left[(\lambda_1 - \lambda_{23})\cos\beta + 3\lambda_4\sin\beta\right]Z^H_{12} + \alpha_4 v_R Z^H_{13}\right\}, \tag{5.47}$$

where we fix $g_L = g_R = g_2 = 0.663$, $g_{BL} = 0.422$, and $v_R = 6400$ GeV, while the scalar potential parameters are varied as in Eq. (5.45). At the LHC, the heavy Higgs boson h' is predominantly produced via gluon-gluon fusion (ggF), which accounts for approximately 90% of its total

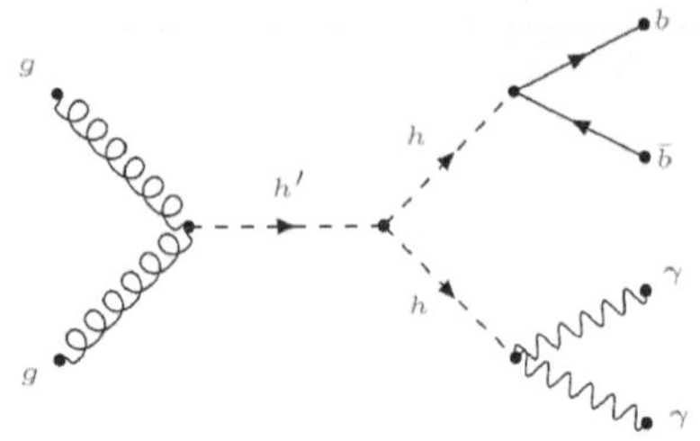

Figure 5.3 Feynman diagram for h' production via ggF and decay into $hh \to b\bar{b}\gamma\gamma$.

production. For $m_{h'} \leq 600$ GeV, the decay branching ratio BR($h' \to hh$) remains sizable, typically exceeding 10%, making this channel promising for discovery.

Among the possible final states, the decay chain $h' \to hh \to b\bar{b}\gamma\gamma$ offers a clean experimental signature and manageable background. The corresponding Feynman diagram is shown in Fig. 5.3. We perform a dedicated analysis for the three benchmark masses of h': 250, 400, and 600 GeV.

Using the narrow width approximation, the total cross-section for the process is:

$$\begin{aligned}\sigma(pp \to h' \to hh \to b\bar{b}\gamma\gamma) &\approx \sigma(pp \to h') \times \text{BR}(h' \to hh) \\ &\times \text{BR}(h \to b\bar{b}) \times \text{BR}(h \to \gamma\gamma).\end{aligned}$$

Table 5.2 summarizes the production cross-section, branching ratio, and total signal rate for the three considered h' masses. To assess the discovery potential, we analyze the signal and background distributions at $\sqrt{s} = 14$ TeV with an integrated luminosity of 300 fb^{-1}. Figure 5.4 shows

TABLE 5.2 Production cross-section and decay branching ratios for different values of $m_{h'}$.

$m_{h'}$ (GeV)	$\sigma(pp \to h')$ (pb)	BR($h' \to hh$)	$\sigma(pp \to h' \to hh \to b\bar{b}\gamma\gamma)$ (fb)
250	12.140	0.30	6.30
400	5.050	0.20	1.01
600	0.504	0.18	0.05

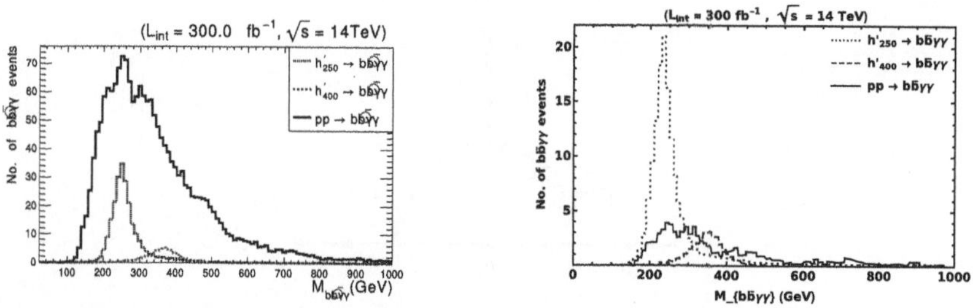

Figure 5.4 Signal event distributions for $h' \to b\bar{b}\gamma\gamma$ at $m_{h'} = 250$ GeV (dotted line) and 400 GeV (dashed line), compared to the SM background (solid line), before (left) and after (right) applying selection cuts at $\sqrt{s} = 14$ TeV and $L_{\text{int}} = 300$ fb^{-1} [94].

the number of signal events for $m_{h'} = 250$ and 400 GeV before and after applying selection cuts, following the procedure of Ref. [94].

5.4.3 $\mu \to e\gamma$ in the TeV-Scale $B - L$ Model with IS

We now examine the charged lepton flavor violating (LFV) process $\mu \to e\gamma$ within the framework of LRIS. A number of experimental efforts have been dedicated to probing such LFV decays, with the $\mu \to e\gamma$ channel receiving particular attention. The most stringent upper limit to date, reported by the MEG collaboration [95], is:

$$\text{BR}(\mu \to e\gamma) < 4.2 \times 10^{-13}. \tag{5.48}$$

Future experiments aim to enhance this sensitivity by up to three orders of magnitude.

It is worth noting that, within the SM augmented by tiny neutrino masses, the branching ratio for this decay is extremely suppressed:

$$\text{BR}(\mu \to e\gamma)^{\text{SM}} \simeq 10^{-55}, \tag{5.49}$$

rendering any experimental signal of this process a definitive indicator of new physics.

The decay amplitude receives contributions from both light and heavy neutrinos in the loop, along with the W boson, as depicted in Fig. 5.5. In the limit $m_e \to 0$, the amplitude can be expressed as:

$$A(\mu \to e\gamma) \simeq \frac{m_\mu G_F}{32\sqrt{2}\pi^2} \sum_{i=1}^{9} V_{\mu i}^* V_{ei} f(r_i) \times \bar{u}(p)[2e(p' \cdot \epsilon)]u(p'), \tag{5.50}$$

where $f(r_i)$ is the loop function and $r_i = m_{\nu_i}^2 / M_W^2$.

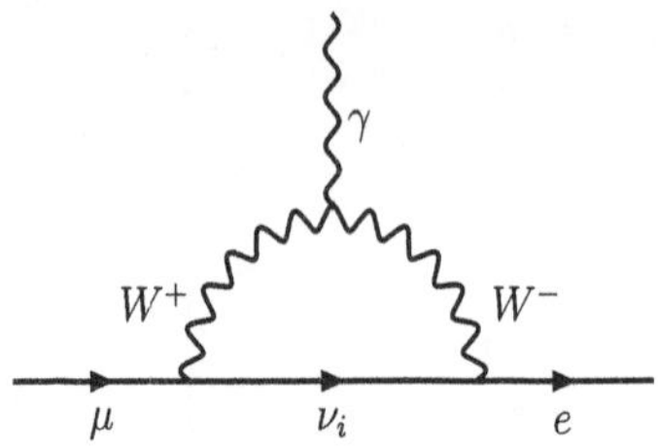

Figure 5.5 Feynman diagram for the dominant contribution to $\mu \to e\gamma$ in the LRIS.

Defining the overall amplitude coefficient as:

$$a = \frac{e m_\mu G_F}{32\sqrt{2}\pi^2} \sum_{i=1}^{9} V_{\mu i}^* V_{ei} f(r_i), \tag{5.51}$$

the decay width becomes:

$$\Gamma(\mu \to e\gamma) = \frac{m_\mu^3}{8\pi} |a|^2. \tag{5.52}$$

Normalizing to the standard muon decay width $\Gamma(\mu \to e\nu\bar{\nu}) \simeq m_\mu^5 G_F^2/(192\pi^3)$, the branching ratio is given by:

$$\text{BR}(\mu \to e\gamma) = \frac{\Gamma(\mu \to e\gamma)}{\Gamma(\mu \to e\nu\bar{\nu})} = \frac{3\alpha}{64\pi} \left| \sum_{i=1}^{9} V_{\mu i}^* V_{ei} f(r_i) \right|^2, \tag{5.53}$$

with $\alpha = e^2/(4\pi) \simeq 1/137$.

From the MEG upper limit in Eq. (5.48), one obtains the constraint:

$$\left| \sum_{i=1}^{9} V_{\mu i}^* V_{ei} f(r_i) \right| < 1.95 \times 10^{-5}. \tag{5.54}$$

In the limit where LH neutrinos are extremely heavy, the mixing matrix $V_{3\times3}$ becomes nearly unitary, and the light neutrino contributions ($i = 1, 2, 3$) to the amplitude are negligible. Additionally, in this regime, the heavy neutrino contributions are highly suppressed due to the smallness of the mixing terms ($V_{\mu i}, V_{ei} \sim m_D/M_N \sim \mathcal{O}(10^{-9})$), ensuring that the bound in Eq. (5.54) is easily satisfied and $\text{BR}(\mu \to e\gamma) \ll 10^{-12}$.

Conversely, in the TeV-scale IS scenario, the lepton mixing matrix is significantly non-unitary, and the mixing between light and heavy neutrinos is no longer suppressed ($m_D/M_N \sim \mathcal{O}(0.1)$). In this context, the constraint can be rewritten as:

$$\left|\frac{10}{3}\sum_{i=1}^{3} V_{\mu i}^{*} V_{ei} + \sum_{j=4}^{9} V_{\mu j}^{*} V_{ej} f(r_j)\right| < 1.95 \times 10^{-5}, \tag{5.55}$$

where $r_j = (m_{\nu_{H_j}}/M_W)^2$ and the mixing elements $V_{e(\mu)j}$ are given by:

$$\begin{aligned} V_{e(\mu)j} &= \left[\left(0, m_D M_N^{-1}\right) V_{6\times 6}\right]_{1(2),j-3} \\ &= \left[\left(0, U_{MNS}\sqrt{m_{\nu_l}^{\text{diag}}}\, R\sqrt{\mu_s^{-1}}\right) V_{6\times 6}\right]_{1(2),j-3}. \end{aligned} \tag{5.56}$$

Assuming $r_j \gg 1$ so that $f(r_j) \simeq 4/3$, the constraint simplifies to:

$$\left|\sum_{i=1}^{3} V_{\mu i}^{*} V_{ei}\right| < 0.8 \times 10^{-5}, \tag{5.57}$$

which imposes the condition $(FF^T)_{21,12} \lesssim 10^{-4}$. This limit is typically satisfied in the IS framework, given the restrictions on the off-diagonal components of the non-unitary U_{MNS} matrix.

CHAPTER 6

$SU(5)$ Grand Unified Theory

The $SU(5)$ theory stands as a widely studied example of a simple Grand Unified Theory (GUT), first proposed by Georgi and Glashow in 1974 [96]. Unlike SM, $SU(5)$ is a rank-four group[1]. Since the SM gauge group, $SU_C(3) \times SU_L(2) \times U_Y(1)$, also has rank four, any viable GUT must have at least this rank. Possible candidate groups with rank four include:

$$[SU(3)]^2, \quad [SU(2)]^4, \quad O(9), \quad O(8), \quad Sp(8), \quad SU(5),$$
$$F_4 \text{ (exceptional group).}$$

After careful examination, only $[SU(3)]^2$ and $SU(5)$ remain viable candidates, as complex representations are required to accommodate chiral fermions. In the SM, left- and RH fermions transform differently, *e.g.*, $e_L^\dagger \neq e_R$. However, $[SU(3)]^2$ is ultimately excluded because it implies that leptons carry color quantum numbers and leads to unphysical flavor assignments, both of which are incompatible with experimental observations. Consequently, $SU(5)$ stands out as the unique minimal candidate for unification.

6.1 $SU(5)$ FERMION CONTENT

A defining feature of GUTs, particularly $SU(5)$, is the unification of the SM matter content into irreducible representations of higher dimensions

[1] *The rank of a group is the number of mutually commuting generators that can be simultaneously diagonalized.*

DOI: 10.1201/9781003457701-6

of the unified group. This must be achieved while preserving the correct transformation properties under the SM subgroup.

Each generation of SM fermions consists of 15 left-handed Weyl spinors. We organize them using the fundamental representations of $SU(5)$, that is, 5 and its conjugate 5^*. These decompose under $SU(3)_C \times SU(2)_L \times U(1)_Y$ as [97]:

$$5 = (3,1)_{-1/3} \oplus (1,2)_{1/2}, \quad 5^* = (3^*,1)_{1/3} \oplus (1,2)_{-1/2}. \tag{6.1}$$

Here, we adopt the convention that the first three components correspond to $SU(3)_C$ and the last two to $SU(2)_L$. The hypercharge is embedded to satisfy the tracelessness condition of $SU(5)$ generators. In particular, a decomposition such as $(3,1)_{2/3} \oplus (1,2)_{1/2}$ would violate tracelessness and is therefore not allowed.

The 5 representation can be populated as:

$$5 \equiv (\psi_i)_R = \begin{pmatrix} d_r \\ d_g \\ d_b \\ e^+ \\ -\nu_e^c \end{pmatrix}_R, \tag{6.2}$$

where $d_{r,g,b}$ are RH down-type quarks, and e^+ and ν_e^c denote the RH charged lepton and neutrino, respectively.

Applying charge conjugation, $(\psi_R)^c = C\gamma_0\psi_R^* \equiv (\psi^c)_L$, we obtain the 5^* representation:

$$5^* \equiv (\psi^i)_L \equiv \psi_L^c = \begin{pmatrix} d_r^c \\ d_g^c \\ d_b^c \\ e^- \\ -\nu_e \end{pmatrix}_L. \tag{6.3}$$

This assignment matches precisely the quantum numbers of SM particles.

To incorporate the remaining left-handed SM fermions, we consider higher-dimensional representations built from tensor products:

$$5 \otimes 5 = 15_S \oplus 10_A, \quad 5 \otimes 5^* = 24 \oplus 1, \tag{6.4}$$

where the subscripts denote symmetric (S) and antisymmetric (A) parts. Their decompositions are:

$$15 = (6,1)_{-4/3} \oplus (3,2)_{1/3} \oplus (1,3)_2, \tag{6.5}$$

$$10 = (3^*,1)_{-4/3} \oplus (3,2)_{1/3} \oplus (1,1)_2. \tag{6.6}$$

The 15 representation contains exotic color sextets not observed experimentally, while the 10 representation beautifully matches the SM content.

Explicitly, the antisymmetric 10 representation, accommodating u_L^c, Q_L, and e_L^+, is given by:

$$10 \equiv (\chi_{ij})_L = \frac{1}{\sqrt{2}} \begin{pmatrix} 0 & u_3^c & -u_2^c & u_1 & d_1 \\ -u_3^c & 0 & u_1^c & u_2 & d_2 \\ u_2^c & -u_1^c & 0 & u_3 & d_3 \\ -u_1 & -u_2 & -u_3 & 0 & e^+ \\ -d_1 & -d_2 & -d_3 & -e^+ & 0 \end{pmatrix}_L . \tag{6.7}$$

Thus, the SM fermions of a single generation neatly fit into the $5^* \oplus 10$ of $SU(5)$.

6.2 CHARGE QUANTIZATION IN *SU*(5)

One of the profound successes of $SU(5)$ GUTs is the natural explanation of electric charge quantization [98]. In this framework, charge quantization arises from two key principles:

- The SM gauge groups $SU(3)_C$, $SU(2)_L$, and $U(1)_Y$ commute, implying that particles differing only by color or weak isospin have identical hypercharges.
- The $SU(5)$ generators are traceless, constraining the embedding of $U(1)_Y$.

Remarkably, the sum of electric charges within a given $SU(5)$ multiplet vanishes. For instance, in the 5^* representation:

$$3q(d^c) + q(e^-) + q(\nu_e) = 0, \tag{6.8}$$

leading to the relation $3(-\frac{1}{3}) + (-1) + 0 = 0$, which fixes the down quark charge as $q(d) = -1/3$. This elegant result is a direct consequence of group theory and provides deep insight into one of the most mysterious features of the SM.

6.3 SU(5) GENERATORS

The group $SU(n)$ is defined as the set of $n \times n$ unitary matrices with determinant one. A general $SU(n)$ transformation is expressed as

$$U = \exp\left(i \sum_{a=1}^{n^2-1} B^a \, T_a \right) = \exp\left(-i \, \mathbf{B} \cdot \mathbf{T}\right), \tag{6.9}$$

where the T_a are the generators of the Lie algebra, satisfying the properties of being Hermitian, traceless, and normalized according to

$$\mathrm{Tr}(\lambda_a \lambda_b) = 2\delta_{ab}, \tag{6.10}$$

with λ_a the analogs of Gell-Mann matrices. For $SU(5)$, there are 24 generators ($a = 1, \ldots, 24$), corresponding to the dimension of the group.

To identify the 24 associated vector bosons, denoted by $\mathbf{A}_\mu = A^a_\mu \, T_a = A^a_\mu \lambda_a/2$, we embed the SM gauge group $SU(3)_C \times SU(2)_L \times U(1)_Y$ into $SU(5)$ as follows:

- Eight generators $(T_1, \ldots, T_8)$ are associated with the $SU(3)_C$ subgroup,
- Three generators (T_9, T_{10}, T_{11}) correspond to $SU(2)_L$,
- One generator (T_{12}) is assigned to $U(1)_Y$.

The $SU(3)$ generators act on the first three indices, while the $SU(2)$

generators act on the last two, ensuring their commutativity. Explicitly:

$$T_{1\ldots8}=\frac{1}{2}\begin{pmatrix} & & & 0 & 0\\ & \lambda_{1\ldots8} & & 0 & 0\\ & & & 0 & 0\\ 0 & 0 & 0 & 0 & 0\\ 0 & 0 & 0 & 0 & 0 \end{pmatrix}, \tag{6.11}$$

where $\lambda_{1\ldots8}$ are the usual $SU(3)_C$ Gell-Mann matrices. Similarly, the $SU(2)_L$ generators are embedded as

$$T_{9,10,11}=\frac{1}{2}\begin{pmatrix} 0 & 0 & 0 & 0 & 0\\ 0 & 0 & 0 & 0 & 0\\ 0 & 0 & 0 & 0 & 0\\ 0 & 0 & 0 & \sigma_{1,2,3} & \\ 0 & 0 & 0 & & \end{pmatrix}, \tag{6.12}$$

where $\sigma_{1,2,3}$ are the Pauli matrices.

The $SU(2)_L$ weak isospin third generator, corresponding to T_{11}, is given explicitly by

$$T_{11}=\begin{pmatrix} 0 & 0 & 0 & 0 & 0\\ 0 & 0 & 0 & 0 & 0\\ 0 & 0 & 0 & 0 & 0\\ 0 & 0 & 0 & 1/2 & 0\\ 0 & 0 & 0 & 0 & -1/2 \end{pmatrix}. \tag{6.13}$$

The generator associated with $U(1)_Y$ is chosen to respect commutation with $SU(3)_C$ and $SU(2)_L$ generators, tracelessness, and correct hypercharge normalization:

$$T_{12} = \sqrt{\frac{3}{5}} \begin{pmatrix} -\frac{1}{3} & 0 & 0 & 0 & 0 \\ 0 & -\frac{1}{3} & 0 & 0 & 0 \\ 0 & 0 & -\frac{1}{3} & 0 & 0 \\ 0 & 0 & 0 & \frac{1}{2} & 0 \\ 0 & 0 & 0 & 0 & \frac{1}{2} \end{pmatrix}. \tag{6.14}$$

Thus, the SM gauge group is embedded naturally inside $SU(5)$, and the twelve corresponding gauge bosons are associated with the above generators.

The remaining twelve generators correspond to massive vector bosons X and Y, responsible for transitions between quarks and leptons and mediating proton decay. They are represented as off-diagonal matrices connecting the $SU(3)_C$ triplet and $SU(2)_L$ doublet indices. Schematically, they are of the form:

$$T \sim \frac{1}{\sqrt{2}} \begin{pmatrix} 0_{3\times 3} & \mathbf{M}_{3\times 2} \\ \mathbf{M}^{\dagger}_{2\times 3} & 0_{2\times 2} \end{pmatrix}, \tag{6.15}$$

where $\mathbf{M}$ contains elements corresponding to raising and lowering operators between color and weak indices.

Finally, we note that $SU(5)$ is a rank 4 group, having four diagonal generators. These can be identified as T_3 and T_8 from $SU(3)_C$, T_{11} from $SU(2)_L$, and T_{12} from $U(1)_Y$, consistent with the SM structure.

Charge Operator

In the SM, the Gell-Mann-Nishijima relation expresses the electric charge Q as

$$Q = T_3 + \frac{Y}{2}, \tag{6.16}$$

where I_3 is the third component of weak isospin and Y is the weak hypercharge.

In $SU(5)$, the electromagnetic charge operator Q is expected to be a linear combination of the four diagonal generators. Since Q must commute with all $SU(3)_C$ generators (implying that quarks of the same flavor have the same electric charge irrespective of color), we can express Q as

$$Q = T_{11} + T_{12} \equiv T_{11} + \frac{3}{5}Y, \tag{6.17}$$

where T_{11} is the third generator of $SU(2)_L$ and Y is the properly normalized hypercharge generator defined earlier.

Substituting the explicit matrix forms of T_{11} and T_{12}, the action of the charge operator Q on the fundamental representation 5 yields

$$Q(\psi_i) = \begin{pmatrix} -\frac{1}{3} & 0 & 0 & 0 & 0 \\ 0 & -\frac{1}{3} & 0 & 0 & 0 \\ 0 & 0 & -\frac{1}{3} & 0 & 0 \\ 0 & 0 & 0 & 1 & 0 \\ 0 & 0 & 0 & 0 & 0 \end{pmatrix} \equiv Q_i \, \delta_{ij}. \tag{6.18}$$

Accordingly, the fundamental conjugate representation $\bar{5}$ (i.e., 5^*) transforms with opposite charges:

$$Q(\psi^i) = -Q_i \, \delta_{ij} = \begin{pmatrix} \frac{1}{3} & 0 & 0 & 0 & 0 \\ 0 & \frac{1}{3} & 0 & 0 & 0 \\ 0 & 0 & \frac{1}{3} & 0 & 0 \\ 0 & 0 & 0 & -1 & 0 \\ 0 & 0 & 0 & 0 & 0 \end{pmatrix}. \tag{6.19}$$

To determine the charge assignment for the 10 representation, we recall that it transforms as an antisymmetric tensor:

$$\psi_{ij} \to U_{ik} \, U_{jl} \, \psi_{kl}, \tag{6.20}$$

where $U = e^{i\alpha^a T_a}$ is an element of $SU(5)$. Infinitesimally, the variation under a generator T_a reads

$$\delta\psi_{ij} = i\left((T_a)_{ik}\psi_{kj} + (T_a)_{jl}\psi_{il}\right), \tag{6.21}$$

which implies that the charge operator acts as

$$Q(\chi_{ij}) = (Q_i + Q_j)\ \chi_{ij}. \tag{6.22}$$

Thus, the charge of a component χ_{ij} is simply the sum of the charges Q_i and Q_j corresponding to the fundamental indices i and j.

For example, taking $i = 1, j = 2$ yields

$$Q(\chi_{12}) = Q_1 + Q_2 = -\frac{1}{3} - \frac{1}{3} = -\frac{2}{3}, \tag{6.23}$$

which matches the charge of an up-type quark.

Explicitly, the full charge matrix for the 10 is

$$\begin{pmatrix} 0 & -\frac{2}{3} & -\frac{2}{3} & -\frac{2}{3} & -\frac{1}{3} \\ -\frac{2}{3} & 0 & -\frac{2}{3} & \frac{2}{3} & -\frac{1}{3} \\ -\frac{2}{3} & -\frac{2}{3} & 0 & \frac{2}{3} & -\frac{1}{3} \\ -\frac{2}{3} & \frac{2}{3} & \frac{2}{3} & 0 & 1 \\ -\frac{1}{3} & -\frac{1}{3} & -\frac{1}{3} & 1 & 0 \end{pmatrix}. \tag{6.24}$$

An important consequence of charge quantization in $SU(5)$ arises from the tracelessness condition imposed on the generators. Specifically, for the quarks and leptons, the condition

$$3Q_d + Q_{e^+} = 0 \tag{6.25}$$

must hold, where Q_d denotes the down-type quark charge and Q_{e^+} the positron charge. This naturally explains why quarks carry electric charge 1/3 of that of leptons in grand unified theories [96–98].

6.4 GAUGE BOSONS

The $SU(5)$ group possesses $5^2 - 1 = 24$ gauge bosons. Twelve of these correspond to the well-known gauge bosons of the SM, including the photon, $W^{\pm}$, Z^0, and the eight gluons. The remaining twelve are exotic gauge bosons capable of transforming quarks into leptons and vice versa, a distinctive feature of grand unification [96].

This distinction becomes apparent when decomposing the $SU(5)$ adjoint representation under the SM subgroup:

$$24 = (8, 1)_0 \oplus (3, 2)_{-5/3} \oplus (\bar{3}, 2)_{5/3} \oplus (1, 3)_0 \oplus (1, 1)_0. \tag{6.26}$$

Here, the $(8,1)_0$ corresponds to the eight gluons of $SU(3)_C$, the $(1,3)_0$ corresponds to the three weak gauge bosons $W^{\pm}$ and W^3, and the $(1,1)_0$ is the $U(1)_Y$ hypercharge boson B. The $(3,2)_{-5/3}$ and $(\bar{3},2)_{5/3}$ components represent the twelve new heavy gauge bosons, often called X and Y bosons, which mediate baryon and lepton number-violating interactions.

Following the construction of the $SU(5)$ generators, these 24 gauge fields can be organized in matrix form as:

$$\mathbf{A}_\mu = A_\mu^a T_a = \frac{1}{\sqrt{2}} \begin{pmatrix} & & & \bar{X}_\mu^r & \bar{Y}_\mu^r \\ & G_\mu^8 & & \bar{X}_\mu^g & \bar{Y}_\mu^g \\ & & & \bar{X}_\mu^b & \bar{Y}_\mu^b \\ X_\mu^r & X_\mu^g & X_\mu^b & W_\mu^3 & W_\mu^+ \\ Y_\mu^r & Y_\mu^g & Y_\mu^b & W_\mu^- & -\frac{1}{\sqrt{2}} W_\mu^3 \end{pmatrix} + \sqrt{\frac{3}{5}} \begin{pmatrix} -\frac{1}{3} & 0 & 0 & 0 & 0 \\ 0 & -\frac{1}{3} & 0 & 0 & 0 \\ 0 & 0 & -\frac{1}{3} & 0 & 0 \\ 0 & 0 & 0 & \frac{1}{2} & 0 \\ 0 & 0 & 0 & 0 & \frac{1}{2} \end{pmatrix} B_\mu, \tag{6.27}$$

where $G_\mu^8 = A_\mu^{1\ldots 8} T_{1\ldots 8}$ are the gluons, $W_\mu^{\pm} = \frac{1}{\sqrt{2}}(A_\mu^9 \mp i A_\mu^{10})$ are the charged weak bosons, $W_\mu^3 = A_\mu^{11} T_{11}$, and $B_\mu = A_\mu^{12} T_{12}$ corresponds to hypercharge.

The twelve new gauge bosons are defined via linear combinations, for example:

$$\bar{X}_\mu^r = \frac{1}{\sqrt{2}}(A_\mu^{13} - i A_\mu^{14}), \qquad X_\mu^r = \frac{1}{\sqrt{2}}(A_\mu^{13} + i A_\mu^{14}), \tag{6.28}$$

and similarly for the other color indices g and b.

The charges of these gauge bosons, obtained analogously using Eq. (6.17), are:

$$\begin{pmatrix} 0 & 0 & 0 & -\frac{4}{3} & -\frac{1}{3} \\ 0 & 0 & 0 & -\frac{4}{3} & -\frac{1}{3} \\ 0 & 0 & 0 & -\frac{4}{3} & -\frac{1}{3} \\ \frac{4}{3} & \frac{4}{3} & \frac{4}{3} & 0 & 1 \\ \frac{1}{3} & \frac{1}{3} & \frac{1}{3} & -1 & 0 \end{pmatrix}. \tag{6.29}$$

Eqs. (6.27) and (6.29) reveal that the X and Y gauge bosons transform under the SM gauge group as $(\bar{3}, 2)_{-5/3}$, and not as their conjugate representations.

These new bosons can be organized as doublets in color space:

$$\begin{pmatrix} X \\ Y \end{pmatrix} : \quad (\bar{3}, 2)_{-5/3}, \qquad \begin{pmatrix} \bar{X} \\ \bar{Y} \end{pmatrix} : \quad (3, 2)_{5/3}. \tag{6.30}$$

The existence of these twelve heavy gauge bosons leads to distinct new interactions. As previously noted, in $SU(5)$, quarks and leptons are unified into common multiplets [98]. Thus, the X and Y bosons can mediate transitions between quarks and leptons, or between two quarks, thereby enabling baryon and lepton number-violating processes. In particular, the combination of two such vertices can induce proton decay, a key prediction of $SU(5)$ grand unified theories [99], which will be discussed in detail in the next section.

6.5 INTERACTIONS

The Lagrangian for the $SU(5)$ gauge field theory incorporates the dynamics of fundamental particles and their interactions. It includes terms governing gauge fields, matter fields, and their mutual couplings. A simplified expression for the $SU(5)$ interaction Lagrangian reads [96, 98]:

$$\mathcal{L} = \mathcal{L}_{\text{gb}} + \mathcal{L}_f + \mathcal{L}_\Phi + \mathcal{L}_Y, \tag{6.31}$$

where each term corresponds to a distinct sector of the theory. We now describe these contributions in more detail.

Gauge Boson Sector

The kinetic term for the $SU(5)$ gauge bosons is governed by the field strength tensor $F^a_{\mu\nu}$. The gauge boson Lagrangian is given by

$$\mathcal{L}_{\text{gb}} = -\frac{1}{4}F^a_{\mu\nu}F^{a\mu\nu}, \tag{6.32}$$

where the field strength tensor is defined as

$$F^a_{\mu\nu} = \partial_\mu A^a_\nu - \partial_\nu A^a_\mu + g f^{abc} A^b_\mu A^c_\nu, \tag{6.33}$$

with f^{abc} denoting the structure constants of $SU(5)$. The structure constants are given by

$$f_{abc} = \frac{1}{4i}\text{Tr}\left(\lambda_a\left[\lambda_b, \lambda_c\right]\right), \tag{6.34}$$

where λ_a are the generators of $SU(5)$ in the fundamental representation.

The covariant derivative acting on a field in the fundamental representation is

$$D_\mu = \partial_\mu - ig_G \sum_{a=1}^{24} A^a_\mu \frac{\lambda_a}{2} \equiv \partial_\mu - ig_G\, \mathbf{A}_\mu, \tag{6.35}$$

where g_G is the unified gauge coupling constant, and $\mathbf{A}_\mu$ collectively denotes the gauge fields.

Explicitly, the action of the covariant derivative on a fundamental field ψ_p reads

$$D_\mu\psi_p = \partial_\mu\psi_p - ig_G(A_\mu)_{pq}\psi_q, \tag{6.36}$$

whereas for a two-index tensor field χ_{pq} (such as the antisymmetric **10** representation), the covariant derivative is

$$D_\mu\chi_{pq} = \partial_\mu\chi_{pq} - ig_G(A_\mu)^{pr}\chi_{rq} - ig_G(A_\mu)^{qs}\chi_{ps}. \tag{6.37}$$

These structures form the basis for understanding the dynamics and interactions of matter and gauge fields within the $SU(5)$ grand unified theory framework.

Fermion Sector

The kinetic term for the matter fields, denoted by $\mathcal{L}_f$, describes the propagation and gauge interactions of the fermions ψ_i. It incorporates the covariant derivative D_μ to ensure local $SU(5)$ gauge invariance. The fermion Lagrangian reads

$$\mathcal{L}_f = i(\overline{\psi}^p)_L\gamma^\mu D_\mu(\psi_p)_L + i(\overline{\chi}^{pq})_L\gamma^\mu D_\mu(\chi_{pq})_L, \tag{6.38}$$

where ψ_p and χ_{pq} represent the left-handed Weyl fermions in the $\mathbf{5}^*$ and $\mathbf{10}$ representations, respectively.

Substituting the explicit forms of the covariant derivatives from Eqs. (6.36) and (6.37), the Lagrangian becomes

$$\begin{aligned}\mathcal{L}_f &= i\overline{\psi}^p\gamma^\mu\partial_\mu\psi_p + g_G\overline{\psi}^p\gamma^\mu(A_\mu)_{pq}\psi_q + i\,\overline{\chi}^{pq}\gamma^\mu\partial_\mu\chi_{pq}\\ &\quad - ig_G(A_\mu)^{pr}\overline{\chi}^{pq}\gamma^\mu\chi_{rq} - ig_G(A_\mu)^{qs}\overline{\chi}^{pq}\gamma^\mu\chi_{ps},\end{aligned} \tag{6.39}$$

clearly exhibiting both the kinetic terms (independent of g_G) and the interaction terms between gauge bosons and fermions.

In particular, the interactions involving the heavy X and Y gauge bosons with the SM fermions are given by

$$\begin{aligned}\mathcal{L}_{(X,Y)} = \frac{g_G}{\sqrt{2}}&\Big[\overline{X}^\alpha_\mu\left(\overline{d}_{\alpha R}\gamma^\mu e^+_R + \overline{d}_{\alpha L}\gamma^\mu e^+_L + \epsilon_{\alpha\beta\gamma}\overline{u}^{c\gamma}_L\gamma^\mu u_{\beta L}\right)\\ &+ \overline{Y}^\alpha_\mu\left(-\overline{d}_{\alpha R}\gamma^\mu\nu^c_R + \overline{u}_{\alpha L}\gamma^\mu e^+_L + \epsilon_{\alpha\beta\gamma}\overline{u}^{c\gamma}_L\gamma^\mu d_{\beta L}\right)\Big] + \text{h.c},\end{aligned} \tag{6.40}$$

where α, β, γ are color indices, and $\epsilon_{\alpha\beta\gamma}$ is the totally antisymmetric tensor. These interactions lead to baryon and lepton number violation, providing a framework for proton decay predictions in $SU(5)$ GUTs [98, 99].

Scalar Sector

The scalar sector of $SU(5)$ plays a crucial role in achieving spontaneous symmetry breaking. In particular, scalar fields belonging to the adjoint $\mathbf{24}$ representation and the fundamental $\mathbf{5}$ representation are responsible for breaking $SU(5)$ down to the SM gauge group and further to $U(1)_{\text{EM}}$.

The adjoint scalar field Φ acquires a VEV that triggers the breaking $SU(5) \to SU(3)_C \times SU(2)_L \times U(1)_Y$, while the 5-plet scalar field H_5 is responsible for electroweak symmetry breaking and giving masses to fermions via Yukawa interactions.

The scalar Lagrangian $\mathcal{L}_\Phi$ is given by

$$\mathcal{L}_\Phi = \text{Tr}\left[(D_\mu\Phi)^\dagger(D^\mu\Phi)\right] + (D_\mu H_5)^\dagger(D^\mu H_5) - V(\Phi, H_5), \tag{6.41}$$

where the trace is taken over the group indices and the covariant derivatives are defined as

$$D_\mu \Phi = \partial_\mu \Phi + i g_G [A_\mu, \Phi], \tag{6.42}$$

$$D_\mu H_5 = (\partial_\mu - i g_G \, \mathbf{A}_\mu)\, H_5. \tag{6.43}$$

The scalar potential $V(\Phi, H_5)$ governs the self-interactions and the mixing between Φ and H_5. A simple form of the potential can be written as [96, 97]

$$\begin{aligned} V(\Phi, H_5) &= \mu_{\mathbf{24}}^2 \,\mathrm{Tr}(\Phi^\dagger \Phi) + \mu_{H_5}^2\, (H_5^\dagger H_5) + \lambda_{\mathbf{24}}\, [\mathrm{Tr}(\Phi^\dagger \Phi)]^2 \\ &+ \lambda_{H_5}\, (H_5^\dagger H_5)^2 + \lambda_{\mathbf{24}H_5}\, \mathrm{Tr}(\Phi^\dagger \Phi)\, (H_5^\dagger H_5), \end{aligned} \tag{6.44}$$

where $\mu_{\mathbf{24}}^2$, $\mu_{H_5}^2$ are mass parameters and $\lambda_{\mathbf{24}}$, λ_{H_5}, $\lambda_{\mathbf{24}H_5}$ are dimensionless couplings controlling the strengths of the interactions.

These terms dictate the vacuum structure of the theory and the mass spectra of the resulting scalar and gauge bosons after spontaneous symmetry breaking.

Yukawa sector

In the construction of the SU(5) invariant Yukawa Lagrangian, the tensor product decompositions are given by:

$$\begin{aligned} 5^* \times 10 &= 5 + 45^*, \\ 10 \times 10 &= 5^* + 45 + 50, \\ 5^* \times 5^* &= 10^* + 15^*. \end{aligned}$$

The above decompositions reveal that, in adhering to the scalar representations of the minimal $SU(5)$ theory, namely $H_5 \equiv 5_H$ and $\Phi \equiv 24_H$, only two invariant Yukawa mass terms exist:

$$5^* \otimes 10 \otimes 5_H^*, \tag{6.45}$$

$$10 \otimes 10 \otimes 5_H. \tag{6.46}$$

Consequently, the Yukawa Lagrangian at the GUT scale becomes:

$$\mathcal{L}_Y^G = \Gamma_1\, (5_L^{C\alpha})^T\, C\, 10_L^{\alpha\beta}\, 5_H^{*\beta} + \Gamma_2\, \epsilon_{\alpha\beta\gamma\delta r}\, (10_L^{\alpha\beta})^T\, C\, 10_L^{\gamma\delta}\, 5^r + \text{h.c.}, \tag{6.47}$$

where Γ_1 and Γ_2 denote the Yukawa couplings.

6.6 SPONTANEOUS SYMMETRY BREAKING

As discussed previously, the scalar potential for the adjoint Higgs multiplet Φ in $SU(5)$ is given by [96]:

$$V(\Phi) = m_1^2 \operatorname{Tr}(\Phi^2) + \lambda_1 \left[\operatorname{Tr}(\Phi^2)\right]^2 + \lambda_2 \operatorname{Tr}(\Phi^4), \tag{6.48}$$

where Φ is expanded over the generators T^a of the adjoint representation:

$$\Phi = \sum_{a=1}^{24} \phi^a T^a. \tag{6.49}$$

To achieve symmetry breaking $SU(5) \to SU(3)_c \times SU(2)_L \times U(1)_Y$, we align the VEV of Φ along the T^{24} generator:

$$\langle \Phi \rangle = \phi_{24} T^{24} = \frac{1}{\sqrt{15}} \phi_{24} \operatorname{Diag}(-2, -2, -2, 3, 3). \tag{6.50}$$

This choice leaves the SM subgroup unbroken while breaking the rest of $SU(5)$. Symmetry is preserved for generators T satisfying $[\Phi, T] = 0$, with the transformation $\Phi \to U\Phi U^\dagger$, $U = \exp(-i\underline{\alpha} \cdot \underline{T})$. The VEV $\langle \Phi \rangle$ commutes with the SM generators but not with the X and Y leptoquark generators, indicating partial symmetry breaking [98]. We find:

$$\operatorname{Tr}(\Phi^2) = 2\phi_{24}^2, \quad \left[\operatorname{Tr}(\Phi^2)\right]^2 = 4\phi_{24}^4, \tag{6.51}$$

and

$$\operatorname{Tr}(\Phi^4) = \frac{14}{15} \phi_{24}^4. \tag{6.52}$$

Minimizing the potential yields:

$$\frac{\delta V}{\delta \Phi} = \phi_{24} \left[-4m_1^2 + \left(16\lambda_1 + \frac{56}{15}\lambda_2\right) \phi_{24}^2\right] = 0, \tag{6.53}$$

leading to the VEV:

$$\phi_{24}^2 = v_\Phi^2 = \frac{m_1^2}{4\lambda_1 + \frac{14}{15}\lambda_2}, \tag{6.54}$$

with stability conditions:

$$\lambda_2 > 0, \quad \lambda_1 > -\frac{7}{30}\lambda_2. \tag{6.55}$$

Thus, the explicit form of the VEV is:

$$\langle \Phi \rangle = \frac{v_\Phi}{\sqrt{15}} \begin{pmatrix} -2 & & & & \\ & -2 & & & \\ & & -2 & & \\ & & & 3 & \\ & & & & 3 \end{pmatrix}. \tag{6.56}$$

The VEV generates masses for the X and Y gauge bosons via the kinetic term:

$$\mathcal{L}_{\text{mass}} = -g_G^2 \operatorname{Tr}([A_\mu, \Phi]^2), \tag{6.57}$$

leading to:

$$\mathcal{L}_{\text{mass}} = \frac{20}{3} g_G^2 v_\Phi^2 \sum_{i=1}^{3} \left(\overline{X}_i^\mu X_{i\mu} + \overline{Y}_i^\mu Y_{i\mu} \right), \tag{6.58}$$

with gauge boson masses:

$$m_X^2 = m_Y^2 = \frac{20}{3} g_G^2 v_\Phi^2. \tag{6.59}$$

Thus, X and Y bosons acquire superheavy masses of order M_{GUT} and decouple from low-energy physics [97, 98].

In the second stage, the scalar field H_5 acquires a VEV, breaking $SU(3)_c \times SU(2)_L \times U(1)_Y$ down to $SU(3)_c \times U(1)_{\text{em}}$. The scalar potential reads:

$$V(H_5) = \frac{m_2^2}{2} H_5^\dagger H_5 + \frac{\lambda_3}{4} (H_5^\dagger H_5)^2. \tag{6.60}$$

Choosing the VEV in the neutral component:

$$\langle H_5 \rangle = \begin{pmatrix} 0 \\ 0 \\ 0 \\ 0 \\ v_H \end{pmatrix}, \tag{6.61}$$

and minimizing the potential gives:

$$v_H^2 = -\frac{m_2^2}{\lambda_3}. \tag{6.62}$$

This mechanism is analogous to electroweak symmetry breaking in the SM, generating $W^{\pm}$ and Z^0 masses.

The fermion mass terms arise from Yukawa couplings:

$$\begin{aligned}\mathcal{L}_m &= Y_1(\bar{d}_L^C C d_L + \bar{e}_L^T C e_L^C)\frac{v_H^*}{\sqrt{2}} + 4Y_2(\bar{u}_L^T C u_L + \bar{u}_L^{CT} C u_L)\frac{v_H}{\sqrt{2}} + \text{h.c.} \\ &= (Y_1 \bar{d}_R d_L - Y_1^T \bar{e}_R e_L)\frac{v_H^*}{\sqrt{2}} + 4(Y_2 \bar{u}_R u_L - Y_2^T \bar{u}_R u_L)\frac{v_H}{\sqrt{2}}. \end{aligned} \tag{6.63}$$

At tree level, the minimal $SU(5)$ model predicts:

$$m_d = m_e, \tag{6.64}$$

$$m_s = m_\mu, \tag{6.65}$$

$$m_b = m_\tau, \tag{6.66}$$

at the unification scale [96]. However, renormalization effects modify these relations. Notably, at low energies:

$$m_b \approx 3m_\tau, \tag{6.67}$$

consistent with observations [100]. Yet, discrepancies arise for lighter generations:

$$m_s \approx 3m_\mu, \tag{6.68}$$

$$\frac{m_\mu}{m_e} \approx \frac{m_s}{m_d}. \tag{6.69}$$

The ratios $m_d/m_s \simeq 1/24$ and $m_e/m_\mu \simeq 1/207$ are starkly different, posing a major flaw in minimal $SU(5)$ [101].

Possible remedies include: (*i*) extending the Higgs sector by adding a 45 representation [102]; (*ii*) introducing higher-dimensional operators suppressed by the Planck scale [100].

6.7 DOUBLET-TRIPLET SPLITTING PROBLEM

The coupling terms between the 24-Higgs scalar representation and the 5-Higgs scalar representation, denoted as $V(\Phi, H)$, are given by:

$$\begin{aligned}V(\Phi, H) &= m_1^2 \text{Tr}\Phi^2 + \lambda_1[\text{Tr}(\Phi^2)]^2 + \lambda_2 \text{Tr}\Phi^4 + \frac{m_2^2}{2} H_5^\dagger H_5 \\ &+ \frac{\lambda_3}{4}(H_5^\dagger H_5)^2 + \underbrace{\alpha H^\dagger H \text{Tr}\Phi^2 + \beta H^\dagger \Phi^2 H}, \end{aligned} \tag{6.70}$$

where the underlined terms yield coupling between h_3 (triplet colored Higgs scalars) and h_2 (SM Higgs doublet) with 24-Higgs. Consequently, the 5-Higgs scalars acquire mass as follows:

$$(2\alpha + \frac{4}{15}\beta)v_\Phi^2 h_3^\dagger h_3 + (2\alpha + \frac{9}{15}\beta)v_\Phi^2 h_2^\dagger h_2. \quad (6.71)$$

However, this leads to a problem: the mass of both h_3 (Higgs triplet) and h_2 (Higgs doublet) would be of the order of $v_\Phi^2 \approx 10^{15}$ GeV. While this may be acceptable for h_3, as it can contribute to proton decay, the doublet Higgs h_2 should ideally have a mass of around 100 GeV to facilitate electroweak symmetry breaking. This discrepancy is known as the doublet-triplet splitting problem [96, 103].

Nevertheless, it is possible to fine-tune the parameters in the equation to make the second term zero, which yields:

$$\alpha = -\frac{9}{30}\beta \qquad \text{and} \qquad \beta < 0. \quad (6.72)$$

The doublet-triplet splitting problem has motivated the development of various mechanisms to naturally generate the splitting without extreme fine-tuning. One of the most notable proposals is the so-called "missing VEV mechanism" [104], originally introduced in the context of supersymmetric GUTs. More sophisticated solutions often involve introducing extended symmetries, additional Higgs fields, or extra dimensions [105]. Resolving this issue remains essential for ensuring the consistency and viability of realistic GUT models.

6.8 GAUGE COUPLING UNIFICATION IN *SU*(5)

The possibility of unifying the gauge couplings within a GUT was first proposed in the seminal work of Georgi and Glashow [96]. In the SM, there are three gauge couplings corresponding to the SM gauge groups. The renormalization group equations (RGEs) describe how the coupling constants of a quantum field theory evolve with the energy scale at which the theory is probed. The RGEs for the gauge couplings in the SM, $\alpha_i = g_i^2/4\pi$ with $i = 1, 2, 3$ referring to $U(1)_Y$, $SU(2)_L$, and $SU(3)_C$, respectively, can be written as:

$$\frac{d\alpha_i}{dt} = \frac{\alpha_i^2}{2\pi} b_i, \quad (6.73)$$

where $t = \ln \mu$ and μ is the renormalization scale. The solution to these equations is given by

$$\alpha_i^{-1}(\mu) = \alpha_i^{-1}(M_X) + \frac{b_i}{2\pi} \ln \left(\frac{M_X}{\mu} \right). \tag{6.74}$$

Here, the running is from the low scale μ (e.g., M_Z) up to a high energy scale M_X. The beta function coefficients, b_i, in the SM with n_f families are given by:

$$b_1 = \frac{1}{10} + \frac{4}{3} n_f, \tag{6.75}$$

$$b_2 = -\frac{43}{6} + \frac{4}{3} n_f, \tag{6.76}$$

$$b_3 = -11 + \frac{4}{3} n_f. \tag{6.77}$$

For $n_f = 3$, one obtains $b_1 = \frac{41}{10}$, $b_2 = -\frac{19}{6}$, and $b_3 = -7$. Since b_1 is positive, α_1 increases with energy, while α_2 and α_3 decrease. However, α_3 decreases more rapidly than α_2. Thus, at some scale higher than the Fermi scale, the strength of the strong interactions may become comparable to the strength of the electroweak interactions.

The evolution of the SM gauge couplings is illustrated in Fig. 6.1, emphasizing that there is no high-energy scale M_X where the gauge

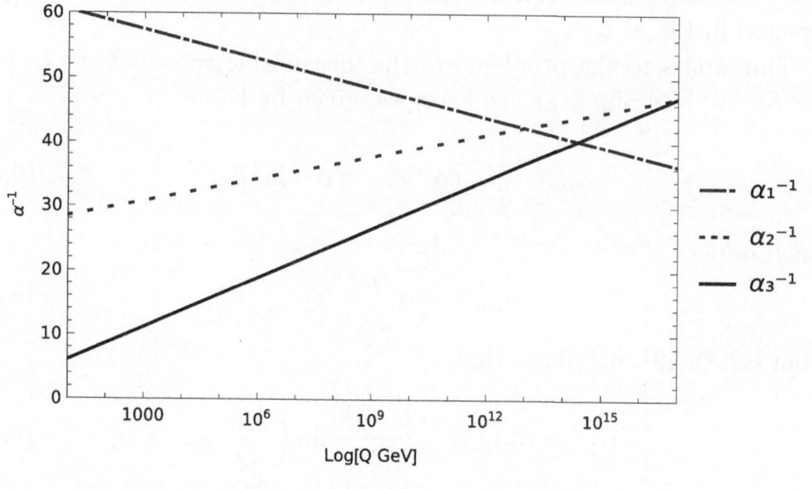

Figure 6.1 Evolution of the inverse of the SM three coupling constants.

couplings meet at a single point. Consequently, their unification within the framework of the SM is not achieved [98].

In the context of the SU(5) GUT, the primary goal is to consolidate all fundamental forces under a single gauge coupling at a high energy scale conventionally denoted as M_X. At energy scales $Q > M_X$, the theory predicts a unique coupling constant g_G associated with the unified gauge group. Thus, at these energies, the strong, weak, and electromagnetic interactions are governed by identical couplings:

$$g_G = g_3 = g_2 = g_1. \tag{6.78}$$

This unification provides an elegant framework for understanding the fundamental forces within a single theoretical structure.

Given the normalization condition imposed on λ_{12}, one has $Y = \sqrt{\frac{5}{3}}\lambda_{12}$, leading to $g' = \sqrt{\frac{3}{5}}g_1$. Consequently, one finds that the Weinberg mixing angle at the unification scale satisfies:

$$\sin^2\theta_W = \frac{g'^2}{g'^2 + g^2} = \frac{3}{8}, \tag{6.79}$$

which holds at M_X. For scales $Q < M_X$, the running of the couplings follows

$$\alpha_i^{-1}(Q) = \alpha_5^{-1} + \frac{b_i}{2\pi}\ln\left(\frac{M_X}{Q}\right), \tag{6.80}$$

where α_5 denotes the unified coupling at the scale M_X. This running is depicted in Fig. 6.2.

The inputs to the problem are the measurable values of the $U(1)_{\rm em}$ and $SU(3)_C$ couplings at low energies, given by [106]:

$$\alpha_{em} = \frac{3}{5}\alpha_1\cos^2\theta_W = \alpha_2\sin^2\theta_W, \tag{6.81}$$

which implies

$$\alpha_{em}^{-1} = \frac{5}{3}\alpha_1^{-1} + \alpha_2^{-1}. \tag{6.82}$$

From Eq. (6.80), it follows that

$$\alpha_1^{-1} - \alpha_2^{-1} = \frac{b_1 - b_2}{2\pi}\ln\left(\frac{M_X}{Q}\right), \tag{6.83}$$

$$\alpha_2^{-1} - \alpha_3^{-1} = \frac{b_2 - b_3}{2\pi}\ln\left(\frac{M_X}{Q}\right). \tag{6.84}$$

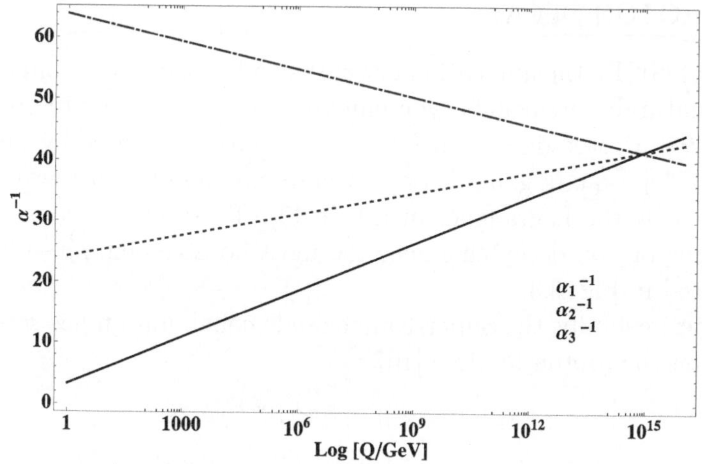

Figure 6.2 The predicted unification scale by the minimal-SU(5) theory for the SM gauge coupling constants.

For $Q = M_Z$, one finds

$$\alpha_{em}^{-1} = \frac{8}{3}\alpha_3^{-1} + \frac{22}{2\pi} \ln\left(\frac{M_X}{M_Z}\right), \tag{6.85}$$

which leads to $M_X \simeq 8 \times 10^{14}$ GeV. Furthermore, the unified coupling is given by $\alpha_5^{-1} \simeq 41.5$.

Additionally, one can estimate $\sin^2\theta_W(M_Z)$ as follows:

$$\begin{aligned} \sin^2\theta_W(M_Z) &= \frac{\alpha_{em}(M_Z)}{\alpha_2(M_Z)} \\ &= \alpha_{em}(M_Z)\left[\frac{1}{\alpha_3(M_Z)} + \frac{22}{12\pi}\ln\left(\frac{M_X}{M_Z}\right)\right] \\ &= \frac{5}{9}\frac{\alpha_{em}(M_Z)}{\alpha_3(M_Z)} + \frac{1}{6} \simeq 0.201, \end{aligned} \tag{6.86}$$

which deviates significantly from the experimental value

$$[\sin^2\theta_W(M_Z)]_{\text{exp}} = 0.23120 \pm 0.00015$$

reported in [106]. Thus, minimal SU(5) with the SM particle content does not successfully achieve gauge coupling unification [98].

6.9 PROTON DECAY

In $SU(5)$ GUTs, the unification of quarks and leptons into common multiplets naturally predicts baryon number violation, leading to proton decay [96,98]. Proton decay processes are primarily mediated by the heavy X and Y leptoquark gauge bosons, whose interactions with fermions are described by the Lagrangian in Eq. (6.40). These interactions give rise to various proton decay channels, the most notable being $p \to e^+\pi^0$, as illustrated in Fig. 6.3.

Experimentally, the Super-Kamiokande collaboration has set a lower bound on the proton lifetime [107]:

$$\tau_p(p \to e^+\pi^0) \gtrsim 5 \times 10^{33} \text{ years.} \tag{6.87}$$

On the theoretical side, the model-independent estimate for the proton decay width due to dimension-6 operators is given by

$$\Gamma_p \approx \alpha_{GUT}^2 \frac{m_p^5}{M_X^4}, \tag{6.88}$$

where α_{GUT} is the unified coupling constant, m_p is the proton mass, and M_X is the mass scale of the heavy leptoquark gauge bosons.

This leads to a lower bound on the M_X scale, assuming $\alpha_{GUT} \sim 1/40 - 1/25$:

$$M_X \gtrsim 3.2 \times 10^{15} \text{ GeV.} \tag{6.89}$$

However, in minimal $SU(5)$ models, gauge coupling unification predicts $M_X \sim 10^{14}$ GeV [108], which is in tension with the experimental lower limits on proton lifetime. This discrepancy highlights a major challenge for minimal $SU(5)$ GUTs, motivating extensions such as supersymmetric unification, missing partner mechanisms, or higher-dimensional constructions [109].

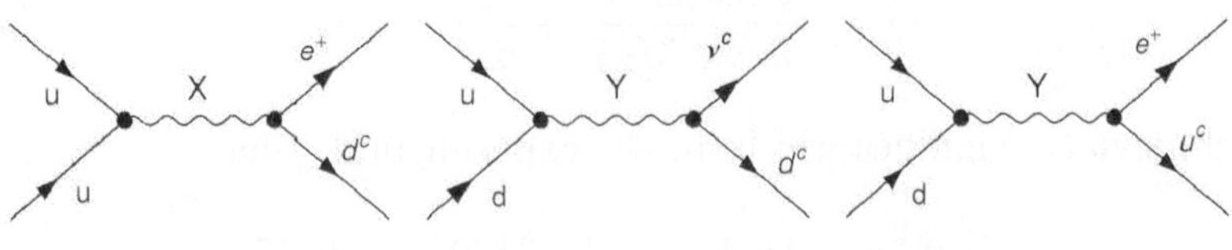

Figure 6.3 Baryon number violating processes at the lowest order via X and Y boson exchange.

6.10 SU(5) ASSESSMENT

6.10.1 Advantages

1. $SU(5)$ is the only GUT with rank 4. While other GUTs, such as $SO(10)$, have higher ranks, which provide greater flexibility in their breaking patterns, they may sacrifice predictability.
2. A distinctive feature of $SU(5)$ is the quantization of electric charge (Q), where $\text{Tr}Q|_5 = 0$ implies $Q(d) = \frac{1}{3}Q(e^-)$.
3. The prediction for the weak mixing angle, $\sin^2\theta_W$, is in excellent agreement with current experimental data.
4. $SU(5)$ predicts a ratio $m_b/m_\tau \simeq 3$ at low energy, which is consistent with experimental observations.
5. The simplest version of $SU(5)$, with a single Higgs 5-plet, avoids tree-level FCNCs.
6. The model is free of gauge anomalies, making it a consistent quantum field theory.
7. $SU(5)$ explains the baryon asymmetry by enabling baryon number violating processes required for baryogenesis; proton decay is a low-energy manifestation of such processes.
8. The minimal $SU(5)$ model makes unique, testable predictions regarding the weak mixing angle $\sin^2\theta_W$ and nucleon lifetime, both of which can be probed experimentally.

6.10.2 Disadvantages

1. The $SU(5)$ model faces significant challenges from new experimental constraints on proton decay, especially the $P \to e^+\pi^0$ channel. If the proton lifetime bound of 10^{32} years is confirmed, the minimal $SU(5)$ model may be ruled out, requiring extensions that reduce predictability.
2. In the minimal version of the model, which only includes Higgs 5-plets, the ratio m_e/m_μ is predicted to be equal to m_d/m_s, which does not match the experimentally observed mass hierarchy of leptons and quarks.

3. Fermions in the $SU(5)$ model are assigned to reducible representations $(5^* + 10)$, which complicates the understanding of their properties.

4. $SU(5)$ does not provide a clear explanation for the masses of fermions. The Yukawa couplings and the structure of the Higgs potential must be fine-tuned to fit experimental data.

5. There exists a significant gap (the "great desert") between the W boson mass (M_W) and the X boson mass (M_X), with no natural explanation for the intermediate mass scale.

6. The Higgs scalar potential introduces considerable arbitrariness due to undetermined parameters in the Higgs potential and the Yukawa Lagrangian.

7. Despite the proximity of the GUT scale M_X to the Planck scale M_P, gravity remains excluded from the unified gauge group picture in $SU(5)$.

8. The gauge hierarchy problem remains unsolved within the $SU(5)$ framework, requiring additional mechanisms such as supersymmetry or extra dimensions.

CHAPTER 7

Non-Minimal $SU(5)$ Model

In this chapter, we discuss the non-minimal $SU(5)$ GUT in which the Higgs sector is extended by the inclusion of a scalar field in the 45 representation [102]. The 45 arises from the tensor product of the antisymmetric 10 representation and the conjugate fundamental 5^* representation, as follows:

$$5 \times 5 = 10_A \oplus 15_S, \tag{7.1}$$

$$10_A \times 5^* = 45 \oplus 5. \tag{7.2}$$

The 45 representation consists of two upper indices from the 10 and a lower index corresponding to the 5^*. To isolate the 45 component from a general tensor $T_\gamma^{\alpha\beta}$, we need to construct a combination that is antisymmetric in α and β, traceless, and satisfies the condition of vanishing total antisymmetrization over all three indices. This leads to the projected form

$$45_\gamma^{\alpha;\,\beta} = 45_\gamma^{\alpha\beta} + 45_\alpha^{\gamma\beta} - 45_\alpha^{\beta\gamma} - 45_\beta^{\alpha\gamma}, \tag{7.3}$$

which ensures the proper mixed symmetry required for the irreducible 45 representation. Therefore, the 45 Higgs scalars field satisfy the following conditions:

$$45_\gamma^{\alpha\beta} = -45_\gamma^{\beta\alpha}, \quad \sum_\alpha^5 (45)_\alpha^{\alpha\beta} = 0, \tag{7.4}$$

with

$$\sum_{i=1}^3 < 45 >_i^{i5} = - < 45 >_4^{45}$$

DOI: 10.1201/9781003457701-7

and

$$v_{45} =< 45 >_1^{15} =< 45 >_2^{25} =< 45 >_3^{35},$$

where the VEVs are defined as v_{45}.

The SU(5)-invariant Higgs potential for the 45_H representation is given by [110]:

$$\begin{aligned} V(45_H) = & -\frac{1}{2}\mu_{45}^2 (45_\gamma^{\alpha\beta} 45_{\alpha\beta}^\gamma) + \lambda_1 (45_\gamma^{\alpha\beta} 45_{\alpha\beta}^\gamma)^2 \\ & + \lambda_2 45_\gamma^{\alpha\beta} 45_{\alpha\beta}^\delta 45_\delta^{k\lambda} 45_{k\lambda}^\gamma \\ & + \lambda_3 45_\gamma^{\alpha\beta} 45_{\alpha\beta}^\delta 45_\lambda^{k\gamma} 45_{k\delta}^\lambda + \lambda_4 45_\beta^{\alpha\delta} 45_{\alpha\gamma}^\beta 45_\lambda^{k\gamma} 45_{k\delta}^\lambda \\ & + \lambda_5 45_\delta^{\alpha\gamma} 45_{\gamma e}^\beta 45_\alpha^{k\delta} 45_{k\beta}^e + \lambda_6 45_\delta^{\alpha\gamma} 45_{\gamma e}^\beta 45_\alpha^{ke} 45_{k\beta}^\delta \\ & + \lambda_7 45_\delta^{\alpha\gamma} 45_{\gamma e}^\beta 45_\beta^{k\delta} 45_{k\alpha}^e + \lambda_8 45_\delta^{\alpha\gamma} 45_{\gamma e}^\beta 45_\beta^{ke} 45_{k\alpha}^\gamma . \end{aligned} \tag{7.5}$$

The SM decomposition of the 45 Higgs representation is given by [97]:

$$\begin{aligned} 45_H = & (8,2)_{1/2} \oplus (6^*,1)_{-1/3} \oplus (3,3)_{-1/3} \oplus (3^*,2)_{-7/6} \oplus (3,1)_{-1/3} \\ & \oplus (3^*,1)_{4/3} \oplus (1,2)_{1/2}. \end{aligned} \tag{7.6}$$

This decomposition follows directly from the tensor product relations in Eq. (7.1) and Eq. (7.2). These components correspond to the fields defined in the spectrum table (Table 7.1).

TABLE 7.1 45-Higgs scalars spectrum.

$(6^*,1,-1/3)_k^{ij}$	$\equiv$	ϕ_k^{ij}	$= \frac{1}{6}\, \epsilon^{ijl}\phi_{lk}$
$(3^*,2,-7/6)_c^{ij}$	$\equiv$	ϕ_c^{ij}	$= \frac{1}{6}\, \epsilon^{ijl}\phi_{lc} = \frac{1}{6}\, \epsilon^{ijl} 45_{lc}$
$(3^*,1,4/3)_k^{ab}$	$\equiv$	ϕ_k^{ab}	$= \frac{1}{2}\, \delta^{ab} 45_k$
$(3,1,1/3)_c^{ib}$ $(3,1,1/3)_k^{ij}$	$\equiv$	T_1^i	$= \frac{1}{2}\, \delta_c^b 45^i = -\frac{1}{2}\, \delta_k^j 45^i$
$(3,3,-1/3)_c^{ib}$	$\equiv$	T_3^i	$= 45_c^{ib} - \frac{1}{2}\delta_c^b 45_d^{id}$
$(1,2,1/2)_c^{ab}$ $(1,2,1/2)_k^{ib}$	$\equiv$	D^b	$= \frac{1}{2}\, \delta_c^a 45^b = -\frac{1}{2}\, \delta_k^i 45^b$

The color octet scalars are defined as:

$$(8,2,1/2)^{ia}_j \equiv S^{ia}_j = (45_H)^{ia}_j - \frac{1}{3}\delta^i_j (45_H)^{ma}_m. \tag{7.7}$$

7.1 LOW ENERGY IMPLICATIONS OF 45-HIGGS

7.1.1 Fermion Masses

From the above mentioned decompositions, the SU(5) invariant Yukawa Lagrangian can be extended to include the 45-Higgs representation as follows [102, 111, 112]:

$$\begin{aligned} L^G_Y =& \Gamma_1\,(5^{C\alpha}_L)^T C\; 10^{\alpha\beta}_L\; 5^{*\beta}_H + \Gamma_2\,(5^{C\gamma}_L)^T C\; 10^{\alpha\beta}_L\; 45^{*\alpha\beta}_\gamma \\ &+ \epsilon_{\alpha\beta\gamma\delta r}\left[\Gamma_3\,(10^{\alpha\beta}_L)^T C\; 10^{\gamma\delta}_L\; 5^r + \Gamma_4\,(10^{\alpha\beta}_L)^T C\; 10^{m\gamma}_L\; 45^{\delta r}_m\right]. \end{aligned} \tag{7.8}$$

After electroweak symmetry breaking, the fermion masses are given by:

$$\begin{aligned} L_m =& (Y_1\,\bar{d}_{Ri}\,d_{Li} + Y_1^T\,\bar{e}_R\,e_L)\frac{\upsilon_5^*}{\sqrt{2}} + (2Y_2\,\bar{d}_{Ri}\,d_{Li} - 6Y_2^T\,\bar{e}_R\,e_L)\frac{\upsilon_{45}^*}{\sqrt{2}} \\ &+ [4(Y_3+Y_3^T)\frac{\upsilon_5}{\sqrt{2}} - 8(Y_4 - Y_4^T)\frac{\upsilon_{45}}{\sqrt{2}}]\bar{u}_{Ri}u_{Li}. \end{aligned} \tag{7.9}$$

The fermion masses are then:

$$\begin{aligned} M_E &= Y_1^T\frac{\upsilon_5^*}{\sqrt{2}} - 6Y_2^T\frac{\upsilon_{45}^*}{\sqrt{2}}, \\ M_D &= Y_1\frac{\upsilon_5^*}{\sqrt{2}} + 2Y_2\frac{\upsilon_{45}^*}{\sqrt{2}}, \\ M_U &= 4(Y_3+Y_3^T)\frac{\upsilon_5}{\sqrt{2}} - 8(Y_4 - Y_4^T)\frac{\upsilon_{45}}{\sqrt{2}}, \end{aligned} \tag{7.10}$$

where $\upsilon^2 = \upsilon_5^2 + \upsilon_{45}^2$. This formulation corrects the fermion mass relations from the previous chapter.

7.1.2 Nucleon Decay Constraints on 45-Higgs Masses

Nucleon decay can be induced by dimension-five operators involving color-triplet Higgs multiplets, where the decay rate is proportional to the square of the Higgs masses ($\tau^{-1}_{n.p} \propto M_\Phi^{-2}$). Therefore, matter stability provides a strong experimental constraint on the masses of new scalars in GUT theories, particularly those carrying color quantum numbers.

The 45-Higgs fields contributing to nucleon decay are derived from the GUT Yukawa Lagrangian:

$$L_Y^{45_H} = \Gamma_2\, (5_L^{C\gamma})^T C\, 10_L^{\alpha\beta}\, 45_\gamma^{\alpha\beta\,*} + \epsilon_{\alpha\beta\gamma\delta r}\, \Gamma_4\, (10_L^{\alpha\beta})^T C\, 10_L^{m\gamma}\, 45_m^{\delta r}. \quad (7.11)$$

Starting with the colored octet scalars, S_j^{ia}, we have:

$$S_j^{ia} = \begin{pmatrix} S_j^{i4} \\ S_j^{i5} \end{pmatrix} = \begin{pmatrix} S_j^{i+} \\ S_j^{i0} \end{pmatrix} \equiv S^A T^A,$$

where $A = 1,\ldots,8$ and T^A are the $SU(3)$ generators. The relevant Yukawa interactions are:

$$\begin{aligned} L_Y^S &= 2Y_2\, (5_L^{Cj})^T C\, 10_L^{ia}\, S^* + 4Y_4 \Big[\epsilon_{klbia}\, (10_L^{kl})^T C\, 10_L^{jb} \\ &\quad + \epsilon_{kblia}\, (10_L^{kb})^T C\, 10_L^{jl}\Big] S + \text{h.c.} \\ &= 2Y_2\, \bar{d}_{Ri}\, S^\dagger\, Q_{Lj} - 4\epsilon_{\alpha\beta}(Y_4 - Y_4^T)\, \bar{u}_{Ri}\, Q_{Lj}^\alpha\, S^\beta + \text{h.c.}. \end{aligned} \quad (7.12)$$

For the 45-Higgs doublet, D^a:

$$D^a = \begin{pmatrix} D^+ \\ D^0 \end{pmatrix}, \quad (7.13)$$

the Yukawa interactions are:

$$L_Y^D = Y_2^{eT}\, \bar{e}_R\, D^\dagger\, l_L + Y_2^d\, \bar{d}_R\, D^\dagger\, Q_L + 2\epsilon_{\alpha\beta} Y_4\, \bar{u}_R\, Q_L^\alpha\, D^\beta + \text{h.c.}. \quad (7.14)$$

Table 7.2 summarizes the interactions of the 45-Higgs scalar multiplets with matter, showing which fields contribute to nucleon decay [111]. Experimental lower bounds on the proton decay rate, such as $\tau(p \to k^+\bar{\nu}) > 2.3 \times 10^{33}$ years [113], require the scalars responsible for nucleon decay to be superheavy.

7.1.3 Gauge Coupling Unification

The minimal SU(5) model fails to achieve gauge coupling unification at a consistent scale and is incompatible with current experimental limits on nucleon decay. To address these issues, we propose an effective SU(5) framework featuring two Higgs doublets, H_5 and D_{45}, at the electroweak scale [114]. In addition, we introduce color-octet scalars with

TABLE 7.2 The interactions of 45-Higgs scalars with matter and their contributions to proton decay.

45_H Scalar	$Y^2_{AB}5^*_A 10_B 45^*$	$Y^4_{AB}10_A 10_B 45$	Interactions	Proton Decay
$(6^*, 1, -1/3)$	$2Y^2_{AB}d^{cT}_{Ak}Cu^c_{Bl}\phi^{lk*}$	$4Y^4_{AB}(u^{lT}_A Cd^k_B - d^{lT}_A Cu^k_B)\phi_{lk}$	$\bar{U}\bar{D}$, UD	No
$(3^*, 2, -7/6)$	$2Y^2_{AB}e^T_A Cu^c_{Bi}\phi^{i1*}$	$4(Y^4_{AB} - Y^4_{BA})u^{iT}_A Ce^c_B\phi_{i1}$	$L\bar{U}, Q\bar{E}$	No
	$-2Y^2_{AB}\nu^T_A Cu^c_{Bi}\phi^{i2*}$	$4(Y^4_{AB} - Y^4_{BA})d^{iT}_A Ce^c_B\phi_{i2}$		
$(3^*, 1, 4/3)$	$2Y^2_{AB}d^{cT}_{Ak}Ce^c_B\phi^{k*}$	$4(Y^4_{AB} - Y^4_{BA})\epsilon^{mnk}u^{cT}_{Am}Cu^c_{Bn}\phi_k$	$\bar{E}\bar{D}, \bar{U}\bar{U}$	Yes
$(3, 1, -1/3)$	$2Y^2_{AB}e^T_A Cu^i_B T^*_i$ $-2Y^2_{AB}\nu^T_A Cd^i_B T^*_i$ $2Y^2_{AB}\epsilon_{lkj}d^{cT}_{Al}Cu^c_{Bk}T^*_j$	$4(Y^4_{AB} - Y^4_{BA})u^{cT}_{Ai}Ce^c_B T^i$	$LQ, \bar{U}\bar{D}$ $\bar{U}\bar{E}$	Yes
$(3, 3, -1/3)$	$-2Y^2_{AB}\nu^T_A Cu^i_B T^*_{i1}$	$4(Y^4_{AB} - Y^4_{BA})\epsilon_{jki}d^T_{Aj}Cd_{Bk}T^{i1}$	LQ, QQ	Yes
	$Y^2_{AB}e^T_A Cu^i_B T^*_{i2}$ $-Y^2_{AB}\nu^T_A Cd^i_B T^*_{i2}$	$2(Y^4_{AB} - Y^4_{BA})\epsilon_{jki}u^T_{Aj}Cu_{Bk}T^{i2}$		
	$2Y^2_{AB}e^T_A Cd^i_B T^*_{i3}$	$4(Y^4_{AB} - Y^4_{BA})\epsilon_{jki}d^T_{Aj}Cu_{Bk}T^{i3}$		

masses in the 1–2 TeV range, and color-triplet scalars T^i_3 with masses around 10^8 GeV. The remaining components of the 45-dimensional Higgs multiplet acquire GUT-scale masses via their couplings with the adjoint 24_H scalar.

The one-loop b_i-coefficients in these effective models are:

$$b^{EW}_i = \left(\frac{21}{5}, -3, -7\right), b^S_i = \left(5, -\frac{5}{3}, -5\right), b^T_i = \left(\frac{26}{5}, \frac{1}{3}, -\frac{9}{5}\right). \quad (7.15)$$

Thus, the renormalization group equations (RGEs) become:

$$\alpha^{-1}(M_X) = \alpha_1^{-1}(100) - \frac{1}{2\pi}\Big[b_1^{(SM+2HD)}\ ln\frac{O_S}{Q_{EW}} + b_1^{(SM+2HD+S)}\ ln\frac{O_T}{Q_S} + b_1^{(SM+2HD+S+\Phi)}\ ln\frac{M_X}{Q_T}\Big], \quad (7.16)$$

$$\alpha^{-1}(M_X) = \alpha_2^{-1}(100) - \frac{1}{2\pi}\Big[b_2^{(SM+2HD)}\ ln\frac{O_S}{Q_{EW}} + b_2^{(SM+2HD+S)}\ ln\frac{O_T}{Q_S} + b_2^{(SM+2HD+S+\Phi)}\ ln\frac{M_X}{Q_T}\Big], \quad (7.17)$$

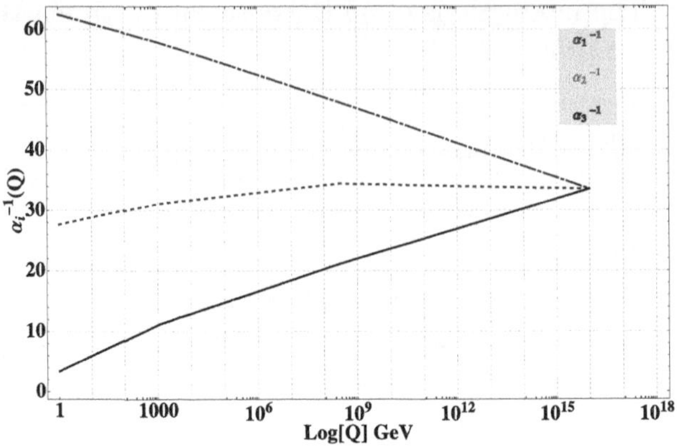

Figure 7.1 Gauge coupling running in the effective $SU(5)$ model with 2HD, TeV-scale octet, and intermediate-scale triplet scalars.

$$\alpha^{-1}(M_X) = \alpha_3^{-1}(100) - \frac{1}{2\pi}\left[b_3^{(SM+2HD)}\, ln\frac{O_S}{Q_{EW}} + b_3^{(SM+2HD+S)}\, ln\frac{O_T}{Q_S} + b_3^{(SM+2HD+S+\Phi)}\, ln\frac{M_X}{Q_T}\right]. \tag{7.18}$$

Solving these equations at $Q_S \sim 1-2$ TeV and $Q_T \sim 10^8$ GeV results in a unification at $M_X \sim 10^{16}$ GeV, as shown in Fig. 7.1 [115], consistent with grand unification scale predictions.

7.2 EFFECTIVE *SU*(5)TWO-HIGGS-DOUBLET MODEL

After the breaking of the $SU(5)$ GUT to the SM, the scalar sector contains two electroweak doublets: one from the $\mathbf{5}_H$ and one from the $\mathbf{45}_H$ representations. These doublets are defined as

$$H \equiv (1,2)_{1/2} = \begin{pmatrix} H^+ \\ H^0 \end{pmatrix}, \quad D \equiv (1,2)_{1/2} = \begin{pmatrix} D^+ \\ D^0 \end{pmatrix}. \tag{7.19}$$

Electroweak symmetry breaking induces vacuum expectation values (vevs),

$$\langle H \rangle = \begin{pmatrix} 0 \\ v_5 \end{pmatrix}, \quad \langle D \rangle = \begin{pmatrix} 0 \\ -3v_{45} \end{pmatrix}, \tag{7.20}$$

where we define $v_1 \equiv v_5$, $v_2 \equiv -3v_{45}$, and $v = \sqrt{v_1^2 + v_2^2}$. The angle $\tan\beta = v_2/v_1$ parameterizes the ratio of vevs. The effective scalar potential at low energies takes the form [114, 116]:

$$V(H, D) = -\mu_H^2 H^\dagger H + \lambda_1 (H^\dagger H)^2 - \mu_D^2 D^\dagger D + \lambda_2 (D^\dagger D)^2 + \lambda_3 (H^\dagger H)(D^\dagger D) + \lambda_4 |H^\dagger D|^2 + \left[\frac{\lambda_5}{2}(H^\dagger D)^2 + \text{h.c.}\right], \tag{7.21}$$

where λ_4 and λ_5 encode the mixing between the doublets. The minimization conditions yield:

$$\mu_H^2 = \lambda_1 v_1^2 + \tfrac{1}{2}(\lambda_3 + \lambda_4 + \lambda_5) v_2^2, \tag{7.22}$$

$$\mu_D^2 = \lambda_2 v_2^2 + \tfrac{1}{2}(\lambda_3 + \lambda_4 + \lambda_5) v_1^2. \tag{7.23}$$

Expanding around the vevs,

$$H = \begin{pmatrix} H^+ \\ v_1 + h_1 + i a_1 \end{pmatrix}, \quad D = \begin{pmatrix} D^+ \\ v_2 + h_2 + i a_2 \end{pmatrix}, \tag{7.24}$$

the CP-even Higgs mass matrix is

$$\mathcal{M}_R^2 = \begin{pmatrix} 2\lambda_1 v_1^2 & (\lambda_3 + \lambda_4 + \lambda_5) v_1 v_2 \\ (\lambda_3 + \lambda_4 + \lambda_5) v_1 v_2 & 2\lambda_2 v_2^2 \end{pmatrix}. \tag{7.25}$$

This matrix is diagonalized by the mixing angle α, defined by

$$\tan 2\alpha = \frac{(\lambda_3 + \lambda_4 + \lambda_5) v_1 v_2}{\lambda_1 v_1^2 - \lambda_2 v_2^2}. \tag{7.26}$$

The mass eigenstates h and H are given by:

$$\begin{pmatrix} H \\ h \end{pmatrix} = \begin{pmatrix} \cos\alpha & \sin\alpha \\ -\sin\alpha & \cos\alpha \end{pmatrix} \begin{pmatrix} h_1 \\ h_2 \end{pmatrix}, \tag{7.27}$$

with masses

$$M_{H,h}^2 = \lambda_1 v_1^2 + \lambda_2 v_2^2 \pm \sqrt{(\lambda_1 v_1^2 - \lambda_2 v_2^2)^2 + ((\lambda_3 + \lambda_4 + \lambda_5) v_1 v_2)^2}. \tag{7.28}$$

This effective 2HDM structure emerges naturally in non-minimal SU(5) extension and plays a key role in scalar sector phenomenology, including potential implications for flavor, CP violation, and collider signatures.

The effective Yukawa couplings of the SM-like Higgs doublet arise from interactions involving the 5_H and 45_H Higgs representations. After symmetry breaking, the relevant Yukawa Lagrangian takes the form:

$$\begin{aligned}\mathcal{L}_Y = \bar{e}_R\left(Y_1^{eT}H^\dagger + Y_2^{eT}D^\dagger\right)l_L + \bar{d}_R\left(Y_1^d H^\dagger + Y_2^d D^\dagger\right)Q_L \\ + \epsilon_{\alpha\beta}\bar{u}_R\left(Y_3' H^\beta + Y_4' D^\beta\right)Q_L^\alpha + \text{h.c.},\end{aligned} \tag{7.29}$$

where $Y_3' = 4(Y_3 + Y_3^T)$ and $Y_4' = 2Y_4$. The couplings Y_1 and Y_2 can be related to fermion mass matrices as:

$$Y_1 = \frac{3M_D + M_E}{4v_5}, \quad Y_2 = \frac{M_D - M_E}{8v_{45}}. \tag{7.30}$$

In the physical basis, assuming flavor diagonal charged leptons and up-quarks, and that the down-quark mass matrix is diagonalized by left-handed rotations only ($V_L^d = V_{CKM}, V_R^d = I$), the couplings of the SM-like Higgs to fermions are:

$$Y_{huu} = -Y_3' \sin\alpha + Y_4' \cos\alpha = -\frac{m_U}{v}\frac{\sin\alpha}{\cos\beta} + Y_4' \cos\alpha, \tag{7.31}$$

$$Y_{hdd} = -\left(\frac{3m_D + m_E V_{CKM}}{4v_5}\right)\sin\alpha + \left(\frac{m_D - m_E V_{CKM}}{4v_{45}}\right)\cos\alpha, \tag{7.32}$$

$$Y_{hee} = -\left(\frac{3m_D V_{CKM} + m_E}{4v_5}\right)\sin\alpha + \left(\frac{m_D V_{CKM} - m_E}{4v_{45}}\right)\cos\alpha. \tag{7.33}$$

The up-quark mass relation $m_U = Y_3' v_5 - 8(Y_4^T - Y_4)v_{45}$ implies that Y_4' vanishes if Y_4 is symmetric.

The couplings of the SM-like Higgs to electroweak gauge bosons arise from the kinetic terms of H and D:

$$\mathcal{L}_{\text{kin}} = (D^\mu H)^\dagger(D_\mu H) + (D^\mu D)^\dagger(D_\mu D). \tag{7.34}$$

After electroweak symmetry breaking and field redefinitions, the couplings to massive vector bosons are:

$$g_{hW^+W^-} = gM_W \sin(\beta - \alpha), \tag{7.35}$$

$$g_{hZZ} = \frac{gM_Z}{\cos\theta_W}\sin(\beta - \alpha). \tag{7.36}$$

The effective Yukawa couplings of the SM-like Higgs doublet arise from interactions involving the 5_H and 45_H Higgs representations. After symmetry breaking, the relevant Yukawa Lagrangian takes the form:

$$\begin{aligned}\mathcal{L}_Y = \bar{e}_R\left(Y_1^{eT}H^\dagger + Y_2^{eT}D^\dagger\right)l_L + \bar{d}_R\left(Y_1^dH^\dagger + Y_2^dD^\dagger\right)Q_L \\ + \epsilon_{\alpha\beta}\bar{u}_R\left(Y_3'H^\beta + Y_4'D^\beta\right)Q_L^\alpha + \text{h.c.},\end{aligned} \tag{7.37}$$

where $Y_3' = Y_3 + Y_3^T$ and $Y_4' = Y_4 - Y_4^T$. In the up-quark sector, the contraction $\epsilon_{\alpha\beta}\bar{u}_RD^\beta Q_L^\alpha$ is antisymmetric in SU(2) indices. Therefore, only the antisymmetric part of Y_4 contributes to this interaction, and if Y_4 is symmetric, the contribution vanishes.

The couplings Y_1 and Y_2 are related to fermion mass matrices as:

$$Y_1 = \frac{3M_D + M_E}{4v_5}, \quad Y_2 = \frac{M_D - M_E}{8v_{45}}. \tag{7.38}$$

In the physical basis, assuming flavor-diagonal up-type quarks and charged leptons, and that the down-type quark mass matrix is diagonalized by a left-handed CKM rotation only (i.e., $V_L^d = V_{\text{CKM}}$, $V_R^d = \mathbb{1}$), the couplings of the SM-like Higgs to fermions become:

$$Y_{huu} = -Y_3'\sin\alpha + Y_4'\cos\alpha = -\frac{m_U}{v}\frac{\sin\alpha}{\cos\beta} + Y_4'\cos\alpha, \tag{7.39}$$

$$Y_{hdd} = -\left(\frac{3m_D + m_EV_{\text{CKM}}}{4v_5}\right)\sin\alpha + \left(\frac{m_D - m_EV_{\text{CKM}}}{8v_{45}}\right)\cos\alpha, \tag{7.40}$$

$$Y_{hee} = -\left(\frac{3m_DV_{\text{CKM}} + m_E}{4v_5}\right)\sin\alpha + \left(\frac{m_DV_{\text{CKM}} - m_E}{8v_{45}}\right)\cos\alpha. \tag{7.41}$$

Here, m_U, m_D, m_E are the diagonal mass matrices for up-type quarks, down-type quarks, and charged leptons, respectively.

Note that the up-quark mass relation

$$m_U = Y_3'v_5 + Y_4'v_{45} \tag{7.42}$$

implies that $Y_4' = 0$ if the original Y_4 is symmetric, since $Y_4' = 2Y_4$ and antisymmetric parts cancel out due to fermion symmetries.

The couplings of the SM-like Higgs to the electroweak gauge bosons originate from the kinetic terms of the scalar fields:

$$\mathcal{L}_{\text{kin}} = (D^\mu H)^\dagger(D_\mu H) + (D^\mu D)^\dagger(D_\mu D). \tag{7.43}$$

After electroweak symmetry breaking and rotation to the Higgs mass eigenstates, the trilinear couplings to massive vector bosons are given by:

$$g_{hW^+W^-} = gM_W \sin(\beta - \alpha), \tag{7.44}$$

$$g_{hZZ} = \frac{gM_Z}{\cos\theta_W} \sin(\beta - \alpha), \tag{7.45}$$

where α is the mixing angle between the CP-even Higgs bosons, and $\tan\beta = v_5/v_{45}$ is the ratio of vacuum expectation values.

7.3 CHARGED HIGGS IN NON-MINIMAL *SU*(5)

In the non-minimal $SU(5)$, the physical charged Higgs field emerges as the orthogonal combination to the Goldstone boson eaten by the $W^\pm$ bosons:

$$H^\pm = -\sin\beta\, H_5^\pm + \cos\beta\, D^\pm, \quad \tan\beta = \frac{v_{45}}{v_5}. \tag{7.46}$$

From the $SU(5)$-invariant scalar potential, one obtains the following charged Higgs mass matrix in the $(H^\pm, D^\pm)$ basis is then given by

$$\mathcal{M}^2_{\text{charged}} = \begin{pmatrix} \mu_5^2 + \lambda_5 v_5^2 & \mu_{54} v \\ \mu_{54} v & \mu_{45}^2 + \lambda_{45} v_{45}^2 \end{pmatrix}, \tag{7.47}$$

where λ_5, λ_{45} are effective quartic couplings. Diagonalizing this matrix yields one Goldstone boson and a physical charged Higgs boson $H^\pm$ with mass:

$$m_{H^\pm}^2 \simeq \mu_{45}^2 + \lambda_{45} v^2 + \Delta m^2, \tag{7.48}$$

where Δm^2 encodes contributions from the mixing parameter μ_{54} and possible loop corrections.

The charged Higgs couplings to fermions depend on both Y_2 and Y_3 and can introduce non-trivial flavor structures:

$$\mathcal{L}_{H^\pm} \supset H^+ \bar{u}_i \left[V_{ij} \left(A_{dj} P_R + A_{uj} P_L\right)\right] d_j + H^+ \bar{\nu}_i A_{eij} P_R e_j + \text{h.c.}, \tag{7.49}$$

with A_{fij} determined by the underlying Yukawa structure and $\tan\beta$. These interactions can lead to flavor-violating processes not present in standard 2HDMs, such as $H^\pm \to \mu\nu$, $H^\pm \to c\bar{s}$, or $t\bar{b}$ with non-standard rates.

7.4 COLORED OCTET SCALARS

In our SU(5) effective model, the color octet scalars arise from the 45_H representation as defined above, with

$$S_j^{ia} = (45_j^{ia})_H - \frac{1}{3}\delta_j^i(45_m^{ma})_H, \qquad S_j^{ia} \equiv (8,2)_{1/2} = S^A T^A, \tag{7.50}$$

where $i, j = 1, 2, 3$, $A = 1, \ldots, 8$, and T^A are the SU(3) generators. These scalars preserve $SU(3)_C$ due to their vanishing VEVs.

The scalar potential involving the octets can be derived from the SU(5)-invariant potential terms [114, 116]. After electroweak symmetry breaking, the mass terms become:

$$m_{S^\pm}^2 = -m_S^2 + \lambda' v_{45}^2, \qquad m_{S^0_{R,I}}^2 = -m_S^2 + \lambda' v_{45}^2, \tag{7.51}$$

$$m_{S^\pm,5}^2 = -m_S^2 + \lambda_3 v_5^2, \qquad m_{S^0_{R,I},5}^2 = -m_S^2 + (\lambda_3 + \lambda_4 \pm \lambda_5) v_5^2, \tag{7.52}$$

where $\lambda' = \lambda_2' + \lambda_3' + \lambda_4' - \lambda_6'$. The VEV v_5 induces a mass splitting between the charged and neutral components.

The octet scalars couple to fermions via SU(5) Yukawa interactions:

$$\mathcal{L}_Y^S = 2Y_2\, \bar{d}_{Ri} Q_{Lj} S^\dagger - 4\epsilon_{\alpha\beta}(Y_4 - Y_4^T)\bar{u}_{Ri} Q_{Lj}^\alpha S^\beta + \text{h.c.} \tag{7.53}$$

Assuming that Y_4 is symmetric, then the couplings to up-quarks vanish. Using Eq. (7.38), the interaction simplifies to:

$$\mathcal{L}_Y^S = \frac{1}{4v_{45}} \bar{d}_{Ri}(M_D - M_E^T) S^\dagger Q_{Lj} + \text{h.c.} \tag{7.54}$$

In the physical basis and assuming $V_R^d = \mathbf{I}$, $V_L^d = V_{\text{CKM}}$, we find:

$$\mathcal{L}_Y^S = \frac{1}{4v_{45}} \Big[\bar{d}_{Ri}(M_D^{\text{diag}} V_{\text{CKM}}^\dagger - M_E^{\text{diag}}) u_{Lj} S^- + \bar{d}_{Ri}(M_D^{\text{diag}} - M_E^{\text{diag}} V_{\text{CKM}}) d_{Lj} S^0 \Big] + \text{h.c.} \tag{7.55}$$

This leads to flavor-violating couplings of the neutral octet scalar due to the misalignment of M_D and M_E [114, 116]. While the minimal flavor violation hypothesis suppresses these effects, we consider Y_2 as a generic matrix without assuming MFV.

The kinetic term $\text{Tr}[(D_\mu 45_H)^\dagger D^\mu 45_H]$ induces interactions between octet scalars and gluons. The covariant derivative acting on the octet scalar S_j^{ia} is:

$$D_\mu S_j^{ia} = \partial_\mu S_j^{ia} - i\frac{g_s}{2}(G_\mu^\alpha T^\alpha)^{ik} S_j^{ka} - i\frac{g_s}{2}(G_\mu^\alpha T^\alpha)_j^n S_n^{ia} + \cdots, \tag{7.56}$$

resulting in gluon interactions:

$$\mathcal{L}_{Sgg} = ig_s\,\mathcal{F}^{ABD}\left(S^{A-}G^{\mu B}\partial_\mu S^{D+} + S^{A0}_{R,I}G^{\mu B}\partial_\mu S^{D0}_{R,I}\right)$$
$$+\,g_s^2\,\mathcal{F}^{ABCD}\left(S^{A-}G^{\mu B}G^C_\mu S^{D+} + S^{A0}_{R,I}G^{\mu B}G^C_\mu S^{D0}_{R,I}\right) + \text{h.c.}, \tag{7.57}$$

with structure constants $\mathcal{F}^{ABD} = \text{Tr}(T^A T^B T^D)$ and $\mathcal{F}^{ABCD} = \text{Tr}(T^A T^B T^D t^E)$. The SU(3) generators satisfy:

$$\{t^A, t^B\} = \frac{1}{N}\delta^{AB} + d^{ABC}t^C, \qquad [t^A, t^B] = if^{ABC}t^C, \tag{7.58}$$

where d^{ABC} and f^{ABC} are the symmetric and antisymmetric structure constants, respectively.

7.5 SCALAR LEPTOQUARK IN NON-MINIMAL *SU*(5)

As emphasized in Table 7.2 that while the scalar triplets ϕ_k^{ab} and ϕ_c^{ib} contribute to the proton decay and they must be superheavy, the scalar triplet ϕ_c^{ij} does not. It has no interaction terms that would cause proton decay. By writing ϕ_c^{ij} as $(\phi_1^i, \phi_2^i)^T$, one can demonstrate that the scalar triplet has the following peculiar interactions:

$$\mathcal{L} = 2Y^2_{AB}e^T_A C u^c_{Bi}\phi^{i1*} + 4(Y^4_{AB} - Y^4_{BA})u^{iT}_A C e^c_B \phi_{i1}$$
$$-\,2Y^2_{AB}\nu^T_A C u^c_{Bi}\phi^{i2*} + 4(Y^4_{AB} - Y^4_{BA})d^{iT}_A C e^c_B \phi_{i2}. \tag{7.59}$$

The first two interaction terms would imply the decay of $b \to s\ell^+\ell^-$ through scalar triplet leptoquark ϕ^{i1} mediation, while the last two interaction terms clearly account for the decay $b \to c\tau\nu$ via scalar triplet leptoquark ϕ^{i2} mediation. These terms can be written as

$$\mathcal{L} = 2Y^2_{AB}\bar{u}_{Bi}P_L\nu_A\phi^{i2*} - 4Y^{4'}_{AB}\bar{e}_B P_L d^i_A \phi_{i2} + h.c., \tag{7.60}$$

where we used $C^T = -C$ and $\bar{\Psi} = \Psi^{cT}_L$, and define $Y^{4'}_{AB} \equiv (Y^4_{AB} - Y^4_{BA})$. In the mass eigenstate basis, where

$$d_A \to V^{CKM}_{AB}d_B, \quad \nu_A \to V^{\text{PMNS}}_{AB}\nu_B, \quad u_A \to u_A, \quad e_A \to e_A,$$

the above Lagrangian takes the form:

$$\mathcal{L} = 2Y^2_{AB}\bar{u'}_{Bi}P_L V^{\text{PMNS}}_{AK}\nu'_k\phi^{i2*} - 4Y^{4'}_{AB}\bar{e'}_B P_L V^{\text{CKM}}_{AK}d'_K\phi_{i2}$$
$$+\,h.c. \tag{7.61}$$

In this regards, the amplitude of $b \to c\tau\nu$ transition is given by

$$\mathcal{M} = -\frac{8Y^{4'}_{13}V^{\text{CKM}}_{13}}{M^2_\phi}\Big[\frac{1}{2}(\bar{u}_\tau P_L v_{\nu_\tau})(\bar{u}_C P_L u_b) + \frac{1}{8}(\bar{u}_\tau \sigma^{\mu\nu} P_L v_{\nu_\tau})(\bar{u}_C P_L \sigma^{\mu\nu} u_b) \times \Big(Y^2_{12}V^{\text{PMNS}}_{13} + Y^2_{22}V^{\text{PMNS}}_{23} + Y^2_{32}V^{\text{PMNS}}_{33}\Big)\Big] + \Big(Y^{4'}_{13}V^{\text{CKM}}_{13} \to Y^{4'}_{23}V^{\text{CKM}}_{23}\Big). \tag{7.62}$$

Because V^{CKM}_{13} and V^{CKM}_{23} are so small (10^{-3} and 10^{-2}, respectively), the amplitude of $b \to c\tau\nu$ is essentially determined by the leptoquark masses $M_\phi, Y^2_{22}, Y^2_{32}$, and $Y^{4'}_{13}$.

CHAPTER 8

Supersymmetry: Central Candidate for BSM

Supersymmetry (SUSY) is a theoretical framework that extends the symmetries of space-time. At its core is the Super Poincaré algebra [117,118], which includes:

- P_μ (Translation): This represents the symmetry of translations in space and time.
- $M_{\mu\nu}$ (Rotation and Lorentz Transformation): These are the symmetries of rotations and Lorentz transformations, crucial for maintaining the consistency of relativistic theories.
- Q_α (SUSY Transformation): The SUSY transformation introduces a symmetry between fermions and bosons, extending the standard symmetries of space-time.

SUSY plays a critical role in addressing some of the fundamental challenges in modern theoretical physics:

1. Stability of the Hierarchy: SUSY provides a mechanism that ensures the stability of the hierarchy between the weak scale and the Planck scale. Without SUSY, the vast difference between these scales could be destabilized by quantum corrections, leading to a severe fine-tuning problem in the SM.
2. Unified Theory BSM: Supersymmetric theories are promising candidates for a unified theory that extends beyond the SM. They offer a framework that could potentially integrate all fundamental forces under a single theoretical umbrella.

DOI: 10.1201/9781003457701-8

3. Natural Electroweak Symmetry Breaking: SUSY makes the mechanism of electroweak symmetry breaking more natural. In the SM, the Higgs boson mass is sensitive to quantum corrections, leading to a fine-tuning problem. SUSY alleviates this issue by introducing superpartners that cancel out the problematic corrections, thereby stabilizing the Higgs boson mass.

Overall, Supersymmetry offers a compelling extension to our understanding of the universe, providing solutions to some of the most profound challenges in high-energy physics [118].

8.1 HIERARCHY PROBLEM AND SUPERSYMMETRY

One of the significant challenges in modern physics is the hierarchy problem, which arises when attempting to unify gravity with the other fundamental forces. Theories like string theory and GUTs suggest that the fundamental forces should unify at a very high energy scale, close to the Planck scale (10^{19} GeV). However, this presents a dilemma when considering the vastly lower energy scale associated with the electroweak interactions, particularly the mass of the Higgs boson.

The main issue is that quantum corrections to the Higgs boson mass are extremely sensitive to the highest energy scales, leading to potential divergences that require severe fine-tuning. Specifically, without any additional mechanisms, the Higgs boson mass would need to be fine-tuned to an accuracy of $\mathcal{O}(10^{34})$, which is an extraordinarily unnatural requirement.

SUSY offers a compelling solution to this problem by introducing a symmetry between fermions and bosons. In SUSY, the quantum corrections to the Higgs mass from standard particles are exactly canceled by the corresponding contributions from their superpartners. This cancellation occurs because the loop diagrams that are quadratically divergent in the SM cancel out, term by term, against equivalent diagrams involving the superpartners. This mechanism avoids the problematic fine-tuning and stabilizes the Higgs boson mass at a natural scale.

For SUSY to effectively address the hierarchy problem, the superpartner masses should lie around $\mathcal{O}(1)$ TeV, especially given the observed Higgs boson mass of $m_H \approx 125$ GeV. This prediction implies that some of these superpartners should be detectable at the LHC, providing a testable aspect of the theory. Through this mechanism, SUSY provides a natural framework for solving the hierarchy problem, making it a central candidate for BSM [118].

8.2 GAUGE COUPLING UNIFICATION

A compelling motivation for low-scale SUSY, particularly around the TeV scale, stems from the idea of gauge coupling unification. In the SM, the gauge couplings, α_1 ($\mathrm{U}(1)_Y$), α_2 ($\mathrm{SU}(2)_L$), and α_3 ($\mathrm{SU}(3)_C$), evolve differently with energy due to quantum corrections, as governed by the renormalization group equations (RGEs).

The one-loop RGE for the gauge coupling α_i is:

$$\frac{d\alpha_i(t)}{dt} = \frac{b_i}{2\pi}\alpha_i^2(t), \quad t = \ln(\mu), \tag{8.1}$$

where b_i are beta function coefficients that depend on the field content of the theory. In the SM, the beta function coefficients for $n_g = 3$ generations and $n_H = 1$ Higgs doublet are:

$$b_1^{\mathrm{SM}} = \frac{41}{10}, \tag{8.2}$$

$$b_2^{\mathrm{SM}} = -\frac{19}{6}, \tag{8.3}$$

$$b_3^{\mathrm{SM}} = -7. \tag{8.4}$$

These couplings do not unify at a single scale, suggesting that the SM alone does not naturally lead to unification.

However, in the Minimal Supersymmetric Standard Model (MSSM), the particle content is extended to include superpartners and typically $n_H = 2$ Higgs doublets. The beta function coefficients become:

$$b_1^{\mathrm{MSSM}} = \frac{33}{5}, \tag{8.5}$$

$$b_2^{\mathrm{MSSM}} = 1, \tag{8.6}$$

$$b_3^{\mathrm{MSSM}} = -3. \tag{8.7}$$

With these values, the RG evolution of the gauge couplings leads to convergence at a common unification scale, around $M_{\mathrm{GUT}} \sim 10^{16}$ GeV [119–121]. This elegant unification strongly supports the MSSM and motivates the presence of superpartners near the TeV scale, potentially within reach of collider experiments such as the LHC.

8.3 WHAT IS SUPERSYMMETRY?

SUSY is a theoretical framework that postulates a fundamental symmetry between two distinct classes of particles: bosons (integer spin) and fermions (half-integer spin). In SUSY, each bosonic state has a corresponding fermionic partner and vice versa. This symmetry is encapsulated by the action of fermionic operators Q_α and their Hermitian conjugates $\bar{Q}_{\dot{\alpha}}$, which satisfy:

$$Q_\alpha|\text{Boson}\rangle = |\text{Fermion}\rangle, \quad Q_\alpha|\text{Fermion}\rangle = |\text{Boson}\rangle. \tag{8.8}$$

Since Q_α carries spin-$\frac{1}{2}$, SUSY extends the symmetries of space-time, making it a space-time symmetry rather than a purely internal one.

Supersymmetry was first formulated as an extension of the Poincaré group, which underlies special relativity, by Golfand and Likhtman in 1971 [122], and later developed into a full quantum field theoretic framework in 1973 [123]. In quantum field theory, symmetries are generally classified as:

1. External (Space-Time) Symmetries: These include translations $x^\mu \to x^\mu + a^\mu$ and Lorentz transformations $x^\mu \to \Lambda^\mu{}_\nu x^\nu$.

2. Internal Symmetries: These act on the fields themselves, e.g., $\phi^a(x) \to M^a{}_b\phi^b(x)$, and include gauge symmetries like $U(1)$, $SU(2)$, and $SU(3)$.

According to Noether's theorem, every continuous symmetry corresponds to a conserved quantity. In particle physics, conserved quantities such as energy, momentum, electric charge, and color charge are associated with these fundamental symmetries.

Efforts to unify internal and space-time symmetries into a larger algebraic structure encountered a major obstacle: the Coleman-Mandula theorem [124]. It states that under certain reasonable assumptions, the most general symmetry group of the S-matrix is a direct product of the Poincaré group and an internal symmetry group, implying no non-trivial unification is possible.

SUSY evades this restriction by introducing fermionic symmetry generators that transform as spinors, rather than scalars. These additional generators extend the Poincaré algebra to the so-called Super-Poincaré algebra, enabling a consistent and non-trivial unification of space-time and internal symmetries.

SUSY Algebra and Mass Degeneracy

The complete set of SUSY generators includes not only those of the Poincaré group but also two additional spinor generators, Q and $\bar{Q}$, which act on the fields. These spinor generators introduce a symmetry between bosons and fermions and satisfy specific algebraic relations:

$$\{Q_\alpha, Q_\beta\} = 0, \quad \{\bar{Q}_{\dot{\alpha}}, \bar{Q}_{\dot{\beta}}\} = 0, \quad \{Q_\alpha, \bar{Q}_{\dot{\beta}}\} = 2\sigma^\mu_{\alpha\dot{\beta}} P_\mu, \tag{8.9}$$

where $\sigma^\mu \equiv (1, \sigma^i)$ and $\bar{\sigma}^\mu \equiv (1, -\sigma^i)$. Here, $\alpha, \beta, \dot{\alpha}, \dot{\beta} = 1, 2$ are spinor indices, and P_μ represents the momentum operator [123, 125].

These generators also obey specific commutation relations with the generators of the Poincaré group:

$$\begin{aligned} &[P^\mu, Q_\alpha] = \left[P^\mu, \bar{Q}_{\dot{\alpha}}\right] = 0, \quad [M^{\mu\nu}, Q_\alpha] = -i(\sigma^{\mu\nu})_\alpha^\beta Q_\beta, \\ &\left[M^{\mu\nu}, \bar{Q}^{\dot{\alpha}}\right] = -i(\bar{\sigma}^{\mu\nu})^{\dot{\alpha}}_{\dot{\beta}} \bar{Q}^{\dot{\beta}}. \end{aligned} \tag{8.10}$$

The commutators of Q_α and $\bar{Q}_{\dot{\alpha}}$ with internal symmetry generators T_i generally vanish, indicating that SUSY transformations do not mix with internal symmetries.

The momentum squared operator $P^2 = P_\mu P^\mu$ serves as a Casimir operator in the SUSY algebra, meaning it is invariant under SUSY transformations:

$$\left[Q_\alpha, P^2\right] = 0. \tag{8.11}$$

This invariance implies that particles related by SUSY transformations, such as a fermion $|F\rangle$ and a boson $|B\rangle$, must have the same mass. Specifically, if a fermion state $|F\rangle$ is obtained from a bosonic state $|B\rangle$ through the action of the SUSY generator Q_α, then:

$$\begin{aligned} m_F|F\rangle &= P^2|F\rangle = P^2 Q_\alpha|B\rangle = Q_\alpha P^2|B\rangle = m_B Q_\alpha|B\rangle = m_B|F\rangle \\ \Longrightarrow & \quad m_F = m_B. \end{aligned} \tag{8.12}$$

This relation confirms that particles within the same irreducible representation of the SUSY algebra, such as a boson and its corresponding fermion partner, share the same mass, a defining prediction of supersymmetric theories [122, 125].

8.4 SUPERSPACE AND SUSY TRANSFORMATIONS

Supersymmetric theories are naturally formulated in *superspace*, an extension of spacetime incorporating both bosonic coordinates x^μ and anticommuting fermionic coordinates θ^α, $\overline{\theta}^{\dot{\alpha}}$. Fields in this space are represented by superfields $\Phi(x, \theta, \overline{\theta})$, which encapsulate entire SUSY multiplets.

Supersymmetry extends spacetime translations by introducing fermionic generators Q_α and $\overline{Q}_{\dot{\alpha}}$. A finite SUSY transformation induces specific shifts in superspace coordinates:

$$\begin{aligned} x^\mu &\to x^\mu + y^\mu + i\eta\sigma^\mu\overline{\theta} - i\theta\sigma^\mu\overline{\eta}, \\ \theta^\alpha &\to \theta^\alpha + \eta^\alpha, \\ \overline{\theta}^{\dot{\alpha}} &\to \overline{\theta}^{\dot{\alpha}} + \overline{\eta}^{\dot{\alpha}}, \end{aligned} \tag{8.13}$$

where η^α, $\overline{\eta}^{\dot{\alpha}}$ are Grassmann-valued parameters. These transformations unify bosonic and fermionic degrees of freedom through geometry.

Expanding an infinitesimal transformation allows identification of the SUSY generators:

$$\begin{aligned} P_\mu &= i\partial_\mu, \\ Q_\alpha &= i\left(\partial_{\theta^\alpha} + i\sigma^\mu_{\alpha\dot{\alpha}}\overline{\theta}^{\dot{\alpha}}\partial_\mu\right), \\ \overline{Q}_{\dot{\alpha}} &= i\left(\partial_{\overline{\theta}^{\dot{\alpha}}} - i\theta^\alpha\sigma^\mu_{\alpha\dot{\alpha}}\partial_\mu\right). \end{aligned} \tag{8.14}$$

To construct SUSY-invariant actions, one introduces covariant derivatives that anticommute with the supercharges:

$$\begin{aligned} D_\alpha &= \partial_{\theta^\alpha} + i\sigma^\mu_{\alpha\dot{\alpha}}\overline{\theta}^{\dot{\alpha}}\partial_\mu, \\ \overline{D}_{\dot{\alpha}} &= -\partial_{\overline{\theta}^{\dot{\alpha}}} - i\theta^\alpha\sigma^\mu_{\alpha\dot{\alpha}}\partial_\mu, \end{aligned} \tag{8.15}$$

with the algebra:

$$\{D_\alpha, \overline{D}_{\dot{\beta}}\} = 2i\sigma^\mu_{\alpha\dot{\beta}}\partial_\mu, \quad \{D_\alpha, D_\beta\} = \{\overline{D}_{\dot{\alpha}}, \overline{D}_{\dot{\beta}}\} = 0. \tag{8.16}$$

This framework forms the basis for constructing supersymmetric field theories and ensures manifest SUSY invariance in superspace [117, 118, 125].

8.5 SUSY LAGRANGIANS

Supersymmetric theories unify bosons and fermions through superfields. Two essential types are the chiral and vector superfields. A chiral superfield Φ satisfies $\overline{D}_{\dot{\alpha}}\Phi = 0$, depending only on the shifted coordinate $y^\mu = x^\mu + i\theta\sigma^\mu\bar{\theta}$. Its component expansion includes a complex scalar ϕ, a Weyl fermion ψ, and an auxiliary field F [117, 126]. Chiral superfields are closed under multiplication, enabling the construction of supersymmetric superpotentials.

Vector superfields V, satisfying $V = V^\dagger$, describe gauge bosons. Gauge redundancies are removed via SUSY gauge transformations $V \to V + \Lambda + \Lambda^\dagger$, with Λ chiral. In the Wess-Zumino (WZ) gauge, V contains only the physical components: gauge field v_μ, gauginos $\lambda, \bar{\lambda}$, and auxiliary field D [117].

A general renormalizable SUSY Lagrangian in superspace reads:

$$\mathcal{L} = \Phi_i^\dagger \Phi^i|_{\theta^2\bar{\theta}^2} + W(\Phi_i)|_{\theta^2} + \bar{W}(\Phi_i^\dagger)|_{\bar{\theta}^2}, \tag{8.17}$$

where the superpotential W is:

$$W(\Phi_i) = \frac{1}{2}M_{ij}\Phi^i\Phi^j + \frac{1}{3}Y_{ijk}\Phi^i\Phi^j\Phi^k. \tag{8.18}$$

Expanding Φ in components gives:

$$\Phi(x,\theta,\bar{\theta}) = \phi + \sqrt{2}\theta\psi + \theta\theta F + i\theta\sigma^\mu\bar{\theta}\partial_\mu\phi \tag{8.19}$$

$$- \frac{i}{\sqrt{2}}\theta\theta\sigma^\mu\bar{\theta}\partial_\mu\psi + \frac{1}{4}\theta\theta\bar{\theta}\bar{\theta}\Box\phi. \tag{8.20}$$

The kinetic term is:

$$\Phi_i^\dagger\Phi^i|_{\theta^2\bar{\theta}^2} = \partial_\mu\phi_i^\dagger\partial^\mu\phi_i + i\bar{\psi}_i\bar{\sigma}^\mu\partial_\mu\psi_i + F_i^*F_i. \tag{8.21}$$

The full component Lagrangian becomes:

$$\mathcal{L} = \partial_\mu\phi_i^\dagger\partial^\mu\phi_i + i\bar{\psi}_i\bar{\sigma}^\mu\partial_\mu\psi_i + F_i^*F_i \tag{8.22}$$

$$+ \left[M_{ij}(\phi_iF_j - \frac{1}{2}\psi_i\psi_j) + Y_{ijk}(\phi_i\phi_jF_k - \psi_i\psi_j\phi_k) + \text{h.c.}\right]. \tag{8.23}$$

Eliminating the auxiliary field via its equation of motion:

$$F_i^* = -\frac{\partial W}{\partial \phi_i}, \tag{8.24}$$

yields the on-shell Lagrangian:

$$\mathcal{L} = \partial_\mu\phi_i^\dagger\partial^\mu\phi_i + i\bar{\psi}_i\bar{\sigma}^\mu\partial_\mu\psi_i - \left|\frac{\partial W}{\partial\phi_i}\right|^2 \tag{8.25}$$

$$- \left(\frac{1}{2}M_{ij}\psi_i\psi_j + Y_{ijk}\psi_i\psi_j\phi_k + \text{h.c.}\right). \tag{8.26}$$

The scalar potential is:

$$V = \sum_i \left|\frac{\partial W}{\partial\phi_i}\right|^2. \tag{8.27}$$

For Abelian gauge fields, define the field strength superfields:

$$W_\alpha = -\frac{1}{4}\bar{D}\bar{D}D_\alpha V, \quad \bar{W}_{\dot{\alpha}} = -\frac{1}{4}DD\bar{D}_{\dot{\alpha}}V. \tag{8.28}$$

In the WZ gauge:

$$V_{\text{WZ}} = \theta\sigma^\mu\bar{\theta}v_\mu + i\theta\theta\bar{\theta}\bar{\lambda} - i\bar{\theta}\bar{\theta}\theta\lambda + \frac{1}{2}\theta\theta\bar{\theta}\bar{\theta}D, \tag{8.29}$$

and the vector Lagrangian becomes:

$$\int d^2\theta\, W^\alpha W_\alpha = \frac{1}{2}D^2 - \frac{1}{4}v^{\mu\nu}v_{\mu\nu} - i\lambda\sigma^\mu\partial_\mu\bar{\lambda}. \tag{8.30}$$

The potential is:

$$V = \frac{1}{2}D^2. \tag{8.31}$$

In non-Abelian SUSY gauge theories, the chiral superfields transform as $\Phi \to e^{i\Lambda}\Phi$, with $\Lambda = \Lambda^a T^a$, where T^a are Lie algebra generators obeying:

$$[T^a, T^b] = if^{abc}T^c, \quad \text{Tr}(T^aT^b) = \frac{1}{2}\delta^{ab}. \tag{8.32}$$

The vector superfield transforms as:

$$e^V \to e^{i\Lambda^\dagger}e^V e^{-i\Lambda}. \tag{8.33}$$

Define:

$$W_\alpha = -\frac{1}{4}\bar{D}\bar{D}\left(e^{-V}D_\alpha e^V\right), \quad W_\alpha \to e^{i\Lambda}W_\alpha e^{-i\Lambda}. \tag{8.34}$$

The complete non-Abelian SUSY Lagrangian is:

$$\mathcal{L} = \Phi_i^\dagger(e^V)^i_j\Phi^j\big|_{\theta^2\bar{\theta}^2} + \frac{1}{4g^2}\text{Tr}[W^\alpha W_\alpha]\big|_{\theta^2} + W(\Phi^i)\big|_{\theta^2} + \text{h.c.} \tag{8.35}$$

After eliminating auxiliary fields:

$$V_{\text{SUSY}} = \frac{1}{2}D^aD^a + F_i^*F^i, \quad F^i = \frac{\partial W}{\partial\phi_i}, \quad D^a = g^a\phi_i^*T^a\phi_i. \tag{8.36}$$

8.6 SUPERSYMMETRY BREAKING

SUSY, while theoretically well-motivated, cannot be an exact symmetry of nature. If it were unbroken, superpartners of the SM particles, such as electrons and squarks, would have the same masses as their SM counterparts, which is clearly not the case experimentally. The absence of such degenerate states implies that SUSY must be broken. To stabilize the electroweak scale and resolve the hierarchy problem, SUSY breaking should occur at or moderately above the electroweak scale.

As with other symmetries, SUSY can be broken spontaneously, dynamically, or explicitly. The SUSY algebra is given by

$$\{Q_\alpha, \bar{Q}_{\dot{\alpha}}\} = 2\sigma^\mu_{\alpha\dot{\alpha}} P_\mu, \tag{8.37}$$

which, upon contraction with $(\bar{\sigma}^\nu)^{\dot{\beta}\alpha}$, leads to

$$H = \frac{1}{4}\left(\bar{Q}_1 Q_1 + Q_1 \bar{Q}_1 + \bar{Q}_2 Q_2 + Q_2 \bar{Q}_2\right) \geq 0. \tag{8.38}$$

This implies a semi-positive definite Hamiltonian. For a supersymmetric vacuum where $Q_\alpha |0\rangle = \bar{Q}_{\dot{\alpha}} |0\rangle = 0$, the vacuum energy vanishes:

$$E_{\text{vac}} = \langle 0| H |0\rangle = 0. \tag{8.39}$$

Conversely, if the vacuum is not annihilated by some SUSY generators, then $E_{\text{vac}} > 0$, signaling spontaneous SUSY breaking.

Spontaneous SUSY breaking implies that the variation of some field under SUSY transformations acquires a non-zero vacuum expectation value (VEV). For a chiral superfield,

$$\langle 0|\delta_\xi \psi(x)|0\rangle \propto \langle 0|F|0\rangle \neq 0, \tag{8.40}$$

and for a vector superfield,

$$\langle 0|\delta_\xi \lambda(x)|0\rangle \propto \langle 0|D|0\rangle \neq 0. \tag{8.41}$$

While spontaneous SUSY breaking is conceptually attractive, constructing realistic models remains challenging. Alternatively, SUSY can be broken explicitly by introducing terms in the Lagrangian that violate SUSY but do not reintroduce quadratic divergences. These are the so-called soft SUSY breaking terms, classified in [127]:

- Scalar mass terms: $\tilde{m}^2_{ij}\phi^*_i\phi_j$,
- Gaugino mass terms: $\frac{1}{2}M_a\lambda^a\lambda^a + \text{h.c.}$,

- Bilinear scalar terms: $B_{ij}\phi_i\phi_j + \text{h.c.}$,
- Trilinear scalar terms: $A_{ijk}\phi_i\phi_j\phi_k + \text{h.c.}$.

These soft terms are essential for phenomenology: they shape the SUSY mass spectrum and contribute to radiative electroweak symmetry breaking [118]. Ultimately, a full understanding of SUSY breaking would allow all soft terms to be derived from a fundamental theory, such as gravity mediation, gauge mediation, or anomaly mediation [118].

8.7 MINIMAL SUPERSYMMETRIC STANDARD MODEL

The Minimal Supersymmetric Standard Model (MSSM) is the simplest supersymmetric extension of the SM that retains its gauge group $SU(3)_C \times SU(2)_L \times U(1)_Y$ while introducing a superpartner for each SM particle [118, 126]. The MSSM arranges fields into supermultiplets, pairing SM particles with superpartners differing in spin by 1/2:

Supermultiplet	**SM Particle**	**SUSY Partner**	$SU(3)_C \times SU(2)_L \times U(1)_Y$
Q_L	$(u_L, d_L)^T$	$(\tilde{u}_L, \tilde{d}_L)^T$	$(3, 2, 1/6)$
U_L^c	u_L^c	$\tilde{u}_L^c$	$(\bar{3}, 1, -2/3)$
D_L^c	d_L^c	$\tilde{d}_L^c$	$(\bar{3}, 1, 1/3)$
L_L	$(\nu_L, e_L)^T$	$(\tilde{\nu}_L, \tilde{e}_L)^T$	$(1, 2, -1/2)$
E_L^c	e_L^c	$\tilde{e}_L^c$	$(1, 1, 1)$
H_u	(H_u^+, H_u^0)	$(\tilde{H}_u^+, \tilde{H}_u^0)$	$(1, 2, 1/2)$
H_d	(H_d^0, H_d^-)	$(\tilde{H}_d^0, \tilde{H}_d^-)$	$(1, 2, -1/2)$

Unlike the SM, the MSSM requires two Higgs doublets. This stems from the requirement that the superpotential be analytic in chiral superfields, thus excluding conjugate fields like H^* [117]. Furthermore, anomaly cancellation demands both H_u and H_d, since their fermionic superpartners (Higgsinos) contribute oppositely to the gauge anomalies [118]. The MSSM includes gauge superfields for each SM interaction:

- $(B_\mu, \tilde{B})$ for $U(1)_Y$,
- $(W_\mu^a, \tilde{W}^a)$ for $SU(2)_L$,
- $(G_\mu^a, \tilde{g}^a)$ for $SU(3)_C$.

The full MSSM Lagrangian is

$$\mathcal{L}_{\text{MSSM}} = \mathcal{L}_{\text{gauge}} + \mathcal{L}_{\text{matter}} + \mathcal{L}_{\text{superpotential}} + \mathcal{L}_{\text{soft}}. \tag{8.42}$$

The gauge and matter sectors follow from supersymmetric field theory [117]. The superpotential encodes Yukawa interactions and Higgs mixing:

$$W = Y_u Q U^c H_u + Y_d Q D^c H_d + Y_e L E^c H_d + \mu H_u H_d. \tag{8.43}$$

An explicit example involving the top quark superfield illustrates the structure of SUSY interactions:

$$\begin{aligned}\mathcal{L}_{\text{int}} =& -\frac{Y_t}{2}\left(t_L H_u t_L^c + \tilde{t}_L \tilde{H}_u t_L^c + t_L \tilde{H}_u \tilde{t}_L^c + \text{h.c.}\right)\\ &-Y_t^2\left(|H_u \tilde{t}_L^c|^2 + |H_u \tilde{t}_L|^2 + |\tilde{t}_L \tilde{t}_L^c|^2\right).\end{aligned} \tag{8.44}$$

In addition to the standard top Yukawa interaction, SUSY introduces new interactions involving the scalar superpartners (stops) and the Higgsino. These additional terms are a distinctive feature of SUSY and play a crucial role in its phenomenology.

The Higgs and lepton doublets share gauge quantum numbers, allowing for lepton and baryon number-violating terms:

$$W' = \lambda_{ijk} L_i L_j E_k^c + \lambda'_{ijk} L_i Q_j D_k^c + \lambda''_{ijk} D_i^c D_j^c U_k^c + \mu'_i L_i H_u, \tag{8.45}$$

which can mediate rapid proton decay [118, 128]. For instance:

$$\frac{\lambda''_{211}\lambda'_{112}}{m_{\tilde{s}}^2}(u^c d^c)(ue), \tag{8.46}$$

contributes to $p \to e^+\pi^0$, with decay rate:

$$\Gamma \sim \lambda'_{112}\lambda''_{211}\frac{m_p^5}{m_{\tilde{s}}^4}, \quad \tau_p \sim 6 \times 10^{-13}\,\text{s}\left(\frac{m_{\tilde{s}}}{1\ \text{TeV}}\right)^4 (\lambda'_{112}\lambda''_{211})^{-2}. \tag{8.47}$$

To satisfy experimental limits $\tau_p > 1.6 \times 10^{33}\,\text{yr}$, these couplings must be extremely suppressed:

$$\lambda'_{112}\lambda''_{211} < 3 \times 10^{-26}. \tag{8.48}$$

To forbid these terms, R-parity is introduced, defined by:

$$R_P = (-1)^{3B+L+2S}. \tag{8.49}$$

Its consequences are:

1. SUSY particles must be pair-produced.
2. The lightest supersymmetric particle (LSP) is stable and a dark matter candidate.

Soft SUSY-breaking terms introduce masses and trilinear couplings:

$$\mathcal{L}_{\text{soft}} = -\frac{1}{2}M_a\lambda^a\lambda^a - m^2_{\phi_i}\phi_i^*\phi_i - \left(A_fY_f\tilde{f}_L\tilde{f}_RH + B\mu H_uH_d + \text{h.c.}\right). \tag{8.50}$$

These terms, if unconstrained, introduce $\sim$ 100 new parameters. The constrained MSSM (cMSSM) assumes GUT-scale universality:

$$m^2_{\tilde{f}} = m_0^2, \quad m^2_{H_u} = m^2_{H_d} = m_0^2, \tag{8.51}$$

$$A_f = A_0, \quad M_1 = M_2 = M_3 = m_{1/2}, \tag{8.52}$$

reducing the parameter space to $\{m_0, m_{1/2}, A_0, \tan\beta, \text{sign}(\mu)\}$. This setup is often motivated by minimal supergravity (mSUGRA) models.

8.8 SUPERSYMMETRIC SPECTRUM IN THE MSSM

In the MSSM, each SM particle has a corresponding superpartner differing by half a unit of spin. These superpartners, collectively known sparticles, play a crucial role in stabilizing the electroweak scale and providing viable dark matter candidates. The MSSM particle spectrum includes gauginos, Higgsinos, squarks, and sleptons. This section discusses the mass generation, mixing, and phenomenological roles of the key supersymmetric particles: gluinos, charginos, neutralinos, squarks, and sleptons [118].

Gluinos

The gluino ($\tilde{g}$), the fermionic partner of the gluon, acquires mass from the soft SUSY-breaking term:

$$\mathcal{L}_{\text{soft}} \supset -\frac{1}{2}M_3\tilde{g}^a\tilde{g}^a. \tag{8.53}$$

At tree level, the gluino mass is $m_{\tilde{g}} = |M_3(Q)|$. At one-loop, the gaugino mass parameters run as:

$$M_i(Q) = \frac{\alpha_i(Q)}{\alpha_i(M_X)}M_i(M_X), \quad i = 1, 2, 3, \tag{8.54}$$

with universal boundary condition $M_1 = M_2 = M_3 = m_{1/2}$ at the GUT scale. At the electroweak scale:

$$M_3 : M_2 : M_1 \approx 7 : 2 : 1. \tag{8.55}$$

Charginos

Charginos are mass eigenstates formed from charged winos ($\tilde{W}^{\pm}$) and Higgsinos ($\tilde{H}_u^+$, $\tilde{H}_d^-$). The mass terms arise from:

$$\mathcal{L} \supset -\begin{pmatrix} \tilde{W}^- & \tilde{H}_d^- \end{pmatrix} \mathcal{M}_C \begin{pmatrix} \tilde{W}^+ \\ \tilde{H}_u^+ \end{pmatrix} + \text{h.c.}, \tag{8.56}$$

with

$$\mathcal{M}_C = \begin{pmatrix} M_2 & \sqrt{2} M_W \sin\beta \\ \sqrt{2} M_W \cos\beta & \mu \end{pmatrix}. \tag{8.57}$$

This matrix is diagonalized by unitary matrices U and V to yield the physical chargino masses $m_{\tilde{\chi}_1^{\pm}}$, $m_{\tilde{\chi}_2^{\pm}}$, where the lighter mass is often near the electroweak scale.

Neutralinos

Neutralinos are mixtures of bino ($\tilde{B}$), wino ($\tilde{W}^3$), and neutral Higgsinos ($\tilde{H}_u^0$, $\tilde{H}_d^0$). The mass matrix is:

$$\mathcal{M}_N = \begin{pmatrix} M_1 & 0 & -M_Z s_W c_\beta & M_Z s_W s_\beta \\ 0 & M_2 & M_Z c_W c_\beta & -M_Z c_W s_\beta \\ -M_Z s_W c_\beta & M_Z c_W c_\beta & 0 & -\mu \\ M_Z s_W s_\beta & -M_Z c_W s_\beta & -\mu & 0 \end{pmatrix}. \tag{8.58}$$

Diagonalization yields four neutralino mass eigenstates. The lightest, χ_1^0, is a dark matter candidate. Its composition is:

$$\chi = N_{11}\tilde{B} + N_{12}\tilde{W}^3 + N_{13}\tilde{H}_d^0 + N_{14}\tilde{H}_u^0. \tag{8.59}$$

The gaugino fraction $f_g = |N_{11}|^2 + |N_{12}|^2$ indicates the neutralino's character: gaugino-like ($f_g > 0.5$) or Higgsino-like ($f_g < 0.5$).

Squarks

Squarks are scalar partners of quarks. Their mass matrices include soft terms, Yukawa interactions, and D-term contributions. For third-generation squarks, mixing is significant due to large Yukawa couplings. The stop mass matrix is:

$$\mathcal{M}_{\tilde{t}}^2 = \begin{pmatrix} m_{\tilde{q}_L}^2 + m_t^2 - \frac{\cos 2\beta}{6}(M_Z^2 - 4M_W^2) & m_t(A_t - \mu \cot\beta) \\ m_t(A_t - \mu\cot\beta) & m_{\tilde{t}_L^c}^2 + m_t^2 + \frac{2}{3}M_Z^2 \sin^2\theta_W \cos 2\beta \end{pmatrix}. \quad (8.60)$$

The lighter stop mass $m_{\tilde{t}_1}$ can be significantly below the TeV scale.

Sleptons

Sleptons are scalar partners of leptons. The stau ($\tilde{\tau}$) mass matrix in mSUGRA takes the form:

$$\mathcal{M}_{\tilde{\tau}}^2 = \begin{pmatrix} m_{\tilde{\ell}_L}^2 + m_\tau^2 + \frac{\cos 2\beta}{2}(M_Z^2 - 2M_W^2) & m_\tau(A_\tau - \mu\tan\beta) \\ m_\tau(A_\tau - \mu\tan\beta) & m_{\tilde{\ell}_L^c}^2 + m_\tau^2 - M_Z^2 \cos 2\beta \sin^2\theta_W \end{pmatrix}. \quad (8.61)$$

For large $\tan\beta$, significant left-right mixing occurs, which can impact LSP coannihilation and collider signals.

8.9 THE HIGGS SECTOR IN THE MSSM

The MSSM requires two Higgs doublets, H_u and H_d, to give mass to both up- and down-type fermions and ensure anomaly cancellation. These doublets acquire vacuum expectation values (VEVs) $v_u = v\sin\beta$ and $v_d = v\cos\beta$, with $v \simeq 246\,\text{GeV}$ and $\tan\beta = v_u/v_d$. The Higgs potential receives contributions from the superpotential, soft SUSY-breaking terms, and D-terms:

$$\begin{aligned} V =& (m_{H_u}^2 + |\mu|^2)H_u^\dagger H_u + (m_{H_d}^2 + |\mu|^2)H_d^\dagger H_d + (B\mu H_u \cdot H_d + \text{h.c.}) \\ &+ \frac{g^2 + g'^2}{8}(H_u^\dagger H_u - H_d^\dagger H_d)^2 + \frac{g^2}{2}|H_u^\dagger H_d|^2. \end{aligned} \quad (8.62)$$

Charged Higgs Bosons

The charged Higgs mass matrix, expressed in the basis $(H_d^\mp, H_u^\pm)$, yields a physical charged Higgs with mass:

$$m_{H^\pm}^2 = m_A^2 + m_W^2, \quad (8.63)$$

where m_A is the mass of the CP-odd Higgs. The light charged Higgs decays predominantly into $\tau^\pm\nu$ and $c\bar{s}$, while heavier states decay into $t\bar{b}$ or $\tau\nu$, depending on phase space.

CP-odd Higgs Boson

The CP-odd Higgs mass matrix in the (a_d, a_u) basis has one zero eigenvalue (Goldstone boson) and a massive pseudoscalar:

$$m_A^2 = \frac{2B\mu}{\sin 2\beta}. \tag{8.64}$$

This state decays primarily into heavy fermions such as $\tau^+\tau^-$ and $b\bar{b}$, with rates sensitive to $\tan\beta$.

CP-even Higgs Bosons

The CP-even Higgs mass matrix in the (h_d, h_u) basis is:

$$\mathcal{M}_H^2 = \begin{pmatrix} m_Z^2\cos^2\beta + m_A^2\sin^2\beta & -(m_Z^2 + m_A^2)\sin\beta\cos\beta \\ -(m_Z^2 + m_A^2)\sin\beta\cos\beta & m_Z^2\sin^2\beta + m_A^2\cos^2\beta \end{pmatrix}. \tag{8.65}$$

The lighter eigenstate satisfies $m_h \leq m_Z|\cos 2\beta|$ at tree level, necessitating radiative corrections to reach the observed value of 125 GeV [129–131]. These loop corrections primarily involve top-stop loops:

$$V^{(1)} = \frac{1}{64\pi^2}\mathrm{Str}\left[M^4(\varphi_i)\left(\ln\frac{M^2(\varphi_i)}{Q^2} - C_s\right)\right], \tag{8.66}$$

where C_s is a spin-dependent constant in the $\overline{\mathrm{DR}}$ scheme.

8.10 SUSY SEARCHES AT THE LHC

The search for SUSY particles at the LHC is a central aspect of the experimental effort to discover new physics beyond the SM. At the LHC, SUSY particle production is predominantly driven by strongly interacting particles, such as squarks and gluinos. These particles have substantial production cross-sections due to their strong interactions, making them the primary focus of SUSY searches [132, 133].

In typical high-mass SUSY scenarios, the production of SUSY particles initiates with squarks and gluinos. These particles then undergo a series of decay steps, eventually leading to lighter SUSY particles and SM particles. The decay chains typically involve:

- Squarks and Gluinos: Decay into quarks, gluons, and lighter SUSY particles such as charginos and neutralinos.
- Charginos and Neutralinos: Further decay into electroweak gauge bosons (W and Z bosons), Higgs bosons, and eventually to the lightest SUSY particle, which is often the neutralino ($\tilde{\chi}_1^0$).

The decay processes result in final states characterized by energetic jets and significant missing transverse energy (E_T^{miss}), as the lightest SUSY particle escapes detection. The general search strategy involves identifying these distinctive signatures in the data collected by the LHC detectors [134, 135].

LHC Constraints on the MSSM Spectrum

The LHC has significantly constrained the parameter space of the MSSM spectrum, particularly in the context of SUSY particle masses, cross-sections, and decay modes [135, 136]. Key constraints include:

- Gluino Mass: Lower bounds of $\sim 2.5\,\text{TeV}$ in simplified models, reaching ~ 2.7–$3.0\,\text{TeV}$ in complex scenarios [132].
- Squark Mass: $\sim 2.0\,\text{TeV}$ for first/second-generation, and up to $\sim 2.5\,\text{TeV}$ for stops/bottoms [135].
- Neutralino and Chargino Masses: Neutralino $\gtrsim 100$–$200\,\text{GeV}$, chargino $\gtrsim 500$–$600\,\text{GeV}$ [133].
- Higgs Sector: Light Higgs mass fixed at $\sim 125\,\text{GeV}$, heavy Higgs and charged Higgs constrained above $\sim 100\,\text{GeV}$ [136].
- Stop Mass: Approx. ~ 1.0–$1.5\,\text{TeV}$, depending on decay channels [134].

Despite no direct SUSY signals yet, the exclusion of parameter space is substantial and continues to shape theoretical developments. Future LHC runs with higher luminosity will improve sensitivity [134].

CHAPTER 9

BSM and Extra Dimensions

9.1 INTRODUCTION

The concept of extra dimensions is a central theme in theories that extend beyond the SM of particle physics. By postulating the existence of additional spatial dimensions beyond the familiar three, these frameworks aim to resolve longstanding theoretical and phenomenological puzzles, such as the hierarchy problem, the unification of forces, and the formulation of a consistent theory of quantum gravity.

One of the earliest and most influential attempts to incorporate extra dimensions was the Kaluza-Klein theory [137, 138], which unified gravity and electromagnetism by introducing a compact fifth spatial dimension. This pioneering idea laid the foundation for modern approaches involving higher dimensions.

A major revival of interest came with the proposal of Large Extra Dimensions in Arkani-Hamed, Dimopoulos, and Dvali (ADD) model [139], which suggested that extra dimensions could be large compared to the Planck scale. This framework offers an elegant resolution to the hierarchy problem by lowering the fundamental scale of gravity. Subsequently, the Randall-Sundrum (RS) models [140, 141] introduced the idea of warped extra dimensions, providing an alternative geometric mechanism to address the same problem.

Beyond addressing the hierarchy problem, extra dimensions are integral to string theory, where consistency requires spacetime to be ten-dimensional [142], or eleven-dimensional in the context of M-theory [143]. These extra dimensions also appear in cosmological models, offering

DOI: 10.1201/9781003457701-9

insights into early-universe inflation [144] and the nature of dark matter and dark energy [145].

9.2 GENERAL RELATIVITY IN 5D SPACETIME

One of the earliest attempts to formulate a unified field theory was introduced by Theodor Kaluza in 1921 [137]. Kaluza extended Einstein's theory of general relativity by proposing a five-dimensional spacetime, integrating both gravitational and electromagnetic forces within the same framework. This innovative approach set the foundation for many modern theories involving extra dimensions. In this five-dimensional model, the spacetime is described by a generalized line element:

$$d\hat{s}^2 = g_{MN} dx^M dx^N, \tag{9.1}$$

where g_{MN} is the metric tensor that encompasses all five dimensions, and M, N represent the spacetime indices ranging from 0 to 4. The five-dimensional metric tensor is structured as follows:

$$g_{MN} = \begin{pmatrix} g_{\mu\nu} - \kappa^2\phi^2 A_\mu A_\nu & -\kappa\phi^2 A_\mu \\ -\kappa\phi^2 A_\nu & -\phi^2. \end{pmatrix} \tag{9.2}$$

Here, $g_{\mu\nu}$ represents the familiar four-dimensional gravitational field, A_μ is the electromagnetic four-potential, and ϕ is a scalar field. Together, they form the five-dimensional gravitational potential, where the electromagnetic and gravitational components are unified.

A key assumption in this model is that the five-dimensional metric is independent of the extra-dimensional coordinate, denoted by y. This assumption is known as the "cylindrical condition":

$$\frac{\partial g_{MN}}{\partial x_4} = 0. \tag{9.3}$$

By adopting this condition, Kaluza was able to reduce the theory to a four-dimensional spacetime with an embedded electromagnetic field. After identifying $g_{44} = -\phi^2 = -1$ and $\kappa = \sqrt{\frac{16\pi G}{c^4}}$, the resulting field equations, $G_{MN} = 0$, naturally split into two familiar sets of equations:

$$\tilde{G}_{\mu\nu} = \frac{8\pi G}{c^4} T_{\mu\nu}, \tag{9.4}$$

$$\nabla^\mu F_{\mu\nu} = 0 \tag{9.5}$$

In these equations, $T_{\mu\nu}$ is the energy-momentum tensor for the electromagnetic field, defined as:

$$T_{\mu\nu} \equiv \frac{1}{2}\left(g_{\mu\nu}F_{\alpha\beta}F^{\alpha\beta} - F_{\mu}{}^{\alpha}F_{\nu\alpha}\right). \tag{9.6}$$

This remarkable result, where the field equations of general relativity yield both Einstein's gravitational equations and Maxwell's equations of electromagnetism, is often referred to as the "Kaluza miracle" [137]. It demonstrated, for the first time, how extra dimensions could provide a path toward unifying the fundamental forces of nature, laying the groundwork for more advanced theories like String Theory [142].

9.3 COMPACTIFIED EXTRA DIMENSION

To justify the cylindrical condition in Kaluza's five-dimensional theory, Oskar Klein proposed a groundbreaking idea in 1926: the extra dimension is not infinite and extended, but rather compactified, curled up into a microscopic, unobservable scale [138]. This compactification of the extra dimension is assumed to be toroidal, meaning that the extra spatial dimension is rolled into a circle with a small radius, R. The periodic nature of the compactified dimension is described as:

$$X(x^{\mu}, x^{4}) = X(x^{\mu}, x^{4} + 2\pi R). \tag{9.7}$$

In this scenario and as shown in Fig. 9.1, the extra dimension $y = x_4$ is periodic, with each point in the dimension returning to itself after traveling a distance of $2\pi R$. This compactification allows the theory to remain consistent with the observation that we experience only four large dimensions while the fifth dimension remains hidden due to its extremely small size.

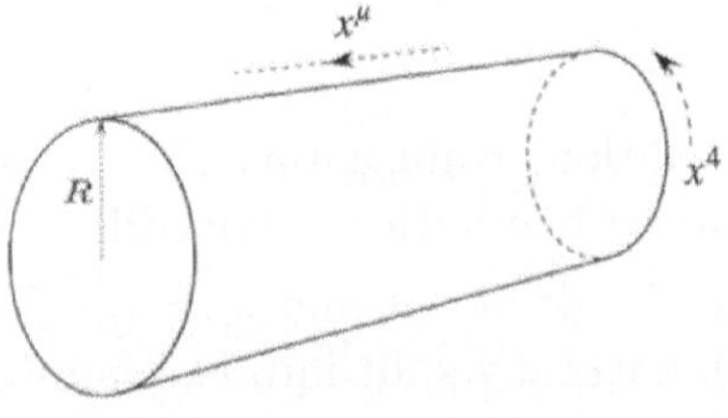

Figure 9.1 Fifth dimension compactification on a circle

Klein's assumption that all fields are periodic in the extra dimension led to the expansion of these fields into Fourier series. This approach allows each field to be decomposed into a series of harmonics, indexed by an integer n, corresponding to different modes of the compactified dimension. The expansions for the metric, electromagnetic potential, and scalar field are given by:

$$g_{\mu\nu}(x,y) = \sum_{n=-\infty}^{+\infty} g_{\mu\nu n}(x) e^{in\cdot y/R}, \tag{9.8}$$

$$A_{\mu}(x,y) = \sum_{n=-\infty}^{+\infty} A_{\mu n}(x) e^{in\cdot y/R}, \tag{9.9}$$

$$\phi(x,y) = \sum_{n=-\infty}^{+\infty} \phi_{n}(x) e^{in\cdot y/R}. \tag{9.10}$$

Here, $g_{\mu\nu n}(x)$, $A_{\mu n}(x)$, and $\phi_n(x)$ represent the Fourier coefficients, corresponding to different Kaluza-Klein modes. These modes reflect the quantized nature of the extra dimension. The harmonics n introduce an entire tower of particles with increasing mass, known as Kaluza-Klein excitations. For consistency, the following relations hold between the Fourier components:

$$g^*_{\mu\nu n}(x) = g_{\mu\nu -n}(x), \quad A^*_{\mu n}(x) = A_{\mu -n}(x), \quad \phi^*_n(x) = \phi_{-n}(x). \tag{9.11}$$

These conditions ensure that the physical fields are real, with their complex conjugates satisfying the necessary symmetries.

The compactification of the extra dimension explains why we do not directly observe the fifth dimension. Since the dimension is microscopic, the higher Kaluza-Klein modes, corresponding to larger values of n, acquire masses that are too large to be detected at low energies, leaving only the zero modes to manifest as observable particles in four-dimensional spacetime. Thus, the compactified extra dimension provides a consistent framework to incorporate higher-dimensional theories while maintaining agreement with experimental observations. This compactification mechanism is crucial not only for Kaluza-Klein theory [137,138] but also for modern approaches such as string theory [142], where additional dimensions are similarly compactified on complex geometrical shapes, like Calabi-Yau manifolds. The notion of compactified extra dimensions has deep implications, providing insight into the structure of spacetime and offering a pathway to unification of the fundamental forces.

The Kaluza-Klein theory describes an infinite number of four-dimensional fields, each corresponding to different modes of a compactified extra dimension. These fields obey the equations of motion derived from the five-dimensional action. The equations of motion for the metric, electromagnetic field, and scalar field in this framework are:

$$(\partial^\mu\partial_\mu - \partial^y\partial_y)g_{\mu\nu}(x,y) = (\partial^\mu\partial_\mu + \frac{n^2}{R^2})g_{\mu\nu n}(x) = 0, \tag{9.12}$$

$$(\partial^\mu\partial_\mu - \partial^y\partial_y)A_\mu(x,y) = (\partial^\mu\partial_\mu + \frac{n^2}{R^2})A_{\mu n}(x) = 0, \tag{9.13}$$

$$(\partial^\mu\partial_\mu - \partial^y\partial_y)\phi(x,y) = (\partial^\mu\partial_\mu + \frac{n^2}{R^2})\phi_n(x) = 0. \tag{9.14}$$

These resemble the Klein-Gordon equation, with an added term that gives rise to a "mass" for each field mode. The Kaluza-Klein modes introduce a mass scale for these fields, given by:

$$m_n \sim \frac{n}{R}. \tag{9.15}$$

Here, n denotes the mode of excitation, and R is the radius of compactification. The higher the mode n, the larger the associated mass of the field. In four-dimensional spacetime, these excited states appear with masses proportional to $O(n/R)$, and since the goal is to unify the electromagnetic interactions with gravity, the natural radius of compactification is of the order of the Planck length, $R = \frac{1}{M_p}$, where the Planck mass: $M_p \sim 10^{18}$ GeV.

The resulting action for the scalar field (often referred to as the "dilaton") is expressed as:

$$S = \int d^4x \left\{ \frac{1}{2}\partial_\mu\phi^{(0)}\partial^\mu\phi^{(0)} + \sum_{n=1}^{\infty}\left[\partial_\mu\phi^{(n)\dagger}\partial^\mu\phi^{(n)} - \frac{n^2}{R^2}\phi^{(n)\dagger}\phi^{(n)}\right]\right\}. \tag{9.16}$$

This action describes both the zero mode and the infinite tower of massive Kaluza-Klein states, where each higher mode contributes a scalar field with a mass proportional to n/R.

From the four-dimensional perspective, the action describes an infinite series of particles, known as the Kaluza-Klein tower, with masses given by:

$$m_{(n)} = \frac{n}{R}. \tag{9.17}$$

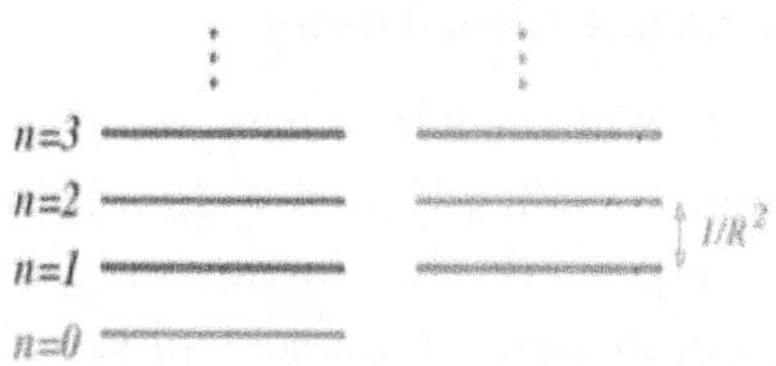

Figure 9.2 KK mass spectrum for a field on the circle

If the original field $\Phi(x^\mu, y)$ in five dimensions has a mass m_0, the corresponding four-dimensional Kaluza-Klein particles will have masses:

$$m_{(n)}^2 = m_0^2 + \frac{n^2}{R^2}. \tag{9.18}$$

This relation introduces a mass spectrum for the Kaluza-Klein modes, where the lowest mode $n = 0$ represents the massless or lightest state, while the higher modes $n > 0$ are increasingly massive. The diagram in Fig. 9.2 illustrates the Kaluza-Klein mass spectrum for a field compactified on a circular dimension:

In this framework, the five-dimensional gauge field $A_M(x^\mu, y)$ has a Fourier decomposition along the compact dimension:

$$A_M(x^\mu, y) = \frac{1}{\sqrt{2\pi R}} \sum_n A_M^{(n)}(x^\mu) e^{i\frac{n}{R}y}. \tag{9.19}$$

The action for the five-dimensional gauge field can be written as:

$$S = \int d^4x\, dy \left[-\frac{1}{4} F_{MN} F^{MN} \right]. \tag{9.20}$$

Upon compactification, the action becomes:

$$S = \int d^4x \Big\{ \Big(-\frac{1}{4} F_{\mu\nu}^{(0)} F^{(0)\mu\nu} + \frac{1}{2} \partial_\mu A_5^{(0)} \partial^\mu A_5^{(0)} \Big) + \sum_{n\geq 1} 2 \Big(-\frac{1}{4} F_{\mu\nu}^{(-n)} F^{(n)\mu\nu} + \frac{1}{2} \frac{n^2}{R^2} A_\mu^{(-n)} A_\mu^{(n)} \Big) \Big\}. \tag{9.21}$$

This shows that the zero modes consist of a four-dimensional gauge field and a real scalar field, while the higher modes include massive vector and scalar fields. These Kaluza-Klein excitations introduce a rich structure of particles that could be relevant in models of unification or physics beyond the SM [137, 138, 142].

Graviton Spectrum in Kaluza-Klein Theory

In the framework of Kaluza-Klein theory, the metric in higher-dimensional spacetime, specifically a five-dimensional (5D) spacetime, has 15 independent components due to the additional dimension. However, not all of these components represent physical degrees of freedom. To reduce this number, gauge fixing is required. The harmonic gauge condition is imposed to simplify the metric components by applying the following constraint:

$$\partial_M g_N^M - \frac{1}{2}\partial_N g_M^M = 0 \tag{9.22}$$

This condition eliminates 5 components, reducing the independent degrees of freedom to 10. Additionally, the residual gauge symmetry

$$g_{MN} \rightarrow g_{MN} + \partial_M \epsilon_N + \partial_N \epsilon_M \tag{9.23}$$

with the condition $\Box\epsilon_N = 0$, allows the imposition of 5 more constraints. Ultimately, the 5D graviton has 5 physical degrees of freedom.

In contrast, a massless graviton in four-dimensional (4D) spacetime possesses only 2 polarization states. This discrepancy indicates that upon compactification, the extra degrees of freedom manifest as additional fields from the 4D perspective, specifically, a vector field and a scalar field. This decomposition of the higher-dimensional graviton forms the essence of the Kaluza-Klein spectrum [146].

These additional fields reflect the influence of the compactified extra dimension and provide the theoretical basis for the Kaluza-Klein spectrum, where the higher-dimensional graviton yields not only the usual graviton but also additional particles in four-dimensional spacetime.

To understand this more concretely, consider the five-dimensional Einstein-Hilbert action:

$$\hat{S} = \frac{1}{2\hat{k}^2}\int d^5\hat{x}\sqrt{-\hat{g}}\hat{R}. \tag{9.24}$$

Upon compactification of the fifth dimension on a circle of radius R, and focusing on the zero modes ($n = 0$), the effective four-dimensional action becomes:

$$S = \frac{1}{2k^2}\int d^4x\sqrt{-g}\left[R - \frac{1}{4}e^{-\sqrt{3}\phi}F_{\mu\nu}^{(0)}F^{\mu\nu(0)} - \frac{1}{2}\partial_\mu\phi^{(0)}\partial^\mu\phi^{(0)}\right]. \tag{9.25}$$

Here, R is the Ricci scalar of the 4D spacetime, $F_{\mu\nu}^{(0)}$ corresponds to the field strength of a $U(1)$ gauge field arising from $g_{\mu 5}$, and $\phi^{(0)}$ is a scalar field (commonly called the dilaton) associated with g_{55}. These fields reflect the zero-mode decomposition of the 5D graviton and are the physical remnants of the higher-dimensional geometry in the compactified theory.

The higher Fourier modes ($n \neq 0$) correspond to massive spin-2 Kaluza-Klein excitations, which appear as a tower of massive gravitons in four-dimensional spacetime with masses quantized as $m_n = n/R$. These modes are generically suppressed at low energies if the compactification radius R is sufficiently small, typically near the Planck scale [147].

9.4 THEORIES WITH LARGE EXTRA DIMENSIONS

A key consideration in formulating theories with higher dimensions is ensuring that the low-energy limit, often referred to as the four-dimensional (4D) effective theory, matches experimental observations in our familiar 4D spacetime. This serves as a critical test of the viability of such theories. A generalization of the Einstein-Hilbert action in $4+n$ dimensions is given by:

$$S_{4+n} = -M_*^{n+2} \int d^{4+n}x \sqrt{-g^{(4+n)}} R^{(4+n)}, \tag{9.26}$$

where M_* is the fundamental mass scale of the higher-dimensional theory. The power $n+2$ ensures that the action remains dimensionless in natural units. In this framework, the extra dimensions are compactified on an n-sphere or torus with a characteristic radius R. The full spacetime metric is then:

$$ds^2 = \eta_{\mu\nu} dx^\mu dx^\nu - R^2 d\Omega_n^2, \tag{9.27}$$

where $\eta_{\mu\nu}$ denotes the flat Minkowski metric in four dimensions, and $d\Omega_n^2$ is the line element on the n-dimensional compact space.

To connect this higher-dimensional theory with 4D gravity, one must match the low-energy limit of the higher-dimensional Einstein-Hilbert action to the standard 4D action. This comparison yields:

$$-M_*^{n+2} \int d^n y \sqrt{g^{(n)}} \int d^4x \sqrt{-g^{(4)}} R^{(4)} = -M_{\rm Pl}^2 \int d^4x \sqrt{-g^{(4)}} R^{(4)}, \tag{9.28}$$

where $M_{\rm Pl} \sim 10^{18}$ GeV is the four-dimensional Planck mass. The integral over the compact space gives its volume:

$$V_n(R) = \int d^n y \sqrt{g^{(n)}} = \frac{\pi^{n/2}}{\Gamma\left(\frac{n}{2}+1\right)} R^n, \tag{9.29}$$

which corresponds to the volume of an n-dimensional ball or torus, depending on the geometry.

Equating both sides, we find the relationship between the Planck scale and the fundamental scale of gravity:

$$M_{\text{Pl}}^2 = M_*^{n+2} V_n(R). \tag{9.30}$$

Solving for the compactification radius gives:

$$R \sim \frac{M_{\text{Pl}}^{2/n}}{M_*^{1+2/n}}. \tag{9.31}$$

This expression reveals how the size of the extra dimensions depends on the number of dimensions n and the fundamental scale M_*. For instance, if $M_* \sim 1$ TeV, a scale motivated by attempts to solve the hierarchy problem, then for $n = 2$, the radius R could be as large as a fraction of a millimeter, potentially within the reach of submillimeter gravitational experiments [139, 148].

Gauge Field in Higher Dimensions

The action for a gauge field in $4+n$ dimensions generalizes the standard Yang-Mills action as:

$$S^{(4+n)} = -\int d^{4+n}x \, \frac{1}{4g_*^2} F_{MN} F^{MN} \sqrt{-g^{(4+n)}}, \tag{9.32}$$

where F_{MN} is the field strength tensor in the higher-dimensional space, and g_* is the higher-dimensional gauge coupling. Assuming that $g_* \sim \mathcal{O}(1)$, we can match the higher-dimensional gauge action to its four-dimensional effective counterpart. This procedure yields the relation:

$$1 \approx R^n M_*^n, \tag{9.33}$$

where R is the radius of the compact extra dimensions and M_* is the fundamental scale of gravity. Combining this with the previous relation between M_{Pl}, M_*, and R,

$$M_{\text{Pl}}^2 = M_*^{n+2} R^n, \tag{9.34}$$

we find that:

$$R \sim \frac{1}{M_{\text{Pl}}}, \tag{9.35}$$

if $M_* \sim M_{\text{Pl}}$, implying extremely small compactification scales consistent with the absence of observed deviations from four-dimensional physics.

Historically, it was widely assumed that any extra dimensions must be compactified at the Planck scale, $R \sim 10^{-33}$ cm, rendering their physical effects unobservable. However, this paradigm was revolutionized in the late 1990s by developments in string theory and brane-world scenarios. In particular, the idea that Standard Model fields may be localized on a four-dimensional submanifold (brane) embedded in a higher-dimensional bulk allowed for a radical rethinking of extra-dimensional geometry.

In the ADD model, only gravity propagates in the higher-dimensional bulk, while SM particles are confined to a 3-brane. This framework permits large compactification radii up to $R \lesssim 0.1\,\text{mm}$, without violating known experimental constraints from tests of Newtonian gravity at short distances.

In this scenario, the fundamental scale of gravity M_* can be as low as the electroweak scale, $M_* \sim 1\,\text{TeV}$, providing a compelling solution to the hierarchy problem. The apparent weakness of gravity at large distances is then explained as a consequence of its dilution into the extra dimensions.

Gravitational Field in Higher Dimensions

We consider a flat five-dimensional metric:

$$ds^2 = \eta_{MN} dx^M dx^N, \quad M, N = 0, 1, 2, 3, 4, \tag{9.36}$$

where η_{MN} is the Minkowski metric in 5D spacetime. In the weak-field limit and for static perturbations, the gravitational field equation generalizes Gauss's law to higher dimensions [139]:

$$\partial_i \partial^i \phi = 4\pi G_D \rho, \tag{9.37}$$

where G_D is the gravitational constant in D-dimensions and ρ is the mass density. Applying the divergence theorem, and defining the gravitational field as $\mathbf{g} \equiv -\nabla\phi$, we obtain:

$$\oint_S (\mathbf{g} \cdot \mathbf{n})\, dS = -4\pi G_D m, \tag{9.38}$$

where S is the surface enclosing a D-dimensional volume, and m is the enclosed mass.

In a space with n extra spatial dimensions (i.e., $D = 4 + n$), this surface becomes a sphere in $n + 3$ spatial dimensions:

$$\oint_{S^{n+2}} (\mathbf{g} \cdot \mathbf{n})\, dA = -4\pi G_D m. \tag{9.39}$$

The surface area of a $(n+2)$-dimensional sphere is related to the volume of the corresponding ball:

$$V_{n+2}(r) = \frac{\pi^{\frac{n}{2}+1}}{\Gamma\left(\frac{n}{2}+2\right)} r^{n+2}. \tag{9.40}$$

From this, the magnitude of the gravitational field is:

$$g = \frac{-4\Gamma\left(\frac{n}{2}+2\right) G_D m}{\pi^{n/2} r^{n+2}}, \tag{9.41}$$

and integrating this expression yields the gravitational potential:

$$\phi(r) = \frac{4\Gamma\left(\frac{n}{2}+2\right) G_D m}{\pi^{n/2}} \int_{\infty}^{r} \frac{1}{r'^{n+2}}\, dr' = \frac{4\Gamma\left(\frac{n}{2}+2\right) G_D m}{\pi^{n/2} r^{n+1}}. \tag{9.42}$$

This result illustrates that gravity becomes stronger at short distances in the presence of extra dimensions due to the modified scaling of the potential, a central feature of models with large extra dimensions [148].

The higher-dimensional Newton constant G_D is related to the fundamental Planck scale $M_D \equiv M_{\mathrm{pl}(4+n)}$ by dimensional analysis. In natural units ($\hbar = c = 1$), its mass dimension is:

$$[G_D] = \frac{1}{{M_D}^{n+2}}. \tag{9.43}$$

This scaling is crucial for understanding how gravity can appear weak in 4D while being fundamentally strong in higher-dimensional theories [139, 148].

Matching to Newtonian Gravity

To recover an effective four-dimensional theory in the limit $R \ll r$, we compare the potential derived from the higher-dimensional theory to the familiar Newtonian potential:

$$\phi^{(4)} \sim \frac{m}{R^n r {M_{\mathrm{pl}(4+n)}}^{2+n}} \approx \frac{m}{M_{\mathrm{pl}}^2 r}. \tag{9.44}$$

This matching condition leads to the relation [139, 148]:

$$R^n = \frac{M_{\text{pl}}^2}{M_{\text{pl}(4+n)}{}^{2+n}}. \tag{9.45}$$

If we assume that the electroweak scale $m_{\text{EW}} \sim 1\,\text{TeV}$ is the fundamental Planck scale, we can choose the radius R of the extra dimensions to reproduce the observed four-dimensional Planck mass. This yields:

$$R \sim 10^{\frac{30}{n}-17}\,\text{cm} \times \left(\frac{1\,\text{TeV}}{m_{\text{EW}}}\right)^{1+\frac{2}{n}}. \tag{9.46}$$

Constraints from Experimental Bounds

For $n = 1$, the radius R of the extra dimension would be approximately $10^{13}\,\text{cm}$, comparable to the size of the solar system, and thus clearly excluded by observations. To remain consistent with current experimental constraints, at least two extra dimensions are required. For $n = 2$, the radius drops to about $100\,\mu\text{m}$, which lies within the current experimental bounds on deviations from Newtonian gravity at submillimeter scales [149].

9.5 RANDALL-SUNDRUM MODEL

Motivated by the large extra dimensions framework proposed by Arkani-Hamed, Dimopoulos, and Dvali (ADD) [139], a variety of models incorporating extra spatial dimensions have emerged. A particularly influential example is the Randall-Sundrum (RS) model, proposed in 1999 by Lisa Randall and Raman Sundrum [140, 141]. The RS model introduces a non-factorizable, warped geometry between our four-dimensional (4D) spacetime and an additional fifth dimension.

Specifically, the five-dimensional (5D) metric is a warped product between the 4D Minkowski space with coordinates x^μ and the extra-dimensional coordinate y, with the 4D geometry multiplied by a scalar warp factor $f(y)$. The general form of a warped product metric is:

$$ds^2 = f(y)\, g_{\mu\nu}(x)\, dx^\mu dx^\nu + g_{ab}(y)\, dy^a dy^b, \tag{9.47}$$

where $g_{\mu\nu}(x)$ is the metric of the 4D spacetime, and $g_{ab}(y)$ represents the geometry of the internal extra-dimensional space.

In the RS scenario, there is only a single extra dimension compactified on a circle S^1 of radius R, subject to a $\mathbb{Z}_2$ symmetry under reflection $y \to -y$, resulting in the orbifold $S^1/\mathbb{Z}_2$. This orbifold structure has strong roots in string-theoretic constructions, particularly in the Horava-Witten scenario [150], where eleven-dimensional M-theory is compactified on $R^{10} \times S^1/\mathbb{Z}_2$, with six dimensions curled up into a Calabi-Yau manifold. Such constructions motivate the effective appearance of five-dimensional physics at low energies.

Warped Metric and Einstein Field Equations

The RS model assumes that the induced 4D metric on the brane is flat, $g_{\mu\nu}(x) = \eta_{\mu\nu}$, while the warp factor depends solely on the extra dimension. The resulting metric becomes:

$$ds^2 = f(y)\,\eta_{\mu\nu}\,dx^\mu dx^\nu - dy^2, \tag{9.48}$$

where $\eta_{\mu\nu}$ is the Minkowski metric and $f(y)$ is to be determined.

To determine the functional form of $f(y)$, we consider the 5D Einstein-Hilbert action with a bulk cosmological constant Λ:

$$S = -\int d^5x\,\sqrt{-g}\left(R + \Lambda\kappa^2\right), \tag{9.49}$$

where $\kappa^2 = 1/M^3$, and M is the 5D Planck mass. The vacuum Einstein field equations take the form:

$$G_{MN} = -\frac{\Lambda}{2M^3} g_{MN}, \tag{9.50}$$

where G_{MN} is the 5D Einstein tensor.

Solving for the Warp Factor

Using the metric in Eq. (9.48), the non-zero components of the Einstein tensor are:

$$\begin{aligned} G_{00} &= -\frac{3}{2}\frac{d^2 f(y)}{dy^2}, \quad G_{11} = G_{22} = G_{33} = \frac{3}{2}\frac{d^2 f(y)}{dy^2}, \\ G_{44} &= \frac{3}{2f(y)^2}\left(\frac{df(y)}{dy}\right)^2. \end{aligned} \tag{9.51}$$

Substituting these into Eq. (9.50), we obtain:

$$\frac{d^2 f(y)}{dy^2} = -\frac{\Lambda\kappa^2}{3} f(y), \tag{9.52}$$

$$\frac{df(y)}{dy} = \pm\sqrt{\frac{-\Lambda\kappa^2}{3}} f(y). \tag{9.53}$$

These equations require $\Lambda < 0$, implying an anti-de Sitter (AdS_5) geometry. Solving Eqs. (9.52) and (9.53) for negative Λ gives:

$$f(y) = e^{\pm 2ky}, \tag{9.54}$$

where $k = \sqrt{-\Lambda\kappa^2/6}$. Imposing the $\mathbb{Z}_2$ symmetry $y \to -y$ selects:

$$f(y) = e^{-2k|y|}. \tag{9.55}$$

The full RS metric is therefore:

$$ds^2 = e^{-2k|y|}\,\eta_{\mu\nu}dx^\mu dx^\nu - dy^2. \tag{9.56}$$

The exponential suppression factor $e^{-2k|y|}$ generates a large hierarchy of scales from modest input parameters, thus providing a geometric solution to the hierarchy problem.

Conformally Flat Form of the RS Metric

For certain applications, it is useful to express the RS metric in a conformally flat form. Defining a new coordinate z via:

$$dy = e^{-k|y|}dz, \tag{9.57}$$

and solving for z, the metric becomes:

$$ds^2 = f(z)\left(\eta_{\mu\nu}dx^\mu dx^\nu - dz^2\right), \tag{9.58}$$

where

$$f(z) = \frac{1}{(k|z|+1)^2}. \tag{9.59}$$

Thus, the RS metric in conformally flat coordinates takes the form:

$$ds^2 = \frac{1}{(k|z|+1)^2}\left(\eta_{\mu\nu}dx^\mu dx^\nu - dz^2\right), \tag{9.60}$$

which is particularly advantageous for calculations involving field propagation in the bulk.

Brane Tension

It is important to note that the metric (9.56) is not fully determined by the bulk field equations alone; rather, it arises from the specific physical setup of the S^1/Z_2 orbifold. To ensure consistency, we must verify that this metric satisfies the Einstein field equations, potentially requiring an additional interpretation of its source terms.

By transforming the RS metric into its conformally flat form (9.60), the Einstein tensor components can be computed straightforwardly. The relevant component for the four-dimensional spacetime is:

$$G_{\mu\nu} = \eta_{\mu\nu} \left(\frac{6k^2}{(k|z|+1)^2} + \frac{3k\,\mathrm{sgn}'(z)}{k|z|+1} \right). \tag{9.61}$$

The first term corresponds to the contribution of the negative bulk cosmological constant, as expected. The second term, however, is singular and proportional to the derivative of the sign function, indicating the presence of localized sources at the orbifold fixed points. Specifically, since $\mathrm{sgn}'(z) = 2\delta(z) - 2\delta(z-b)$, these contributions are delta functions centered at $z = 0$ and $z = b$, the locations of the two 3-branes.

To account for these singularities, one must introduce brane-localized energy-momentum sources, interpreted as the brane tensions. These are incorporated into the total action as:

$$S = S_{\text{gravity}} + S_{\text{vis}} + S_{\text{hid}}, \tag{9.62}$$

where S_{vis} and S_{hid} are the contributions from the visible and hidden branes, respectively. These terms have the form:

$$S_{\text{brane}} = -\int d^4x\,\sqrt{-g_{\text{ind}}}\,\lambda, \tag{9.63}$$

where λ is the brane tension and g_{ind} is the induced 4D metric on the brane.

Matching the coefficients in the Einstein equations yields the fine-tuning condition:

$$\lambda_{\text{vis}} = -\lambda_{\text{hid}} = 24M^3k, \tag{9.64}$$

where M is the 5D Planck mass. This setup is characteristic of the original RS model (RS1) [140], which features a compact extra dimension bounded by two branes, one of which necessarily has negative tension.

The presence of a negative-tension brane in RS1 is often considered a theoretical drawback, as it may lead to instabilities and violates certain energy conditions. This issue is addressed in the second RS model (RS2) [141], where the extra dimension is extended to infinity, and only a single positive-tension brane remains at $z = 0$. In this configuration, the warp factor still localizes gravity near the brane, effectively reproducing four-dimensional Newtonian gravity despite the infinite extra dimension.

9.6 HIERARCHY PROBLEM

To understand the physical implications of the RS model, we now examine how it addresses its original motivation: the hierarchy problem. Specifically, we compute the effective four-dimensional Planck scale and mass parameters in terms of the five-dimensional fundamental scales.

Starting from the metric (9.56), we perform the coordinate transformation $y = r_c\phi$, where r_c is the compactification radius, and $\phi \in [-\pi, \pi]$. In the low-energy limit, we introduce weak-field perturbations $h_{\mu\nu}(x^\mu) \ll 1$, so the metric takes the form:

$$ds^2 = e^{-2kT(x^\mu)|\phi|}\left[\eta_{\mu\nu} + h_{\mu\nu}(x^\mu)\right] dx^\mu dx^\nu + T^2(x^\mu)d\phi^2, \qquad (9.65)$$

where the modulus field $T(x)$ parametrizes the size of the extra dimension and is assumed to be stabilized at its vacuum expectation value $\langle T(x)\rangle = r_c$. The stabilization mechanism, such as the Goldberger–Wise (GW) mechanism [151], is essential to dynamically fix this scale without fine-tuning. For simplicity, we now treat the modulus as a constant, $T = r_c$, yielding the background metric:

$$ds^2 = e^{-2kr_c|\phi|}\left[\eta_{\mu\nu} + h_{\mu\nu}(x^\mu)\right] dx^\mu dx^\nu + r_c^2 d\phi^2. \qquad (9.66)$$

The Ricci scalar in five dimensions includes the four-dimensional curvature term, scaled by the warp factor:

$$R^{(5)} \supset e^{2k|y|}R^{(4)}, \quad \sqrt{-g^{(5)}} \supset e^{-4k|y|}\sqrt{-g^{(4)}}. \qquad (9.67)$$

Using this, the gravitational part of the action becomes:

$$\begin{aligned} S &= -M^3 \int d^5x\sqrt{-g^{(5)}}\left(R^{(5)} + \kappa^2\right) \\ &\supset -M^3 \int dy\, e^{-2k|y|} \int d^4x\sqrt{-g^{(4)}}\left(R^{(4)} + \kappa^2\right), \end{aligned} \qquad (9.68)$$

where M is the 5D Planck scale and κ^2 is the bulk cosmological constant (here treated generically). Since the graviton zero mode is independent of y, the effective four-dimensional Planck scale is:

$$M_{\rm Pl}^2 = M^3 \int_{-b}^{b} dy\, e^{-2k|y|} = \frac{M^3}{k}\left(1 - e^{-2kb}\right). \qquad (9.69)$$

This result shows that $M \sim M_{\text{Pl}}$ even for moderately large values of b, unlike in the ADD model, where a large compactification volume is required. However, this alone does not fully resolve the hierarchy problem, we must still explain the smallness of the weak scale.

Let us now consider a matter field, such as the Higgs field, localized on the brane at $y = b$. The action for the Higgs scalar is:

$$S_H = \int d^4x \sqrt{-g^{\text{ind}}} \left[g^{\mu\nu}_{\text{ind}} D_\mu H D_\nu H - \lambda \left(H^\dagger H - v^2 \right)^2 \right], \tag{9.70}$$

where $g^{\text{ind}}_{\mu\nu} = e^{-2kb} \eta_{\mu\nu}$ is the induced metric on the brane. Substituting this yields:

$$S_H = \int d^4x \, e^{-4kb} \left[e^{2kb} \eta^{\mu\nu} D_\mu H D_\nu H - \lambda \left(H^\dagger H - v^2 \right)^2 \right]. \tag{9.71}$$

We rescale the field to obtain a canonically normalized kinetic term: $\tilde{H} = e^{-kb} H$. The action becomes:

$$S_H = \int d^4x \left[\eta^{\mu\nu} \partial_\mu \tilde{H} \partial_\nu \tilde{H} - \lambda \left(\tilde{H}^\dagger \tilde{H} - (e^{-kb} v)^2 \right)^2 \right]. \tag{9.72}$$

This is the standard 4D Higgs action, but with an effective vacuum expectation value:

$$\tilde{v} = e^{-kb} v. \tag{9.73}$$

Taking $v \sim M_{\text{Pl}}$ as the fundamental input scale, we find that choosing $kb \approx 37$ generates $\tilde{v} \sim$ TeV, providing a natural explanation for the smallness of the electroweak scale relative to the Planck scale. Therefore, the brane at $y = b$ is referred to as the "TeV brane," while the brane at $y = 0$ is the "Planck brane," where physical masses remain at the fundamental scale.

This exponential suppression of energy scales via the warped geometry elegantly resolves the hierarchy problem without requiring a large extra dimension or fine-tuning of fundamental parameters.

Linearized field equations

We now proceed to explore the implications of gravity in the context of the RS model by studying graviton fluctuations around the metric in the conformal gauge. As discussed earlier, the presence of a negative-tension object at the TeV brane introduces nonphysical outcomes. A closer look at Eq. (9.69) suggests a potential alternative approach. Notably, even if the size of the extra dimension approaches the limit $b \to \infty$, the TeV brane can be excluded from the physical setup. This concept forms the basis of the RS2. Although this approach abandons the original solution for the hierarchy problem discussed in the previous section, RS2 focuses on studying the more intriguing aspects of gravity [141, 152].

In this modified model, where only one brane exists, Einstein's equations derived from the action take the form:

$$R_{MN} - \frac{1}{2}g_{MN}R = \Lambda g_{MN} + \sigma g_{\mu\nu}\delta(z)\delta^{\mu}_{M}\delta^{\nu}_{N}, \tag{9.74}$$

where σ is the brane tension. The linearized field equations for a perturbed metric $ds^2 = e^{-A(z)}dx^M dx^N(\eta_{MN} + h_{\mu\nu})$ can be obtained using the relation that connects the Einstein tensor for a manifold with another conformally related manifold, i.e., $g_{MN} = e^{-A(z)}\tilde{g}_{MN}$, where $e^{-A(z)} = \frac{1}{(k|z|+1)^2}$ in our case:

$$\begin{aligned} G_{MN} &= \tilde{G}_{MN} + \frac{d-2}{d}\Big[\frac{1}{2}\tilde{\nabla}_M A\tilde{\nabla}_N A \\ &+ \tilde{\nabla}_M\tilde{\nabla}_N A - \tilde{g}_{MN}\left(\tilde{\nabla}_k\tilde{\nabla}^k A - \frac{d-3}{4}\tilde{\nabla}_k A\tilde{\nabla}^k A\right)\Big] \end{aligned} \tag{9.75}$$

Here, $\tilde{\nabla}$ refers to the covariant derivative with respect to the perturbed metric $\tilde{g}_{MN} = (\eta_{MN} + h_{MN})$.

With the knowledge of $\tilde{G}_{MN}$ and the covariant derivatives of $A(z)$, the linearized Einstein equation takes the form:

$$-\frac{1}{2}\partial^M\partial_M h_{\mu\nu} + \frac{d-2}{4}\partial^M A\partial_M h_{\mu\nu} = 0. \tag{9.76}$$

The linear expansion of the perturbed metric is performed using Eq. (9.76) under the gauge condition $\partial^\mu h_{\mu\nu} = h^\mu_\mu = 0$, known as the RS gauge choice. A detailed discussion of this specific gauge choice can be found in [152]. After performing a separation of variables $h(x,y) = \psi(y)e^{ip\cdot x}$, the resulting equation of motion is:

$$\left[-\frac{m^2}{2}e^{2k|y|} - \frac{1}{2}\partial_y^2 - 2k\delta(y) + 2k^2\right]\psi(y) = 0, \tag{9.77}$$

where m represents the four-dimensional mass eigenvalue, $\Box h_{\mu\nu} = m^2 h_{\mu\nu}$. To cast this equation in a Schrödinger-like form, we perform a change of variables $z \equiv \text{sgn}(y)\frac{e^{k|y|}-1}{k}$, yielding:

$$\left[-\frac{1}{2}\partial_z^2 + V(z)\right]\psi(z) = m^2\psi(z), \tag{9.78}$$

with the potential

$$V(z) = \frac{15k^2}{8(k|z|+1)^2} - \frac{3k}{2}\delta(z). \tag{9.79}$$

This potential forms a "volcano" with a minimum at $z = 0$ due to the delta function. This equation predicts a Yukawa-like correction to the gravitational potential on the brane, similar to the screened Poisson equation:

$$U(r) \sim \frac{G_N M_1 M_2}{r} + \int_0^\infty dm \frac{G_N M_1 M_2}{r} e^{-mr} \psi_m^2(z), \tag{9.80}$$

where the integral is over the massive Kaluza-Klein modes. Equation (9.78) has a solution in terms of Bessel functions:

$$\psi_m = a_m z^{\frac{1}{2}} Y_{\alpha+\frac{1}{2}}(m(z+\frac{1}{k})) + b_m z^{\frac{1}{2}} J_{\alpha+\frac{1}{2}}(m(z+\frac{1}{k})), \tag{9.81}$$

For our purposes, we are primarily interested in the effective four-dimensional theory at the brane, where $\psi_m(0) \sim \sqrt{\frac{m}{k}}$. Thus, the corrected gravitational potential takes the form:

$$U(r) \sim \frac{G_N M_1 M_2}{r} + c\int_0^\infty dm \frac{G_N M_1 M_2}{kr} m e^{-mr} \tag{9.82}$$

$$= G_N \frac{M_1 M_2}{r}\left(1 + \frac{C}{r^2 k^2}\right), \tag{9.83}$$

where C is a constant related to the integration over m. With k of the order of the fundamental Planck scale, the correction to the Newtonian potential is highly suppressed. To detect these higher Kaluza-Klein modes, extraordinarily high energies beyond current experimental reach would be required. Thus, this RS2 model with a semi-infinite extra dimension successfully reproduces the ordinary four-dimensional Newtonian potential [141, 151, 152].

CHAPTER 10

BSM Collider Phenomenology

10.1 INTRODUCTION

Physicists are driven by the goal of understanding the fundamental nature of the universe, which has led them to explore ideas BSM of particle physics. Collider experiments have played a key role in this search, as they allow scientists to probe extremely high energies where new particles or forces might appear. The main goal of BSM collider studies is to search for signs of new particles predicted by theoretical models. These models aim to answer open questions in particle physics and improve our understanding of matter at the smallest scales. High-energy collisions at the Large Hadron Collider (LHC) and future colliders provide a unique opportunity to discover such new physics [153]. The main areas of focus include:

- Supersymmetric Particles: SUSY predicts a partner particle for each SM particle. These superpartners could help resolve major open questions, such as the hierarchy problem, and may provide a framework for unifying the fundamental forces [118].
- Extra Gauge Bosons: Particles like Z' and W' are predicted by extensions of the SM and may indicate new interactions that could help explain observed anomalies in current experimental data [78].
- Extended Higgs Sector: The discovery of additional Higgs bosons could reshape our understanding of electroweak symmetry breaking and the dynamics of the Higgs mechanism [48, 51].

DOI: 10.1201/9781003457701-10

- Dark Matter Candidates: Exploring particles such as Weakly Interacting Massive Particles (WIMPs) and axions may lead to significant breakthroughs in understanding dark matter's elusive nature [32, 33].

In addition to the search for new particles, collider experiments seek to identify signatures of new physics. These signatures offer crucial insights into the existence of BSM phenomena, including:

- Missing Energy: A hallmark of stable or weakly interacting particles that may indicate the production of new physics [154].
- Jet and Lepton Production: Observable events that suggest the involvement of new particles [155].
- Resonance Searches: The detection of specific decay products that could confirm the presence of new particles or interactions [156].

This chapter will explore the experimental strategies employed in probing BSM collider phenomenology. Due to the limited space available in this chapter, we will focus on a select few examples of new particle discoveries at the LHC and will not consider other future colliders.

10.2 SUPERSYMMETRY COLLIDER PHENOMENOLOGY

This section focuses on identifying signatures of SUSY particles, understanding their production mechanisms, and measuring their properties to test SUSY theories. SUSY particles can be produced at colliders through strong interactions, which primarily generate squarks and gluinos due to the strong coupling, as well as electroweak interactions, leading to the production of sleptons, neutralinos, and charginos. Additionally, associated production processes, such as gluino-neutralino or squark-neutralino pairs, contribute to the variety of production mechanisms.

Additionally, SUSY particles decay through cascades involving intermediate states, ultimately producing the lightest supersymmetric particle (LSP). Key observables include missing transverse energy (E_T^{miss}) from the undetectable LSP [157], multiple energetic jets and leptons from squark, gluino, or slepton decays, and potential signatures of long-lived particles, such as displaced vertices or charged tracks, arising from particles like staus or neutralinos [158, 159].

Here, we focus on three fundamental SUSY particles: the Lightest Supersymmetric Particle (LSP), typically the lightest neutralino and a prime candidate for dark matter; the gluino, a key participant in strong interactions and a potential gateway to new physics; and squarks, the superpartners of quarks, which provide insights into flavor physics and the hierarchy problem.

Search for gluinos, first and second generation squarks

As discussed in Chapter 8, within the MSSM, the gluino mass is approximately given by $m_{\tilde{g}} \approx 2.5\, m_{1/2}$, typically resulting in a value of at least a few TeV. The five light-flavored squark species $(\tilde{d}, \tilde{u}, \tilde{s}, \tilde{c}, \tilde{b})$, including both left- and right-chiral components, are generally assumed to be mass-degenerate. In contrast, the third-generation squarks $\tilde{t}_L$ and $\tilde{t}_R$ experience significant left-right mixing, giving rise to two distinct mass eigenstates, $\tilde{t}_1$ and $\tilde{t}_2$. Notably, $\tilde{t}_1$ is considerably lighter than the other squark states, necessitating a dedicated discussion of its production mechanisms and decay channels [118].

At the LHC, squarks and gluinos can be produced in pairs $\{\tilde{g}\tilde{g}, \tilde{q}\tilde{q}, \tilde{q}\tilde{g}\}$ and decay through processes such as $\tilde{q} \to q\tilde{\chi}_1^0$ and $\tilde{g} \to q\bar{q}\tilde{\chi}_1^0$, where $\tilde{\chi}_1^0$ is the lightest neutralino, as depicted in Fig. 10.1. Searches for squarks and gluinos in final states with high transverse momentum jets and missing transverse momentum have been conducted by both the ATLAS and CMS experiments, using center-of-mass energies of $\sqrt{s} = 7, 8, 13$ TeV and a total integrated luminosity of approximately 20 fb^{-1} [160, 161]. No significant excess above the SM predictions was observed. In Fig. 10.2, we show the ATLAS exclusion limits at 95%

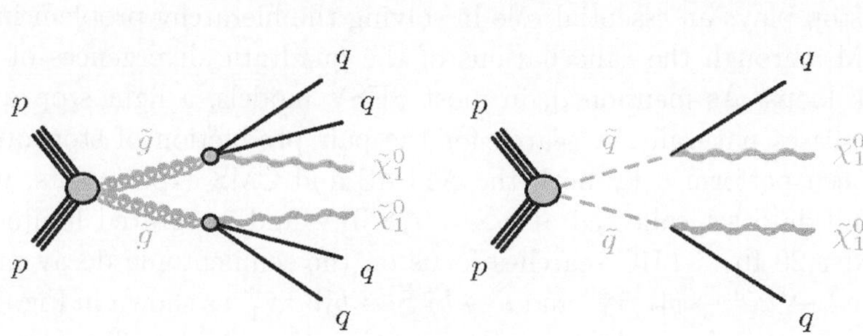

Figure 10.1 Representative diagrams for (left) gluino- and (right) squark-pair production and their direct decays: $\tilde{g} \to q\bar{q}\tilde{\chi}^0$ and $\tilde{q} \to q\tilde{\chi}^0$.

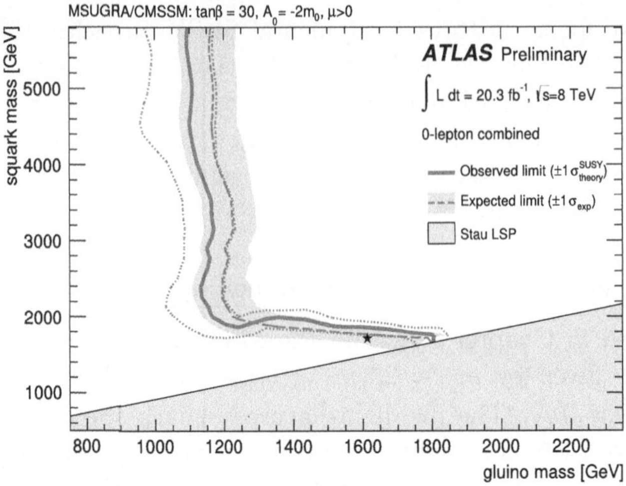

Figure 10.2 ATLAS exclusion limits, at 95% CL, for the MSSM model with $\tan\beta = 30$, $A_0 = -2m_0$ and $\mu > 0$, projected on the $(m_{\tilde{g}} - m_{\tilde{q}})$ plane [160]. CMS limits are quite similar [161].

confidence level for the MSSM model with $\tan\beta = 30$, $A_0 = -2m_0$, and $\mu > 0$, presented (left) in the $(m_{\tilde{g}} - m_{\tilde{q}})$ plane [160]. The CMS limits are quite similar [161]. It is noteworthy that values of $m_{1/2} < 340$ GeV are excluded for $m_0 < 6$ TeV in this scenario, consistent with constraints derived from the measured lightest Higgs mass of $m_h \approx 125$ GeV [162].

Search for stops

The stop plays an essential role in solving the hierarchy problem in the MSSM, through the cancellations of the quadratic divergences of top-quark loops. As mentioned, in most SUSY models, a light stop quark mass arises naturally. A search for the pair production of stop quarks has been performed by both the ATLAS and CMS experiments, using the full data set collected at $\sqrt{s} = 7, 8$ TeV and integrated luminosity of order 20 fb^{-1}. LHC searches focus on the semileptonic decay mode, where $\tilde{t} \to t\tilde{\chi}_1^0 \to bW^{\pm}\tilde{\chi}_1^0$ and $\tilde{t} \to b\tilde{\chi}_1^{\pm} \to bW^{\pm}\tilde{\chi}_1^0$, as shown in Fig. 10.3. Then one $W^{\pm}$ decays hadronically and the other leptonically. Thus, the final state searched for is $4j + \ell +$ missing energy ($\ell = e, \mu$) [163, 164].

An analysis of ~ 13 fb^{-1} and $\sqrt{s} = 13$ TeV at ATLAS [165] and CMS [166] detectors indicated no significant excess over the SM background.

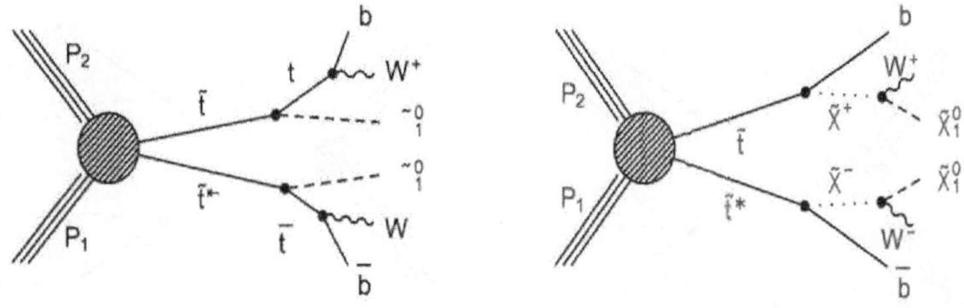

Figure 10.3 Representative diagrams for stop pair production and decay into $b\bar{b}\ W^+W^-\ \tilde{\chi}_1^0\tilde{\chi}_1^0$.

Exclusion limits at 95% CL, shown in Fig. 10.4, are reported. These limits are presented in terms of stop and lightest neutralino masses. In the considered simplified model, where $BR(\tilde{t} \to t\tilde{\chi}^0) = 1$, the stop masses between ~ 310 and ~ 820 GeV are excluded for $m_{\tilde{\chi}^0} < 160$ GeV.

Charginos and neutralinos

The lightest chargino ($\tilde{\chi}_1^{\pm}$) has several leptonic decay modes, producing an isolated lepton and missing energy, due to the undetectable neutrino and the LSP [167,168]:

$$\tilde{\chi}_1^{\pm} \to \tilde{\chi}_1^0\, \ell^{\pm}\, \nu, \tag{10.1}$$

$$\tilde{\chi}_1^{\pm} \to \tilde{\ell}_L^{\pm}\, \nu \to \tilde{\chi}_1^0\, \ell^{\pm}, \tag{10.2}$$

$$\tilde{\chi}_1^{\pm} \to \tilde{\nu}_L\, \ell^{\pm} \to \tilde{\chi}_1^0\, \nu, \tag{10.3}$$

$$\tilde{\chi}_1^{\pm} \to \tilde{\chi}_1^0\, W^{\pm} \to \ell^{\pm}\, \nu. \tag{10.4}$$

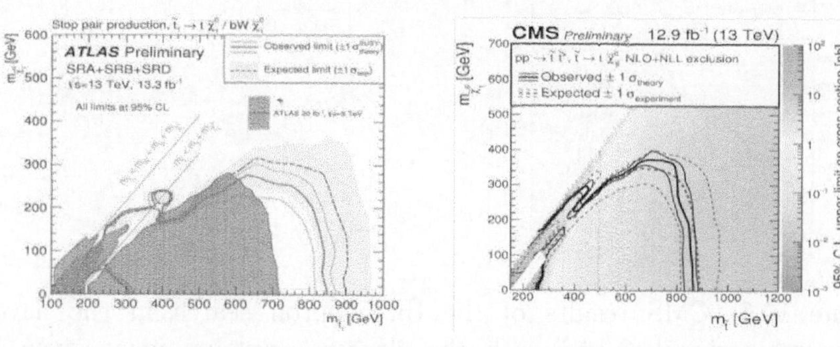

Figure 10.4 ATLAS (left) and CMS (right) exclusion limits at 95% CL for direct stop production, followed by the $\tilde{t} \to t\tilde{\chi}^0$ decay [165, 166].

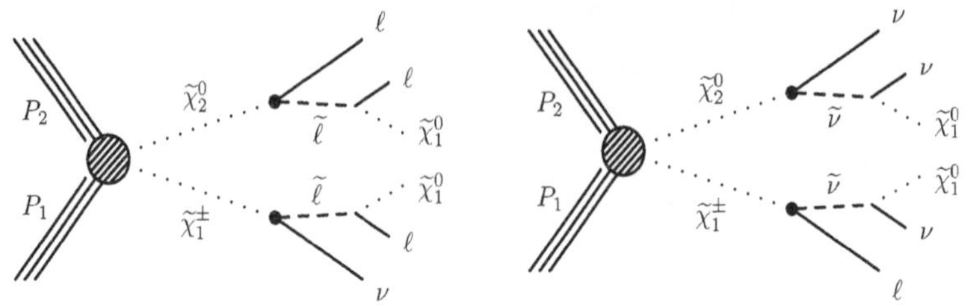

Figure 10.5 Representative diagrams for direct charginos/neutralinos production and decay mediated by sleptons (left) and sneutrinos (right), leading to leptonic final states.

In hadronic collisions, the charginos and neutralinos can be produced directly via 21 different reactions or through the cascade decays of strongly interacting sparticles.

The first searches for the direct electroweak production of supersymmetric charginos and neutralinos at $\sqrt{s} = 13$ TeV were conducted using pp collisions with a luminosity of approximately 13 fb^{-1} in the CMS detector [169]. These searches focus on signatures involving two same-flavor light leptons or three or more leptons, as illustrated in Fig. 10.5. The results are categorized based on the number, sign, and flavor of the leptons, as shown in Fig. 10.6. No significant deviation from Standard Model expectations was observed, and these results were used to set limits on simplified models with large chargino-neutralino pair production

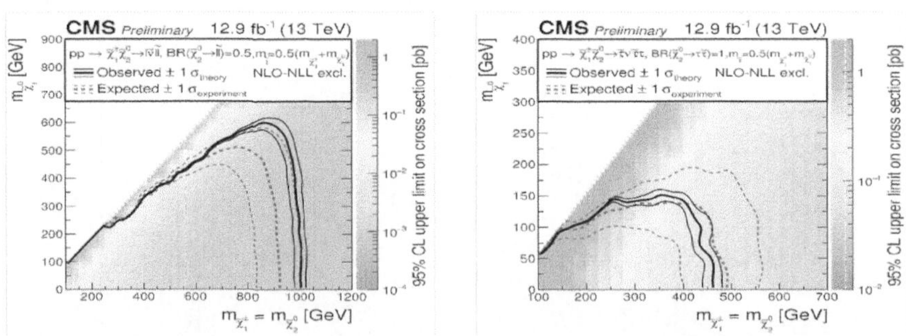

Figure 10.6 CMS results of the three-lepton search in the flavor-democratic signal model with the slepton/sneutrino mass given by $m_{\tilde{\ell}} = m_{\tilde{\nu}} = \frac{1}{2}(m_{\tilde{\chi}^+} + m_{\tilde{\chi}^0})$ [169].

cross-sections. Specifically, they probe chargino and neutralino masses up to 400 – 1000 GeV, depending on the assumed model parameters.

10.3 SEARCH FOR EXTRA GAUGE BOSONS AT THE LHC

The ATLAS and CMS experiments at the Large Hadron Collider (LHC) have conducted extensive searches for extra gauge bosons, such as Z' and W', predicted by various extensions of the SM. These searches are particularly compelling in the leptonic Drell-Yan (DY) channel, where heavy gauge bosons could manifest as resonant peaks in the invariant mass spectrum of lepton pairs (e^+e^- or $\mu^+\mu^-$). DY processes, characterized by the production of lepton pairs through quark-antiquark annihilation ($q\bar{q} \to Z'/W' \to \ell^+\ell^-$), offer a clean experimental signature and serve as powerful tools for discovering or constraining new physics involving heavy spin-1 gauge bosons [170, 171].

Searches for Z'-bosons focus on identifying resonances in the invariant mass spectrum, which could indicate the presence of new physics. These heavy gauge bosons arise in theoretical frameworks such as GUTs, Left-Right Symmetric Models, and models with additional $U(1)$ symmetries. Experimentally, a Z'-boson would appear as a narrow peak above SM backgrounds, with angular distributions and forward-backward asymmetry providing further discriminating power. Recent results from ATLAS and CMS have placed stringent lower bounds on Z'-boson masses, excluding them up to 4.5 TeV depending on the model [170, 171]. Future upgrades at the High-Luminosity LHC are expected to extend the reach of these searches, enabling sensitivity to even heavier Z'-bosons. Enhanced detector resolution and increased luminosity will allow the exploration of rarer events and further tighten constraints, maintaining Z'-searches as a cornerstone of BSM physics exploration at the energy frontier.

Search for Z' at the LHC

The existence of new neutral gauge bosons is a possible signature of GUT models, such as left-right-symmetric (LRS) models and $U(1)'$ extensions of the SM.

Searches for Z' bosons have been extensively conducted by the ATLAS and CMS Collaborations, resulting in stringent bounds on Z' boson production as a function of its mass, $m_{Z'}$, particularly in the $Z' \to \mu^+\mu^-$ and $Z' \to e^+e^-$ decay channels. By combining these two final states

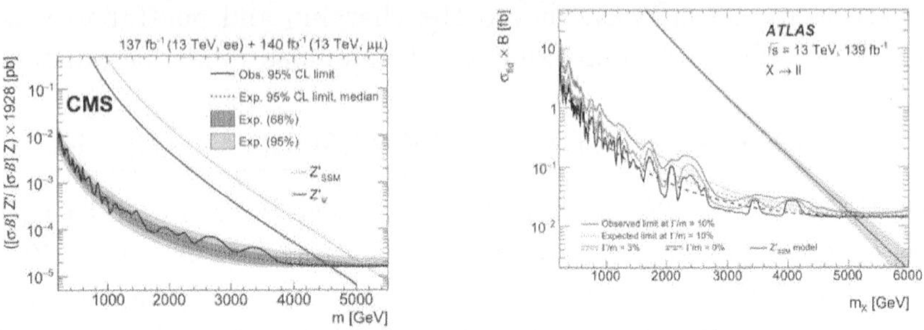

Figure 10.7 CMS and ATLAS Upper Limits on Z' Boson Production: Left: CMS 95% CL upper limits on the normalized cross-section times branching fraction for a spin-1 resonance [172]. Right: ATLAS 95% CL upper limits on the fiducial cross-section times branching ratio for combined dilepton channels with varying resonance widths [173].

and adopting the sequential Standard Model (SSM) framework, the production of an SSM Z' boson has been excluded at the 95% confidence level (CL) for masses below 5.15 TeV (CMS) [172] and 5.10 TeV (ATLAS) [173]. The SSM assumes that the Z' boson has the same couplings to SM particles as the Z boson, specifically with generation-independent couplings to SM fermions.

In Fig. 10.7, the left panel shows the CMS upper limits [172] at the 95% confidence level (CL) on the product of the production cross-section and the branching fraction for a spin-1 resonance with a width equal to 0.6% of the resonance mass. These limits are normalized to the corresponding product for the SM Z boson. Results are provided for the (top left) dielectron channel, (top right) dimuon channel, and (bottom) the combined dilepton channels. The shaded bands represent the 68% and 95% quantiles of the expected limits. Simulated predictions for spin-1 $Z'_{\rm SSM}$ and Z'_ψ resonances are also included for comparison. The right panel presents the ATLAS upper limits [173] at 95% CL on the fiducial cross-section times branching ratio as a function of the pole mass for the combined dilepton channel. The limits are shown for signals with relative widths of zero, 3%, and 10%. Observed limits are depicted as solid lines, while expected limits are indicated by dotted or dashed lines. Theoretical cross-sections for the $Z'_{\rm SSM}$ resonance, with a width-to-mass ratio of $\Gamma/m = 3.0\%$, are overlaid for comparison. These results highlight the combined sensitivity of the CMS and ATLAS searches to Z' boson models with varying resonance widths.

Search for W' at the LHC

Many extensions of the SM of particle physics predict the existence of new, heavy charged gauge bosons, often referred to as W' bosons. These additional particles arise in various theoretical frameworks, such as grand unified theories (GUTs), composite models, and certain models of extra dimensions. The search for these new W' bosons provides an essential tool for probing a wide range of new physics scenarios. By investigating the production and decay of W' bosons, experiments at particle colliders like the LHC can offer insights into the nature of new interactions, the possible structure of underlying physics beyond the SM, and the mass scale of new particles that could explain phenomena unexplained by the SM.

Searches for W' bosons, hypothetical heavier counterparts to the SM W boson, have been extensively conducted by the CMS and ATLAS collaborations. The production of a W' boson in proton-proton collisions at the LHC occurs primarily through quark-antiquark annihilation ($q\bar{q}' \to W'$). The production cross-section is influenced by the parton distribution functions (PDFs) of the quarks and antiquarks in the proton and the coupling strength of the W' boson to quarks. In models like the SSM, the W' boson has couplings similar to the SM W, while other models may propose different coupling patterns. The cross-section is sensitive to the mass of the W' boson, which decreases rapidly with increasing mass due to the behavior of PDFs at high momentum transfer. Experimental searches at the LHC focus on various decay channels, such as leptonic ($W' \to \ell\nu$) and hadronic ($W' \to q\bar{q}'$) decays, with additional sensitivity to the $W' \to tb$ decay mode, especially in models with enhanced couplings to third-generation fermions.

The decay modes of the W' boson provide distinctive signatures for experimental searches. In leptonic decays, the W' boson decays to a charged lepton and a neutrino, yielding a signature of high transverse momentum (p_T) leptons and missing transverse energy (E_T^{miss}), which are prominent in the final states. Hadronic decays lead to high-mass dijet systems, and the $W' \to tb$ decay produces top-bottom quark pairs, which require advanced reconstruction techniques due to the boosted top quarks. Searches by the CMS and ATLAS collaborations have set stringent upper limits on the mass of the W' boson. For example, CMS results exclude W' boson masses up to approximately 6.1 TeV in the leptonic decay channels [174], while ATLAS constraints in the same channels exclude masses up to about 5.9 TeV [170]. Limits for hadronic and

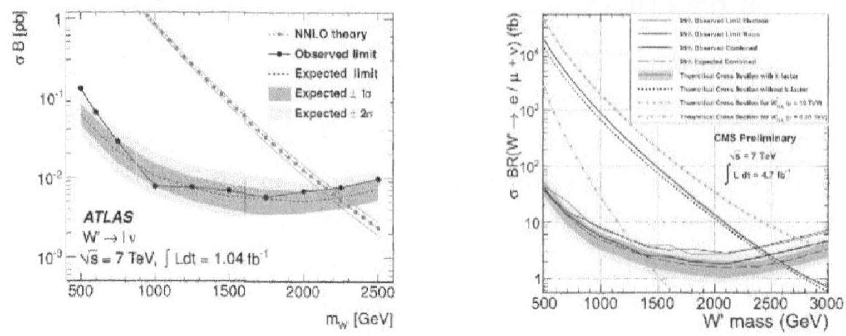

Figure 10.8 ATLAS [170] (left) and CMS [174] (right) cross-section limits on $W' \to \ell\nu$.

heavy-flavor decays extend to approximately 4.5 TeV and 3.8 TeV, respectively.

Both the ATLAS and CMS collaborations have conducted searches for the decay $W' \to \ell\nu$ (where $\ell = e, \mu, \tau$). The ATLAS search sets an SSM mass limit of $m_{W'} > 5.9\,\text{TeV}$, while the CMS search places a slightly stronger limit of $m_{W'} > 6.1\,\text{TeV}$. Fig. 10.8 illustrates the corresponding cross-section limits as a function of W' mass. As typically observed in these analyses, there is good agreement between the observed and expected limits, providing strong constraints on the mass of the W' boson in this decay channel.

10.4 SEARCH FOR HEAVY HIGGS BOSONS AT THE LHC

As discussed in previous chapters, the heavy Higgs boson is a central feature of many BSM theories, including the 2HDM [48], the LRSM) [75], and the MSSM [118]. These models extend the Higgs sector to address unresolved challenges such as the hierarchy problem and the nature of dark matter. Typically possessing a mass in the multi-hundred GeV to TeV range, the heavy Higgs exhibits distinct couplings to fermions and gauge bosons and can decay into exotic particles. Its discovery would mark a breakthrough in the search for new physics, offering crucial insights into extended mechanisms of electroweak symmetry breaking and helping to differentiate between competing BSM frameworks.

The heavy Higgs boson, h', is predominantly produced at the LHC through the gluon-gluon fusion (ggF) process, which accounts for

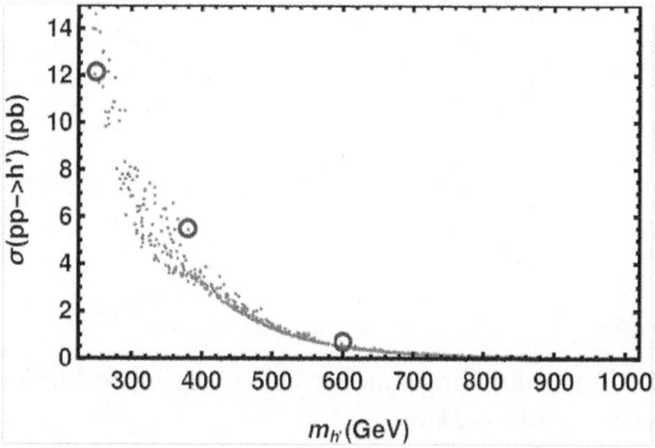

Figure 10.9 Left: The h' production cross-section from ggF as a function of its mass $m_{h'}$. Three benchmark points, under consideration, are surrounded by *dark–shaded region* circles [94].

approximately 90% of its total production cross-section [175]. Other production mechanisms, such as vector boson fusion (VBF), Higgs strahlung, and top-associated Higgs production, contribute to the remaining 10%. Figure 10.9 illustrates the h' ggF production cross-section, $\sigma(pp \to h')$, as a function of its mass, $m_{h'}$, for a range of coupling values. Notably, $\sigma(pp \to h')$ can reach values $\gtrsim 2$ pb for $m_{h'} \lesssim 400$ GeV. In this scenario, an integrated luminosity of $L_{\rm int} = 300$ fb^{-1} is sufficient to probe a relatively light h', while higher luminosities of $L_{\rm int} = 3000$ fb^{-1} are required for heavier h' masses, at a center-of-mass energy of $\sqrt{s} = 14$ TeV.

In Fig. 10.9 (right), we show the relevant h' decay branching ratios as functions of $m_{h'}$. It is remarkable to notice that for $m_{h'} \leq 600$ GeV, the h' decay branching ratio to two SM Higgs bosons is not small, mainly BR$(h' \to hh) \geq 10\%$, which gives hope for probing this heavy Higgs boson through this channel.

One may begin with the decay $h' \to hh \to b\bar{b}b\bar{b}$ for probing h' at the LHC, as the branching ratio BR$(h \to b\bar{b})$ is the largest of h decays. However, this process has a huge background and the signal is much lower than the relevant background, even for a quite heavy Higgs boson. Therefore, the decay process $h' \to hh \to b\bar{b}\gamma\gamma$, given by the Feynman diagram in Fig. 10.10, is considered as the most relevant signal for probing h' [94]. We analyzed this process for $m_{h'} = 250$ GeV, 400 GeV, and 600 GeV.

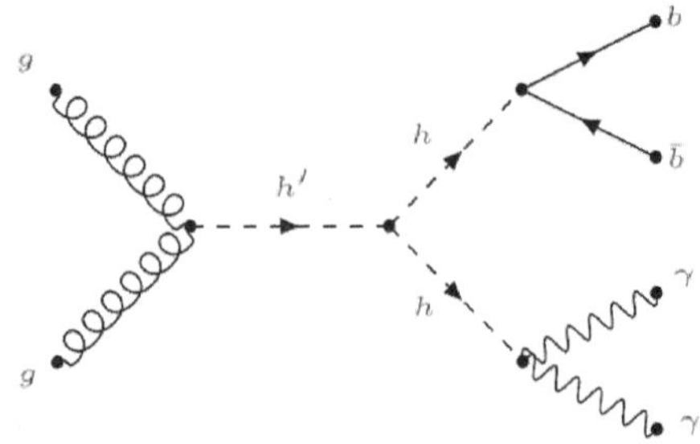

Figure 10.10 Feynman diagram for the h' ggF production and decay process $gg \to h' \to hh \to b\bar{b}\gamma\gamma$.

As the h' decay width $\Gamma_{h'}$ is much smaller than its mass, $\Gamma_{h'}/m_{h'} \ll 1$, the narrow width approximation can be used and the total cross-section $\sigma(pp \to h' \to hh \to b\bar{b}\gamma\gamma)$ can be approximated as

$$\begin{aligned}\sigma(pp \to h' \to hh \to b\bar{b}\gamma\gamma) &\approx \sigma(pp \to h') \times \mathrm{BR}(h' \to hh) \\ &\times \mathrm{BR}(h \to b\bar{b}) \times \mathrm{BR}(h \to \gamma\gamma). \quad (10.5)\end{aligned}$$

The branching ratio of the $h' \to hh$ decay, $\mathrm{BR}(h' \to hh)$, is expressed in terms of the coupling $g_{h'hh}$. Table 10.1 presents the calculated cross-sections and decay branching ratios for the three considered $m_{h'}$ values, following the results from [94].

For potential discovery of h' at the LHC, we analyze both its signal and the corresponding relevant background from the SM processes. These backgrounds can be reduced by appropriate kinematical cuts [94]. Fig. 10.11 shows the number of signal events distributions for $m_{h'} = 250$ GeV and 400 GeV with the relevant irreducible background

TABLE 10.1 $pp \to h'$ production cross-section and its $h' \to hh$ decay branching ratio and the total cross-section for its production and decay process $pp \to h' \to hh \to b\bar{b}\gamma\gamma$ for three different values of $m_{h'} =$ 250 GeV, 400 GeV, and 600 GeV.

$m_{h'}$(GeV)	$\sigma(pp \to h')$(pb)	$\mathrm{BR}(h' \to hh)$	$\sigma(pp \to h' \to hh \to b\bar{b}\gamma\gamma)$(fb)
250	12.140	0.30	6.30
400	5.050	0.20	1.01
600	0.504	0.18	0.05

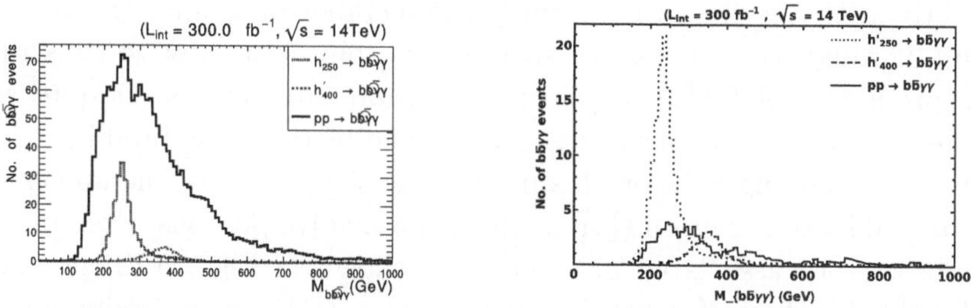

Figure 10.11 Number of signal events for $h' \to b\bar{b}\gamma\gamma$ decays at mass $m_{h'} = 250$ GeV (dotted line) and 400 GeV (dashed line) induced by ggF versus the invariant mass of the final states $b\bar{b}\gamma\gamma$, at $\sqrt{s} = 14$ TeV and $L_{\text{int}} = 300$ fb^{-1} alongside the relevant background events (solid line) before (left) and after (right) applying the cut flow [94]..

before (left) and after (right) applying cuts, respectively. The benchmark point with $m_{h'} = 600$ GeV is not included here as its cross-section is quite tiny with the considered ($L_{\text{int}} = 300$ fb^{-1}).

As can be seen, the application of kinematic cuts effectively suppresses the majority of background events. These backgrounds exhibit broad distributions, as they originate primarily from non-resonant elastic scattering processes rather than resonant production in the signal region. The results clearly establish the $2\gamma + 2b$-jet final state in SM Higgs boson pair production as a distinctive signature for probing the heavy CP-even Higgs (h') in models with enhanced $h'hh$ couplings.

10.5 COLLIDER SEARCHES FOR DARK MATTER

Collider searches for dark matter (DM) focus on the exciting prospect of identifying its particle content, which remains one of the most significant open questions in modern physics. While the existence of dark matter is well established through its gravitational effects on galaxies and galaxy clusters, as well as its impact on the cosmic microwave background, the exact nature of its constituent particles remains elusive. Various theories beyond the Standard Model (SM) suggest potential candidates for dark matter, including weakly interacting massive particles (WIMPs), axions, and dark photons, but none have been conclusively detected.

Colliders such as the LHC provide an ideal environment for probing DM by searching for missing transverse energy (E_T^{miss}) and other signatures indicative of DM production. These searches focus on events where DM is produced in high-energy collisions but does not interact directly with the detector, leaving a detectable imbalance in momentum. Through the analysis of such events, physicists aim to constrain the properties of DM, such as its mass and interactions. Despite the challenges, collider experiments remain one of the most promising avenues for shedding light on this mysterious component of the universe.

Initial state radiation (ISR), which can involve photons, jets, or vector bosons, or associated production as shown in Fig. 10.12, provides valuable signatures for tagging dark matter (DM) pair production in collider experiments. In these searches, the production of a Standard Model (SM) particle(s) X in conjunction with large missing transverse energy (E_T^{miss}) serves as a key indicator of DM interactions. The large E_T^{miss} arises from the undetected dark matter particles, which carry away momentum, resulting in an imbalance that can be measured [176, 177].

The Monojet search is a very general DM search focusing on events with jets and large missing transverse energy (E_T^{miss}). The selection criteria for this search include:

(i) An energetic jet with transverse momentum $p_T > 150\,\text{GeV}$,

(ii) Missing transverse energy $E_T^{\text{miss}} > 200\,\text{GeV}$.

The primary SM backgrounds considered in this analysis include Z+jets, W+jets, diboson, and $t\bar{t}$ events. These backgrounds are estimated using control regions defined by final states with one or two leptons. The background contributions from Z+jets and W+jets processes are predicted with high precision, carrying uncertainties of approximately $\mathcal{O}(1\text{–}2\%)$ [176].

Both ATLAS and CMS employ semi-data-driven techniques in their analyses, utilizing statistically independent signal regions to enhance the robustness of their results. Backgrounds are estimated using control

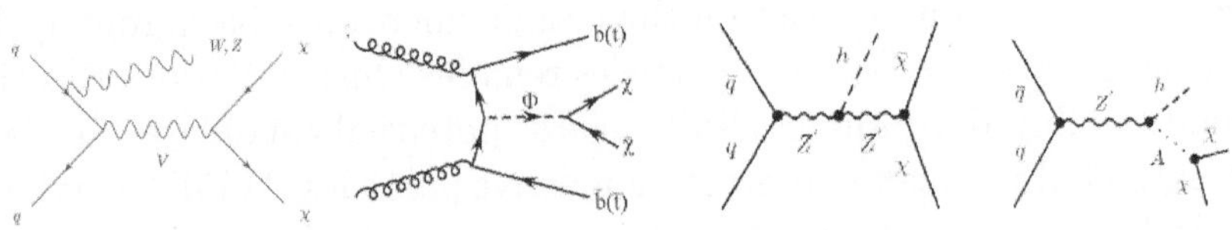

Figure 10.12 Initial state radiation vector boson, Associated production: scalar/pseudo-scalar, and mono-Higgs signatures.

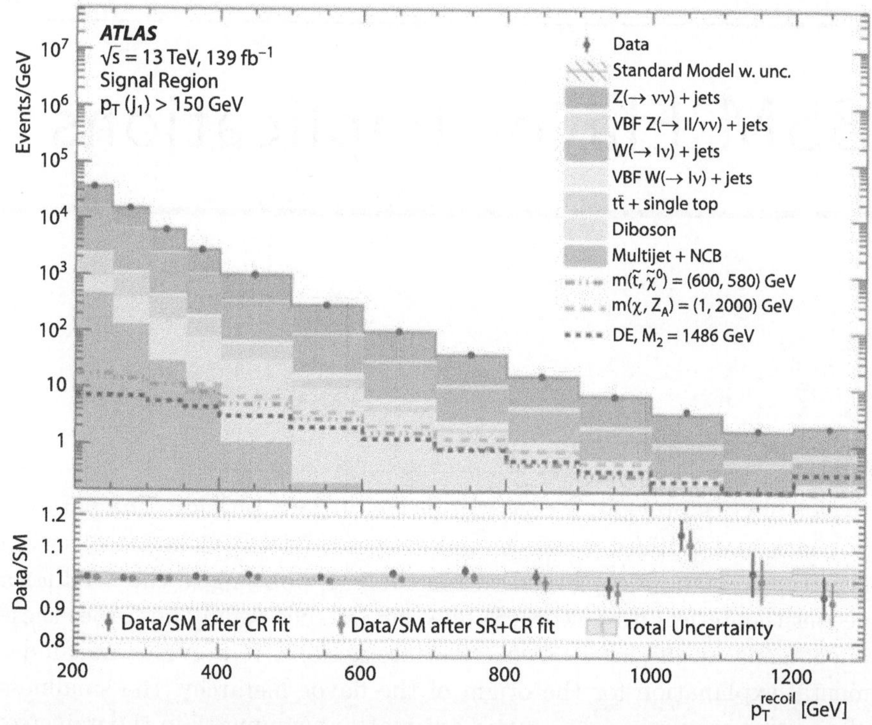

Figure 10.13 ATLAS results for the data and background predictions in the signal region [177]

regions defined by events with one or two leptons, enabling precise validation of Standard Model predictions. Despite the sensitivity of these methods, no significant excess over the expected background has been observed so far.

Figure 10.13 displays the measured distributions of recoil transverse momentum p_T^{recoil} for events with $p_T^{\mathrm{recoil}} > 200$ GeV, compared against SM predictions in the signal region. These distributions are a crucial diagnostic tool for identifying potential deviations from SM expectations, which could indicate DM production or other new physics phenomena. The SM predictions shown include contributions from the dominant backgrounds: Z+jets, W+jets, diboson, and $t\bar{t}$, which are estimated and validated through control regions and precise theoretical calculations [177]. Comparing these distributions enables the identification of excesses or discrepancies in p_T^{recoil}, offering valuable insights into potential new physics.

CHAPTER 11

BSM Flavor Implications

11.1 INTRODUCTION

Flavor physics provides a powerful probe of BSM physics. The patterns of quark and lepton masses, mixing angles, and CP violation observed in nature remain unexplained within the SM, pointing to the need for a deeper theoretical framework. While the SM successfully describes these phenomena at the level of effective interactions, it provides no fundamental explanation for the origin of the flavor hierarchy, the smallness of neutrino masses, or the matter-antimatter asymmetry in the universe. These open questions strongly suggest the presence of new physics at higher energy scales.

BSM theories often address these questions by introducing new particles, symmetries, or dynamics that modify the SM's flavor structure. For example, mechanisms such as the seesaw mechanism for neutrino masses [20], flavor symmetries, or additional Higgs sectors can provide a more fundamental understanding of the observed flavor patterns. However, these modifications must carefully align with the stringent experimental constraints imposed by precision measurements of flavor-changing processes, such as neutral meson oscillations, rare decays, and lepton flavor violations [178, 179]. The absence of significant deviations from SM predictions in these processes places tight restrictions on the parameter space of many BSM models, making flavor physics a key testing ground for new theories.

This chapter focuses on the theoretical implications of BSM physics for flavor phenomena. We explore how various BSM scenarios reshape our understanding of flavor. Particular attention is given to the mechanisms through which these theories generate flavor structures, including

 DOI: 10.1201/9781003457701-11

the role of spontaneous symmetry breaking, alignment mechanisms, and the dynamics of flavor-changing interactions. Additionally, we discuss the potential for new sources of CP violation beyond the SM, which could have profound implications for cosmology and the matter-antimatter asymmetry. Finally, we examine the interplay between flavor physics and other BSM phenomena, such as neutrino masses, dark matter, and the hierarchy problem, highlighting how flavor observables can provide indirect probes of these interconnected aspects of new physics.

11.2 THE FLAVOR PUZZLE AND BSM MOTIVATIONS

The SM of particle physics provides a remarkably successful description of fundamental particles and their interactions. However, when it comes to flavor: the patterns of quark and lepton masses, mixing angles, and CP violation, the SM falls short in providing a fundamental explanation. Key limitations include:

- Unexplained Mass Hierarchies: The masses of the fundamental fermions span many orders of magnitude, from the electron ($m_e \sim 0.5\,\text{MeV}$) to the top quark ($m_t \sim 173\,\text{GeV}$). The SM does not explain why these hierarchies exist or why the Yukawa couplings vary so dramatically.

- Neutrino Masses: The SM predicts neutrinos to be massless, yet neutrino oscillation experiments have conclusively shown that neutrinos have small but non-zero masses [180]. This requires an extension of the SM.

- Flavor Mixing and CP Violation: The observed patterns in the CKM (quark) and PMNS (lepton) mixing matrices are not predicted by the SM. The origin of CP violation in the quark sector (via the CKM phase) and the potential for CP violation in the lepton sector remain unexplained.

- No Fundamental Theory of Flavor: The SM treats Yukawa couplings as free parameters, with no underlying principle to determine their values. This lack of a predictive framework for flavor is a major theoretical gap.

These unresolved issues collectively form the "flavor puzzle," highlighting the need for a deeper theoretical framework beyond the SM [181].

11.3 FLAVOR SYMMETRIES AND MODEL BUILDING

Flavor symmetries are a cornerstone of BSM physics, providing a theoretical framework to explain the observed patterns in fermion masses and mixing angles [182,183]. By imposing symmetry constraints on the Lagrangian, model builders can reduce the number of free parameters and predict specific structures in the fermion mass matrices. This section explores the role of discrete and continuous flavor symmetries in generating fermion masses and mixing patterns, highlighting their potential to address the flavor puzzle.

11.3.1 Discrete Flavor Symmetries

Discrete flavor symmetries are finite groups that act on the flavor space of quarks and leptons. They are particularly attractive because they can naturally explain the observed hierarchies and mixing patterns without introducing excessive complexity [184,185]. As a widely studied example, we consider the A_4 **symmetry**, which is a discrete symmetry group with 12 elements. It is the group of even permutations of four objects and is isomorphic to the symmetry group of a regular tetrahedron. It has four irreducible representations:

- Three singlets: $\mathbf{1}$, $\mathbf{1}'$, and $\mathbf{1}''$.
- One triplet: $\mathbf{3}$.

The multiplication rules for these representations are as follows:

- $\mathbf{1}' \otimes \mathbf{1}' = \mathbf{1}''$
- $\mathbf{1}' \otimes \mathbf{1}'' = \mathbf{1}$
- $\mathbf{1}'' \otimes \mathbf{1}'' = \mathbf{1}'$
- $\mathbf{3} \otimes \mathbf{3} = \mathbf{1} \oplus \mathbf{1}' \oplus \mathbf{1}'' \oplus \mathbf{3}_1 \oplus \mathbf{3}_2$

These rules are essential for constructing invariant terms in the Lagrangian. In A_4 models, the three generations of leptons are typically assigned to the triplet representation $\mathbf{3}$, while the Higgs fields and right-handed neutrinos are assigned to either singlet or triplet representations [186].

We consider the following representations of fields under the discrete flavor symmetry group A_4:

- Left-handed lepton doublets: $L = (L_1, L_2, L_3) \sim \mathbf{3}$
- Right-handed neutrinos: $\nu_R = (\nu_{R1}, \nu_{R2}, \nu_{R3}) \sim \mathbf{3}$
- Standard Model Higgs: $H \sim \mathbf{1}$
- Flavons: $\phi, \chi \sim \mathbf{3}$, $\xi \sim \mathbf{1}$

Therefore, the Lagrangian for the lepton sector includes Yukawa terms and mass terms that are invariant under A_4. These terms are constructed as follows:

Charged Lepton Sector:

The Yukawa terms for charged leptons are constructed using the Higgs field H and the flavon field ϕ, as follows:

$$\mathcal{L}_Y = \frac{y_e}{\Lambda}\bar{L}\phi H e_R + \frac{y_\mu}{\Lambda}\bar{L}\phi H \mu_R + \frac{y_\tau}{\Lambda}\bar{L}\phi H \tau_R + \text{h.c.} \tag{11.1}$$

Here, Λ represents the non-renormalizable suppression scale, and the Yukawa interactions involve both the Higgs field H (for $SU(2)_L$ invariance) and the flavon field ϕ (for A_4 invariance). When the flavon field ϕ and Higgs field H acquire vacuum expectation values, the A_4 and $SU(2)_L$ symmetries are spontaneously broken, generating the charged lepton mass matrix.

After electroweak and flavor symmetry breaking:

$$\begin{aligned}\langle H\rangle &= v,\\ \langle \phi\rangle &= (v_\phi, v_\phi, v_\phi),\\ \langle \chi\rangle &= (v_\chi, v_\chi, v_\chi),\\ \langle \xi\rangle &= v_\xi.\end{aligned}$$

The charged lepton mass matrix is generated and given by:

$$M_\ell = \frac{v_\phi v}{\Lambda}\begin{pmatrix} y_e & 0 & 0\\ 0 & y_\mu & 0\\ 0 & 0 & y_\tau\end{pmatrix}. \tag{11.2}$$

In this context, the A_4 symmetry leads to a diagonal structure for the charged lepton mass matrix M_ℓ. Consequently, the observed hierarchy $m_e \ll m_\mu \ll m_\tau$ can be naturally accommodated by a corresponding hierarchy among the Yukawa couplings y_e, y_μ, y_τ. While the symmetry does not predict the specific values of these couplings, it provides a natural framework where the mass eigenstates coincide with the flavor eigenstates, simplifying the analysis of lepton mixing.

Neutrino Sector:

The neutrino mass terms arise from a combination of Dirac and Majorana terms, constructed to be invariant under A_4 symmetry. The relevant effective Lagrangian is:

$$\mathcal{L}_\nu = \frac{y_D}{\Lambda}(\bar{L}\phi)_{\mathbf{1}}\tilde{H}\nu_R - \frac{1}{2}\nu_R^T C^{-1}(y_1\xi + y_2(\chi)_{\mathbf{3}})\nu_R + \text{h.c.} \tag{11.3}$$

The singlet flavon ξ is not strictly essential, but it is particularly useful. It allows for a simple invariant contraction $(\nu_R\nu_R)_{\mathbf{1}}\xi$ and provides flexibility in generating the structure and scale of the Majorana mass matrix M_R. The presence of both ξ and χ allows for the construction of more realistic models that can reproduce observed neutrino mass hierarchies and mixing angles.

The Dirac mass matrix m_D and Majorana mass matrix M_R are generated as follows:

$$m_D = y_D \frac{v v_\phi}{\Lambda} \begin{pmatrix} 1 & 0 & 0 \\ 0 & 1 & 0 \\ 0 & 0 & 1 \end{pmatrix},$$

$$M_R = y_1 v_\xi \begin{pmatrix} 1 & 0 & 0 \\ 0 & 1 & 0 \\ 0 & 0 & 1 \end{pmatrix} + y_2 v_\chi \begin{pmatrix} 0 & 1 & 1 \\ 1 & 0 & 1 \\ 1 & 1 & 0 \end{pmatrix}$$

The light neutrino mass matrix then follows from the type-I seesaw mechanism:

$$m_\nu = -m_D M_R^{-1} m_D^T \tag{11.4}$$

With appropriate flavon VEV alignments, this framework can accommodate the observed neutrino mixing patterns, such as tri-bimaximal mixing pattern for the PMNS matrix, given by:

$$U_{\text{PMNS}} = \begin{pmatrix} \frac{2}{\sqrt{6}} & \frac{1}{\sqrt{3}} & 0 \\ -\frac{1}{\sqrt{6}} & \frac{1}{\sqrt{3}} & -\frac{1}{\sqrt{2}} \\ -\frac{1}{\sqrt{6}} & \frac{1}{\sqrt{3}} & \frac{1}{\sqrt{2}} \end{pmatrix}. \tag{11.5}$$

This pattern predicts the following mixing angles: $\theta_{13} = 0$, $\theta_{23} = \pi/4$, and $\theta_{12} = \arcsin(1/\sqrt{3})$, which are consistent with early neutrino oscillation data but require modifications to account for the non-zero θ_{13} observed in later experiments. This discrepancy has motivated further refinements of A_4 models, such as the inclusion of additional flavons or small symmetry-breaking effects.

11.3.2 Continuous Flavor Symmetries

The simplest example of continuous symmetries is the $U(1)$ flavor symmetry that assigns different charges to the fermion generations. It is widely used in BSM physics to explain the observed hierarchies in fermion masses and mixing angles [187, 188]. The $U(1)$ symmetry is often spontaneously broken by the vacuum expectation value (VEV) of a scalar field (flavon), leading to hierarchical Yukawa couplings through the Froggatt-Nielsen mechanism.

The $U(1)$ flavor symmetry group consists of all complex numbers of unit modulus under multiplication. In the context of flavor physics, the $U(1)$ symmetry acts on the fermion and scalar fields, assigning them specific charges. Each fermion generation f_i (where $i = 1, 2, 3$) and the flavon field ϕ are assigned $U(1)$ charges:

$$f_i \to e^{iq_i\alpha} f_i, \quad \phi \to e^{iq_\phi\alpha}\phi, \tag{11.6}$$

where q_i and q_ϕ are the $U(1)$ charges of the fermions and flavon, respectively, and α is the transformation parameter.

The Yukawa couplings are constrained by the $U(1)$ symmetry. For example, the Yukawa term for the charged leptons takes the form:

$$\mathcal{L}_Y \sim \bar{L}_i H e_{Rj} \left(\frac{\phi}{\Lambda}\right)^{n_{ij}}, \tag{11.7}$$

where Λ is the cutoff scale of the theory, and n_{ij} are integers determined by the $U(1)$ charge conservation:

$$n_{ij} = q_{L_i} + q_{e_{R_j}} + q_H. \tag{11.8}$$

The Froggatt-Nielsen mechanism explains the hierarchical structure of fermion masses through the spontaneous breaking of the $U(1)$ flavor symmetry. The flavon field ϕ acquires a VEV, breaking the $U(1)$ symmetry. The VEV is typically much smaller than the cutoff scale Λ:

$$\langle \phi \rangle = \epsilon \Lambda, \quad \epsilon \ll 1. \tag{11.9}$$

Thus, the Yukawa couplings are suppressed by powers of the small parameter ϵ:

$$y_{ij} \sim \epsilon^{n_{ij}}. \tag{11.10}$$

This generates a hierarchical structure in the fermion mass matrices. For example, the charged lepton mass matrix takes the form:

$$M_e \sim v_H \begin{pmatrix} \epsilon^{n_{11}} & \epsilon^{n_{12}} & \epsilon^{n_{13}} \\ \epsilon^{n_{21}} & \epsilon^{n_{22}} & \epsilon^{n_{23}} \\ \epsilon^{n_{31}} & \epsilon^{n_{32}} & \epsilon^{n_{33}} \end{pmatrix}, \tag{11.11}$$

where v_H is the VEV of the Higgs field.

One can demonstrate that the $U(1)$ flavor symmetry helps construct the correct fermion masses by generating hierarchical Yukawa couplings. For example, the charged lepton mass matrix is diagonalized by the hierarchy $\epsilon^{n_{11}} \ll \epsilon^{n_{22}} \ll \epsilon^{n_{33}}$, which naturally leads to the observed mass hierarchy $m_e \ll m_\mu \ll m_\tau$.

The quark mass matrices exhibit similar hierarchies, explaining the smallness of the up, charm, and top quark masses relative to each other. For example:

$$M_u \sim v_H \begin{pmatrix} \epsilon^{n_{11}} & \epsilon^{n_{12}} & \epsilon^{n_{13}} \\ \epsilon^{n_{21}} & \epsilon^{n_{22}} & \epsilon^{n_{23}} \\ \epsilon^{n_{31}} & \epsilon^{n_{32}} & \epsilon^{n_{33}} \end{pmatrix}. \tag{11.12}$$

In the neutrino sector, the $U(1)$ symmetry can be combined with the seesaw mechanism to generate small neutrino masses. The Dirac mass matrix M_D and the Majorana mass matrix M_R take the form:

$$M_D \sim v_H \begin{pmatrix} \epsilon^{n_{11}} & \epsilon^{n_{12}} & \epsilon^{n_{13}} \\ \epsilon^{n_{21}} & \epsilon^{n_{22}} & \epsilon^{n_{23}} \\ \epsilon^{n_{31}} & \epsilon^{n_{32}} & \epsilon^{n_{33}} \end{pmatrix}, \quad M_R \sim \Lambda \begin{pmatrix} \epsilon^{m_{11}} & \epsilon^{m_{12}} & \epsilon^{m_{13}} \\ \epsilon^{m_{21}} & \epsilon^{m_{22}} & \epsilon^{m_{23}} \\ \epsilon^{m_{31}} & \epsilon^{m_{32}} & \epsilon^{m_{33}} \end{pmatrix}. \tag{11.13}$$

The effective light neutrino mass matrix M_ν is then given by:

$$M_\nu = M_D^T M_R^{-1} M_D. \tag{11.14}$$

It has been shown that the $U(1)$ flavor symmetry also influences the mixing patterns in both the quark and lepton sectors. For quarks, the hierarchical structure of the mass matrices leads to small mixing angles in the CKM matrix, consistent with experimental observations. In the lepton sector, the $U(1)$ symmetry can predict specific patterns in the PMNS matrix, depending on the charge assignments and symmetry-breaking mechanism, such as generating the large mixing angles observed in neutrino experiments.

While the $U(1)$ flavor symmetry offers a compelling framework for explaining fermion masses and mixing patterns, it faces several challenges. For instance, the introduction of additional parameters, such as charges and vacuum expectation values (VEVs), reduces the model's predictive power. Additionally, precision measurements of flavor-changing processes place strict constraints on the parameter space of $U(1)$ models. Furthermore, the symmetry must align with other BSM phenomena, such as neutrino masses, dark matter, and baryogenesis. Despite these challenges, the $U(1)$ flavor symmetry remains a valuable tool for understanding the origins of fermion masses and mixing. Future experiments, including those studying rare decays and neutrino oscillations, will play a crucial role in testing $U(1)$ models and advancing our understanding of flavor physics.

11.4 FLAVOR AND CP VIOLATION BEYOND THE SM

BSM theories introduce new particles and interactions that can significantly alter the flavor structure and CP-violating mechanisms of the SM. These extensions aim to address unresolved questions in the SM, such as the origin of fermion mass hierarchies, the smallness of neutrino masses, and the matter-antimatter asymmetry in the universe. In this context, we focus on three key examples: the Two-Higgs-Doublet Model (2HDM), described in detail in Chapter 2; the Left-Right Symmetric Model (LRSM), discussed in Chapters 4 and 5; and the Minimal Supersymmetric Standard Model (MSSM), explored in Chapter 8. These models demonstrate how BSM physics can reshape our understanding of flavor and CP violation, offering testable predictions for current and future experiments.

11.4.1 Flavor and CP Violation in 2HDM

The flavor structure of the 2HDM depends on how the Higgs doublets couple to fermions, leading to different types of 2HDMs (Type I, II, and III) [48]. In particular, Type III models allow for tree-level flavor-changing neutral currents (FCNCs), which are tightly constrained by experimental data [189]. In the 2HDM, the Yukawa Lagrangian for the quark sector can be written as:

$$\mathcal{L}_Y = -\bar{Q}_L Y_1^u \tilde{\Phi}_1 u_R - \bar{Q}_L Y_2^u \tilde{\Phi}_2 u_R - \bar{Q}_L Y_1^d \Phi_1 d_R - \bar{Q}_L Y_2^d \Phi_2 d_R + \text{h.c.}, \tag{11.15}$$

where Q_L is the left-handed quark doublet, u_R and d_R are the right-handed up- and down-type quark singlets, Φ_1 and Φ_2 are the two Higgs doublets (with $\tilde{\Phi}_i = i\sigma_2\Phi_i^*$), and Y_1^u, Y_2^u, Y_1^d, and Y_2^d are the Yukawa coupling matrices. In Type I and Type II models, a Z_2 symmetry is imposed to eliminate tree-level FCNCs by coupling each type of fermion to only one Higgs doublet. However, in Type III models, this symmetry is relaxed, allowing both Higgs doublets to couple to all fermions, leading to tree-level FCNCs. The flavor-violating interactions can be parameterized as:

$$\mathcal{L}_{\text{FCNC}} = -\frac{\sqrt{2}}{v}\left(\bar{d}_L X_d d_R \phi + \bar{u}_L X_u u_R \phi + \text{h.c.}\right), \tag{11.16}$$

where ϕ represents the neutral Higgs bosons (h, H, A), X_d and X_u are the flavor-violating coupling matrices for down-type and up-type quarks, respectively, and $v = \sqrt{v_1^2 + v_2^2} \approx 246\,\text{GeV}$ is the SM Higgs VEV. These couplings lead to flavor-violating processes such as $K^0 - \bar{K}^0$ mixing, $B^0 - \bar{B}^0$ mixing, and rare decays like $B \to X_s\gamma$.

The flavor-violating couplings in the 2HDM are tightly constrained by experimental data. For example, the mass difference Δm_K and the

CP-violating parameter ϵ_K in $K^0 - \bar{K}^0$ mixing provide stringent constraints on the off-diagonal elements of X_d [190]. The contribution to Δm_K is given by:

$$\Delta m_K \sim \frac{1}{m_H^2} \left|(X_d)_{12}\right|^2, \tag{11.17}$$

where m_H is the mass of the mediating Higgs boson. Similarly, the mass differences Δm_{B_d} and Δm_{B_s} in $B^0 - \bar{B}^0$ mixing constrain the off-diagonal elements $(X_d)_{13}$ and $(X_d)_{23}$. The contribution to Δm_{B_d} is:

$$\Delta m_{B_d} \sim \frac{1}{m_H^2} \left|(X_d)_{13}\right|^2. \tag{11.18}$$

Rare decays such as $B \to X_s\gamma$ and $B_s \to \mu^+\mu^-$ provide additional constraints. For instance, the branching ratio of $B \to X_s\gamma$ is sensitive to the charged Higgs contribution:

$$\mathrm{BR}(B \to X_s\gamma) \propto \left|\frac{(X_d)_{23}}{m_{H^\pm}^2}\right|^2, \tag{11.19}$$

where $m_{H^\pm}$ is the mass of the charged Higgs boson.

In the alignment limit, the light CP-even Higgs boson h behaves like the SM Higgs, suppressing FCNCs [191]. This limit is achieved when the mixing angle α between the CP-even Higgs states satisfies $\alpha \approx \beta - \pi/2$. In this limit, the flavor-violating couplings are suppressed, and the model becomes consistent with experimental constraints. The charged Higgs bosons $H^\pm$ can also mediate flavor-changing processes, such as $B \to X_s\gamma$. The contribution to the decay amplitude is proportional to:

$$\mathcal{A}(B \to X_s\gamma) \propto \frac{(X_d)_{23}}{m_{H^\pm}^2}. \tag{11.20}$$

This imposes constraints on the charged Higgs mass and the flavor-violating couplings. It is worth stressing that experimental constraints on significant flavor violation in 2HDM Type III necessitate either a precise alignment of the Higgs couplings or severe fine-tuning to suppress tree-level FCNCs.

In addition, CP violation can arise in the 2HDM from complex phases in the Higgs potential, Yukawa couplings, or interactions involving the additional Higgs bosons [192]. These new sources of CP violation can have significant implications for flavor physics, electric dipole moments (EDMs), and cosmology.

In Type III 2HDM, where both Higgs doublets couple to all fermions, complex Yukawa couplings can introduce new sources of CP violation. These phases can contribute to CP-violating observables, such as $B^0-\bar{B}^0$ mixing and electric dipole moments (EDMs). The contribution to the electron EDM, for example, can be written as:

$$d_e \propto \text{Im}\left(\frac{(Y_e)_{11}}{m_H^2}\right), \tag{11.21}$$

where $(Y_e)_{11}$ is the Yukawa coupling of the electron to the Higgs bosons, and m_H is the mass of the mediating Higgs boson. Current experimental limits on EDMs place stringent constraints on the CP-violating phases in the 2HDM.

The 2HDM can also contribute to CP violation in meson mixing, such as $K^0-\bar{K}^0$ and $B^0-\bar{B}^0$ mixing. The contribution to the CP-violating parameter ϵ_K in $K^0-\bar{K}^0$ mixing can be written as:

$$\epsilon_K \propto \text{Im}\left(\frac{(X_d)_{12}^2}{m_H^2}\right). \tag{11.22}$$

The experimental value of the CP-violating parameter is precisely measured, with $|\epsilon_K| \approx 2.228 \times 10^{-3}$. This stringent constraint severely limits the imaginary part of the flavor-changing couplings in the down-quark sector, particularly $\text{Im}[(X_d)_{12}^2]$, thus placing strong bounds on new sources of CP violation within the 2HDM framework. As a result, any viable model must ensure that such contributions remain sufficiently suppressed, either through flavor alignment or by pushing the mass scale m_H to higher values.

11.4.2 Flavor and CP Violation in LRSM

The LRSM introduces new sources of flavor violation through the interactions of the right-handed gauge bosons $W_R^\pm$ and the additional Higgs fields. The right-handed gauge boson $W_R^\pm$ mediates flavor-changing interactions between right-handed quarks, described by the interaction Lagrangian:

$$\mathcal{L}_{W_R} = \frac{g_R}{\sqrt{2}}\bar{u}_R\gamma^\mu V_R d_R W_{R\mu}^+ + \text{h.c.}, \tag{11.23}$$

where g_R is the $SU(2)_R$ gauge coupling, and V_R is the right-handed analog of the CKM matrix [193]. The bidoublet Higgs field Φ can also induce tree-level flavor-changing neutral currents (FCNCs) through its Yukawa couplings. The Yukawa Lagrangian for the quark sector is:

$$\mathcal{L}_Y = \bar{Q}_L(Y_1\Phi + Y_2\tilde{\Phi})Q_R + \text{h.c.}, \tag{11.24}$$

where $\tilde{\Phi} = \sigma_2\Phi^*\sigma_2$, and Y_1, Y_2 are Yukawa coupling matrices. These couplings can lead to flavor-violating processes such as $K^0 - \bar{K}^0$ mixing and $B^0 - \bar{B}^0$ mixing. The LRSM contributions to FCNCs are tightly constrained by experimental data. For example, the contribution to $K^0 - \bar{K}^0$ mixing is given by:

$$\Delta m_K \propto \frac{(V_R)_{12}^2}{m_{W_R}^2}, \tag{11.25}$$

where $(V_R)_{12}$ is the flavor-violating coupling between the first and second generations of right-handed quarks, and m_{W_R} is the mass of the right-handed gauge boson [194].

The LRSM also introduces new sources of CP violation through complex phases in the right-handed CKM matrix V_R and the Higgs sector. The right-handed CKM matrix V_R can have complex phases that contribute to CP-violating observables. For example, the phase of $(V_R)_{13}$ can affect $B^0 - \bar{B}^0$ mixing. The bidoublet Higgs field Φ can have complex Yukawa couplings, leading to CP-violating interactions. The contribution to the CP-violating parameter ϵ_K in $K^0 - \bar{K}^0$ mixing is given by:

$$\epsilon_K \propto \text{Im}\left(\frac{(Y_1)_{12}(Y_2^*)_{12}}{m_H^2}\right), \tag{11.26}$$

where m_H is the mass of the mediating Higgs boson. The LRSM can also contribute to electric dipole moments (EDMs) of particles such as the electron, neutron, and atoms through CP-violating interactions involving the right-handed gauge bosons and Higgs fields. The contribution to the electron EDM is given by:

$$d_e \propto \text{Im}\left(\frac{(Y_e)_{11}}{m_{W_R}^2}\right), \tag{11.27}$$

where $(Y_e)_{11}$ is the Yukawa coupling of the electron to the Higgs fields [190].

11.4.3 Flavor and CP Violation in the MSSM

In the MSSM, the soft SUSY breaking terms, such as sfermion masses and trilinear couplings, can introduce new sources of flavor-changing neutral currents (FCNCs) and CP-violating phases [118]. These terms must be carefully constrained to avoid large contributions to FCNCs and CP-violating observables, which are tightly constrained by precision experimental data.

For example, transitions like $\tilde{d}_L \to \tilde{s}_L$, arising from off-diagonal elements in the down-type squark mass matrix, can contribute to processes such as $K^0 - \bar{K}^0$ mixing [118]. The key observables in the neutral kaon system are the mass difference Δm_K and the CP-violating parameter ϵ_K. It is essential that SUSY contributions to Δm_K and ϵ_K remain smaller than the SM predictions and do not exceed the difference between experimental results and SM predictions. This requirement imposes the following stringent constraint on the off-diagonal elements of the squark mass matrices $(\tilde{m}^2_{Q,D})_{12}$:

$$\left| \frac{(\tilde{m}^2_{Q,D})_{12}}{\tilde{m}^2} \right| \lesssim 10^{-3} \quad \text{for} \quad \tilde{m} \sim 1\,\text{TeV}. \tag{11.28}$$

Moreover, the neutral B-meson system also imposes significant constraints on the SUSY parameter space [118]. The $B^0 - \bar{B}^0$ mixing observables, Δm_{B_d} and Δm_{B_s}, require that the off-diagonal elements in the squark mass matrices $(\tilde{m}^2_{Q,D})_{13}$ and $(\tilde{m}^2_{Q,D})_{23}$ satisfy:

$$\left| \frac{(\tilde{m}^2_{Q,D})_{13}}{\tilde{m}^2} \right|, \left| \frac{(\tilde{m}^2_{Q,D})_{23}}{\tilde{m}^2} \right| \lesssim 10^{-2} \quad \text{for} \quad \tilde{m} \sim 1\,\text{TeV}. \tag{11.29}$$

In addition, rare decays such as $B \to X_s\gamma$ and $B_s \to \mu^+\mu^-$, with their SM predictions, impose stringent constraints on SUSY parameters. Specifically, the trilinear coupling A_t and the charged Higgs mass $m_{H^\pm}$ must satisfy:

$$|A_t| \lesssim 2\,\text{TeV}, \quad m_{H^\pm} \gtrsim 500\,\text{GeV}. \tag{11.30}$$

On the other hand, the CP-violating parameter ϵ_K and $\sin 2\beta$ (B-meson system), with their SM predictions, imply that the off-diagonal elements in the squark mass matrices and the CP-violating phases in SUSY must satisfy:

$$\left| \frac{(\tilde{m}^2_{Q,D})_{12}}{\tilde{m}^2} \right| \lesssim 10^{-3}, \quad \phi_{\text{SUSY}} \lesssim 10^{-2}. \tag{11.31}$$

The Electric Dipole Moments (EDMs) of the electron (d_e), neutron (d_n), and atoms (e.g., $d_{\rm Hg}$) impose stringent constraints on SUSY CP-violating phases. Their current experimental limits are:

$$\begin{aligned} |d_e| &< 1.1 \times 10^{-29}\, e \cdot \mathrm{cm}, && (11.32)\\ |d_n| &< 1.8 \times 10^{-26}\, e \cdot \mathrm{cm}, && (11.33)\\ |d_{\rm Hg}| &< 6.3 \times 10^{-30}\, e \cdot \mathrm{cm}. && (11.34) \end{aligned}$$

The CP-violating phases in the SUSY soft terms (e.g., ϕ_μ, ϕ_{A_t}) must satisfy:

$$\phi_\mu, \phi_{A_t} \lesssim 10^{-2} \quad \text{for} \quad \tilde{m} \sim 1\,\mathrm{TeV}. \tag{11.35}$$

These constraints highlight the stringent limits on SUSY parameter space imposed by flavor and CP-violating observables. Future experiments will further test these constraints and guide the development of SUSY models.

11.5 SCALAR LEPTOQUARK FLAVOR VIOLATION IN $SU(5)$

As advocated above, the scalar triplets contained in the $\mathbf{45}_H$ Higgs representation decompose under the Standard Model gauge group as:

$$\begin{aligned} (3^*, 2)^{ij}_{-7/6} &\equiv (\mathbf{45}_H)^{ij}_c \equiv \Phi^{ij}_c, && (11.36)\\ (3^*, 1)^{ab}_{4/3} &\equiv (\mathbf{45}_H)^{ab}_k \equiv \Phi^{ab}_k, && (11.37)\\ [(3,1) \oplus (3,3)]^{ib}_{-1/3} &\equiv (\mathbf{45}_H)^{ib}_c \equiv \Phi^{ib}_c. && (11.38) \end{aligned}$$

As emphasized in [195], while Φ^{ab}_k and Φ^{ib}_c mediate proton decay and must therefore be superheavy, the scalar triplet Φ^{ij}_c does not couple to baryon number violating operators and hence does not contribute to proton decay. Writing Φ^{ij}_c as a weak doublet $\Phi^{ij}_c = (\phi^i_1, \phi^i_2)^T$, its relevant interactions with matter fields are:

$$\begin{aligned} \mathcal{L} = {} & 2Y^2_{AB}\, e^T_A C u^c_{Bi}\, \phi^{i1*} + 4(Y^4_{AB} - Y^4_{BA})\, u^{iT}_A C e^c_B\, \phi_{i1}\\ & - 2Y^2_{AB}\, \nu^T_A C u^c_{Bi}\, \phi^{i2*} + 4(Y^4_{AB} - Y^4_{BA})\, d^{iT}_A C e^c_B\, \phi_{i2}. \end{aligned} \tag{11.39}$$

The first two terms induce $b \to s\ell^+\ell^-$ transitions via ϕ^{i1} exchange, while the last two account for $b \to c\tau\nu$ through ϕ^{i2} [196]. These interactions can be rewritten using Dirac notation and antisymmetric coupling definitions:

$$\mathcal{L} = 2Y^2_{AB}\, \bar{u}_{Bi} P_L \nu_A\, \phi^{i2*} - 4Y^{4'}_{AB}\, \bar{e}_B P_L d^i_A\, \phi_{i2} + \text{h.c.}, \tag{11.40}$$

where $Y^{4'}_{AB} \equiv Y^4_{AB} - Y^4_{BA}$ and we used the identities $C^T = -C$ and $\bar{\Psi} = \Psi^{cT}_L$. In the mass eigenstate basis:

$$d_A \to V^{\rm CKM}_{AK} d'_K, \quad \nu_A \to V^{\rm PMNS}_{AK} \nu'_K, \quad u_A \to u'_A, \quad e_A \to e'_A,$$

the Lagrangian becomes:

$$\mathcal{L} = 2Y^2_{AB}\, \bar{u}'_{Bi} P_L V^{\rm PMNS}_{AK} \nu'_K\, \phi^{i2*} - 4Y^{4'}_{AB}\, \bar{e}'_B P_L V^{\rm CKM}_{AK} d'_K\, \phi_{i2} + \text{h.c.} \tag{11.41}$$

Consequently, the amplitude for the process $b \to c\tau\nu$ is given by [196]:

$$\begin{aligned}\mathcal{M} = -\frac{8Y^{4'}_{13} V^{\rm CKM}_{13}}{M^2_\phi} \Bigg[\frac{1}{2} (\bar{u}_\tau P_L v_{\nu_\tau}) (\bar{u}_c P_L u_b) + \frac{1}{8} (\bar{u}_\tau \sigma^{\mu\nu} P_L v_{\nu_\tau}) \\ \times (\bar{u}_c P_L \sigma_{\mu\nu} u_b) \left(Y^2_{12} V^{\rm PMNS}_{13} + Y^2_{22} V^{\rm PMNS}_{23} + Y^2_{32} V^{\rm PMNS}_{33}\right) \Bigg] \\ + \left(Y^{4'}_{13} V^{\rm CKM}_{13} \to Y^{4'}_{23} V^{\rm CKM}_{23}\right).\end{aligned} \tag{11.42}$$

Since $V^{\rm CKM}_{13} \sim 10^{-3}$ and $V^{\rm CKM}_{23} \sim 10^{-2}$, the amplitude is predominantly controlled by the leptoquark mass M_ϕ, the couplings Y^2_{22}, Y^2_{32}, and $Y^{4'}_{13}$. The general expression of the effective Hamiltonian for the transition $b \to c\ell\bar{\nu}_\ell$ reads [197]:

$$\begin{aligned}\mathcal{H}_{\rm eff} = \frac{4G_F V_{cb}}{\sqrt{2}} \Big[(1 + g_{V_L}) [\bar{c}\gamma_\mu P_L b][\bar{\ell}\gamma^\mu P_L \nu_\ell] + g_{V_R} [\bar{c}\gamma_\mu P_R b][\bar{\ell}\gamma^\mu P_L \nu_\ell] \\ + g_{S_L} [\bar{c} P_L b][\bar{\ell} P_L \nu_\ell] + g_{S_R} [\bar{c} P_R b][\bar{\ell} P_L \nu_\ell] + g_T [\bar{c}\sigma^{\mu\nu} P_L b][\bar{\ell}\sigma_{\mu\nu} P_L \nu_\ell] \Big],\end{aligned} \tag{11.43}$$

where $P_{L,R} = (1 \mp \gamma_5)/2$ and $g_i = C^{\rm NP}_i / C^{\rm SM}$ with $C^{\rm SM} = \frac{4G_F V_{cb}}{\sqrt{2}}$.

In our model, $g_{V_L} = g_{V_R} = g_{S_R} = 0$, while:

$$g_{S_L} = -\frac{\sqrt{2} Z}{M^2_\phi G_F}, \qquad g_T = \frac{g_{S_L}}{4} = -\frac{Z}{2\sqrt{2} M^2_\phi G_F}, \tag{11.44}$$

with

$$Z = \left(Y^2_{12} V^{\rm PMNS}_{13} + Y^2_{22} V^{\rm PMNS}_{23} + Y^2_{32} V^{\rm PMNS}_{33}\right) \left(\frac{Y^{4'}_{13} V^{\rm CKM}_{13}}{V^{\rm CKM}_{23}} + Y^{4'}_{23} \right). \tag{11.45}$$

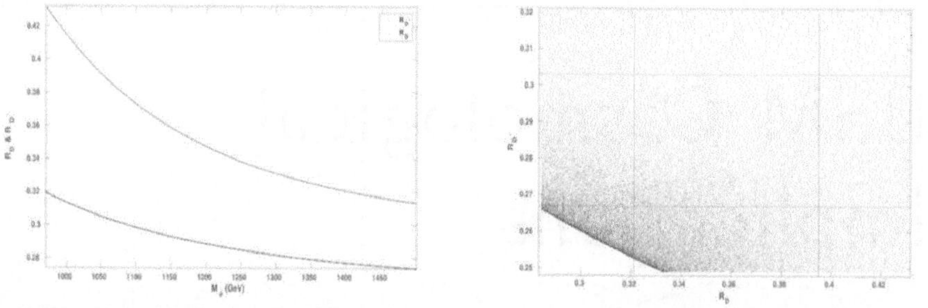

Figure 11.1 R_D and R_{D^*} as function of the Leptoquark mass and The correlation between R_D and R_{D^*}.

Using SM inputs and central form factor values [198], the observables $R(D)$ and $R(D^*)$ are given by:

$$R(D) = R(D)^{\rm SM}\left[1+1.02|g_{S_L}|^2+0.9|g_T|^2+1.49\,{\rm Re}(g^*_{S_L})+1.14\,{\rm Re}(g^*_T)\right] \tag{11.46}$$

$$R(D^*) = R(D^*)^{\rm SM}\left[1+0.04|g_{S_L}|^2+16.07|g_T|^2-0.11\,{\rm Re}(g^*_{S_L})-5.12\,{\rm Re}(g^*_T)\right]. \tag{11.47}$$

A numerical scan shows that for TeV-scale leptoquarks, the model can accommodate the current values of $R(D)$ and $R(D^*)$ [196], while respecting bounds from B_c decays and polarization observables (see Fig. 11.1).

CHAPTER 12

BSM Cosmological Implications

12.1 INTRODUCTION

We now turn our attention to the cosmological implications of BSM physics. As emphasized throughout this work, the SM of particle physics has been extraordinarily successful in describing the fundamental particles and forces that make up the universe. However, we have also highlighted its limitations, gaps that strongly suggest the existence of new physics beyond the SM. These limitations are not just theoretical curiosities; they are intimately tied to some of the most profound questions in cosmology, including the nature of dark matter, the origin of dark energy, the generation of neutrino masses, and the baryon asymmetry of the universe. In this chapter, we will explore how BSM physics addresses these questions and enriches our understanding of the universe [199].

As emphasized earlier, one of the most compelling examples of the SM's limitations is dark matter, which makes up approximately 27% of the universe's energy density. Astronomical observations, such as galaxy rotation curves, gravitational lensing, and the cosmic microwave background (CMB), provide robust evidence for the existence of dark matter. However, the SM lacks a viable candidate particle to account for it. Neutrinos, the only weakly interacting particles in the SM, are far too light to explain the observed dark matter density [200]. This significant gap has driven the search for BSM candidates.

Another profound puzzle is dark energy, the mysterious force responsible for the accelerated expansion of the universe. Observations of distant supernovae [201, 202] and the large-scale structure of the cosmos

DOI: 10.1201/9781003457701-12

have firmly established the presence of dark energy, which constitutes approximately 68% of the universe's total energy content. Despite its dominance, the SM offers no explanation for dark energy. While the cosmological constant in Einstein's equations could, in principle, account for it, the predicted value is orders of magnitude larger than what is observed, a glaring inconsistency known as the "cosmological constant problem" [203,204]. To address this enigma, BSM theories propose alternative explanations, such as dynamic scalar fields (quintessence) [205], modifications to general relativity [206], or contributions from extra dimensions [139].

The baryon asymmetry of the universe, the observed dominance of matter over antimatter, is another profound mystery that the SM cannot resolve. While the SM includes processes that violate baryon number, such as sphaleron transitions, these mechanisms are insufficient to generate the observed asymmetry [207]. Furthermore, the SM fails to satisfy all three Sakharov conditions: baryon number violation, C and CP violation, and departure from thermal equilibrium, required to explain the matter-antimatter imbalance [40]. To address this puzzle, BSM theories propose compelling mechanisms, such as leptogenesis [208,209] and electroweak baryogenesis [210]. These frameworks introduce new sources of CP violation, additional particles, or cosmological phase transitions, providing viable explanations for the origin of the baryon asymmetry and its role in the evolution of the universe. In this context, cosmology plays a pivotal role in testing and constraining BSM theories.

12.2 DARK MATTER AND BSM PHYSICS

Despite profound implications of dark matter (DM) for cosmology and fundamental physics, the nature of DM remains one of the most significant unsolved mysteries in modern science. Although its existence was first suggested over seven decades ago, we still lack a definitive understanding of its composition or interactions beyond gravitational effects. The first strong evidence for DM emerged in 1933 when Swiss astrophysicist Fritz Zwicky studied the Coma galaxy cluster. He observed that the visible, luminous matter in the cluster accounted for only a fraction of the total mass required to explain the gravitational motions of its member galaxies. This discrepancy suggested the presence of an unseen mass component, which he referred to as DM [211]. However, for decades, this idea remained largely speculative. It was not until the 1970s, through the work of Vera Rubin and others on the anomalous rotation curves of spiral galaxies, that dark matter began to be seriously considered as a necessary component of the Universe. The observed rotational speeds of galaxies could not be explained by visible matter alone, reinforcing the hypothesis that a dominant, yet invisible, form of matter must be present to account for these observations [212].

The concept of inferring unseen masses based on gravitational effects is not new. A historical analogy can be drawn from the discovery of Neptune in 1846, which was guided by irregularities in Uranus' orbit. These deviations could not be explained by known celestial bodies, leading astronomers to predict the existence of an unknown planet, later confirmed through direct observation. Similarly, the unexplained gravitational effects in galaxies and clusters suggest the presence of DM, even though it has yet to be directly detected.

While the dark matter hypothesis is widely accepted, a small number of researchers argue that it may not be necessary to explain observed galactic rotation curves. Instead, they propose modifications to Newtonian dynamics at galactic scales, suggesting that deviations from expected motion could result from an incomplete understanding of gravity rather than the presence of an unseen mass component. The most well-known alternative in this regard is Modified Newtonian Dynamics (MOND), which introduces adjustments to Newton's laws at extremely low accelerations to account for the observed rotational speeds of galaxies without invoking DM [213]. However, these approaches are generally considered ad hoc and lack a strong theoretical foundation within a broader physical framework. Moreover, while such modifications may offer an explanation for galaxy rotation curves, they fail to account for the full range of astrophysical and cosmological evidence supporting DM, including gravitational lensing effects, cosmic microwave background anisotropies [31], and large-scale structure formation [214]. The necessity of dark matter extends far beyond individual galaxies, as its presence is also inferred from observations of galaxy clusters and the evolution of the Universe on cosmological scales, reinforcing the view that a fundamental, non-luminous component must exist [215].

12.2.1 What is DM made of?

The nature of DM remains one of the most compelling mysteries in modern physics. Observations from CMB radiation, galaxy rotation curves, and large-scale structure formation strongly suggest the existence of dark matter, which interacts primarily through gravity and possibly weakly via other forces [211, 212]. However, its exact composition is unknown. Constraints from Big Bang nucleosynthesis indicate that DM cannot be baryonic, as the abundance of baryonic matter in the universe is insufficient to account for the observed gravitational effects [216]. A viable DM candidate must satisfy several essential properties:

- It must be stable on cosmological timescales, suggesting either absolute stability or a lifetime much longer than the age of the universe.
- It is non-baryonic, as the observed baryonic matter is insufficient to explain the total gravitational effects attributed to DM.
- It interacts very weakly with SM particles, at least through non-electromagnetic forces, to remain undetected in direct searches.
- It is cold and non-relativistic at the time of structure formation, as required by large-scale structure observations.
- It does not emit, absorb, or reflect light and does not participate in strong interactions.
- It must produce a relic density consistent with cosmological observations ($\Omega_{\rm DM} \approx 0.27$) [31].

The SM of particle physics does not provide a suitable DM candidate, as all known particles either interact too strongly or fail to match the required relic density [217]. Extensions beyond the SM propose new particles, which may fulfill the above criteria. DM is broadly classified based on its velocity dispersion and interaction properties in the early Universe. The main types of dark matter include:

- Cold Dark Matter (CDM), which consists of particles that were non-relativistic (moving at sub-relativistic speeds) at the time of their decoupling from the thermal bath. These particles clump under gravitational attraction, leading to the formation of large-scale structures in the Universe. Weakly Interacting Massive Particles (WIMPs) [218] and axions [219] are prominent CDM candidates.
- Warm Dark Matter (WDM), which had a relativistic speed at early times but became non-relativistic later. Their intermediate velocity suppresses small-scale structure formation compared to CDM,

making them a possible solution to certain discrepancies in standard structure formation models [220]. Sterile neutrinos are a leading candidate for WDM [221].

- Hot Dark Matter (HDM), which consists of particles that remained relativistic for a significant period after their decoupling. Due to their high velocities, HDM particles do not easily cluster, suppressing the formation of small-scale structures. A classic example is the standard neutrino [222].

Each of these DM types has distinct implications for cosmology and structure formation. Observational evidence, such as measurements from the CMB, galaxy distributions, and large-scale structure evolution, strongly favors a cold dark matter-dominated Universe, as predicted by the ΛCDM model [31]. WIMPs are among the most compelling candidates for CDM in the Universe. These hypothetical particles are expected to interact only through the weak nuclear force and gravity, making their direct detection extremely challenging [218].

In the early Universe, WIMPs were in thermal equilibrium with SM particles. Frequent interactions kept their number density comparable to that of photons. As the Universe expanded and cooled, the interaction rate between WIMPs and SM particles diminished. When the temperature dropped below the WIMP mass ($T \ll M_{\text{WIMP}}$), annihilation processes became inefficient at maintaining thermal equilibrium, resulting in the decoupling of WIMPs from the thermal plasma. This freeze-out process left behind a relic abundance of WIMPs, which persists today as a substantial component of the dark matter [215]. The decoupling of WIMPs is a key aspect of their cosmological evolution. At high temperatures ($T \gg M_{\text{WIMP}}$), WIMPs and SM particles maintained equilibrium through frequent interactions. As the Universe cooled, the WIMP number density underwent an exponential suppression, with their abundance freezing out at a characteristic value. Today, these relic WIMPs form a dominant, non-relativistic component of dark matter, with their number density significantly lower than that of photons [217].

This thermal production mechanism, often referred to as the "WIMP miracle," naturally leads to a relic abundance that aligns with observations from CMB measurements and large-scale structure formation [31]. Despite their strong theoretical motivation, WIMPs have yet to be detected experimentally, making their search a major goal of ongoing direct and indirect detection experiments, as well as collider studies [223].

12.2.2 Experimental Evidence for Dark Matter

Multiple lines of astronomical and cosmological observations provide compelling evidence for the existence of DM:

1. Galaxy Cluster Rotation Curves:

   Early signs of dark matter were suggested by mass discrepancies in galaxy clusters, but strong empirical support came in the 1970s through observations of spiral galaxy rotation curves [212]. These studies showed that stars and gas in the outer regions of galaxies orbit at nearly constant velocities, inconsistent with the distribution of visible matter.

   According to Newtonian dynamics, the rotational velocity $v(r)$ at a radius r should satisfy:

   $$M(r) = \frac{Gv(r)^2}{r}, \tag{12.1}$$

   where $M(r)$ is the mass enclosed within radius r, and G is Newton's gravitational constant. If only luminous matter contributed, $v(r)$ would decrease with increasing r. Instead, observations show that $v(r)$ remains flat at large radii, as shown in Fig. 12.1, indicating the presence of additional unseen mass.

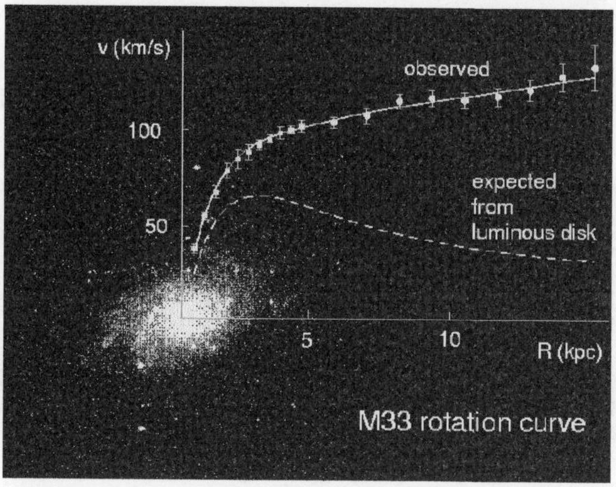

Figure 12.1 Rotation curve of the dwarf spiral galaxy M33 overlaid on its optical image [28].

This phenomenon has been observed in hundreds of spiral galaxies, supporting the idea that they are embedded in extended dark matter halos. These halos dominate the mass distribution at large distances and provide the gravitational influence needed to sustain the flat rotation curves, reinforcing the dark matter hypothesis.

2. Gravitational Lensing:

   When light from distant galaxies passes through a massive foreground object like a galaxy cluster, its path is bent by gravity, a phenomenon known as gravitational lensing. This effect distorts the images of background galaxies, allowing astronomers to map the mass distribution of the lensing object. In many cases, the mass inferred from lensing far exceeds that of visible matter, providing strong evidence for the presence of DM [224].

3. The Bullet Cluster:

   The Bullet Cluster (1E0657-558) [30] offers one of the most compelling pieces of evidence for the existence of dark matter. X-ray observations from the Chandra satellite reveal a shock front in the hot gas resulting from the collision of two galaxy clusters. Meanwhile, weak gravitational lensing maps derived from Hubble Space Telescope data show that most of the mass is located in regions offset from the luminous matter. This spatial separation between the baryonic matter and the gravitational potential strongly indicates the presence of non-luminous, collisionless dark matter halos. The Bullet Cluster thus serves as a striking, direct observational demonstration of DM's gravitational influence (see Fig. 12.2).

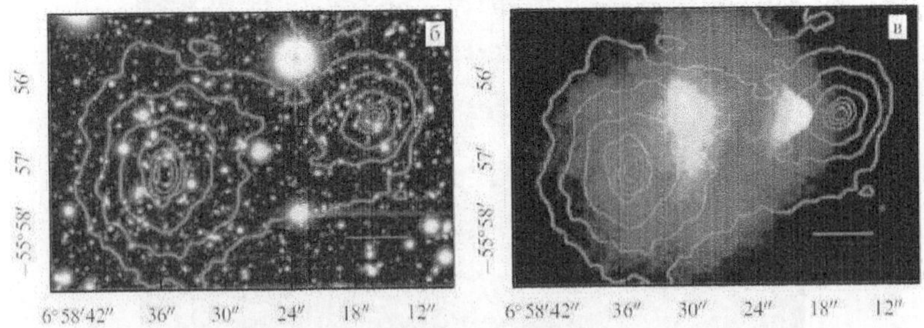

Figure 12.2 Optical (left) and X-ray (right) images of the Bullet Cluster with gravitational lensing contours, showing that most of the mass is offset from the luminous matter [30].

4. Cosmic Microwave Background

While various observations provide compelling evidence for dark matter (DM), they do not directly determine its total abundance or composition. In contrast, the analysis of the Cosmic Microwave Background (CMB) offers crucial insights into these fundamental properties. The CMB, a relic radiation from the early universe, was discovered in 1965 by Arno Penzias and Robert Wilson. Since then, experiments such as WMAP and Planck have mapped its anisotropies with remarkable precision, providing a wealth of cosmological data. In the 2018 data release, the Planck collaboration reported the following values for the total matter and baryonic matter density parameters [31]:

$$\Omega_{\rm m}h^2 = 0.1430 \pm 0.0011, \quad \Omega_{\rm b}h^2 = 0.02237 \pm 0.00015. \tag{12.2}$$

The significant gap between the total matter density $\Omega_{\rm m}$ and the baryonic matter density $\Omega_{\rm b}$ indicates that most of the matter in the universe is non-baryonic, that is, DM. These results are also consistent with predictions from Big Bang nucleosynthesis (BBN), lending additional support to the inferred value of $\Omega_{\rm b}$ [32].

CMB observations have significantly advanced our understanding of DM's role in cosmic evolution. The precise measurements of temperature fluctuations reveal the composition of the universe as follows (see Fig. 12.3):

- 4.9%: Ordinary (baryonic) matter, atoms that make up stars, planets, and visible structures.

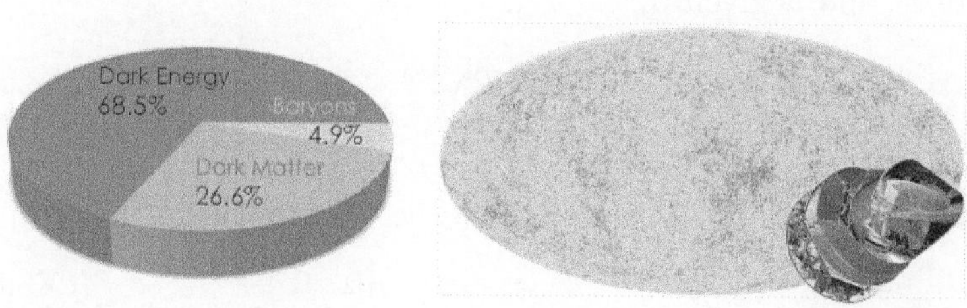

Figure 12.3 Left: Energy density distribution among different components of the universe. Right: Temperature fluctuations in the CMB as observed by the Planck satellite.

- 26.6%: Dark matter, non-luminous matter detectable only through its gravitational effects.
- 68.5%: Dark energy, responsible for the accelerated expansion of the universe.

These results provide one of the most robust confirmations of dark matter's existence, underscoring its essential role in the formation and structure of the universe.

12.2.3 DM Relic Abundance

In this framework, we assume that χ, a DM candidate, was initially in thermal equilibrium with SM particles in the early universe and decoupled when it became non-relativistic. The annihilation rate of χ, given by $\Gamma_\chi = \langle\sigma_\chi^{\rm ann} v\rangle n_\chi$, eventually became smaller than the expansion rate of the universe ($\Gamma_\chi \leq H$), at which point χ ceased to annihilate and fell out of equilibrium [199]. The relic density of χ was then frozen and has remained constant ever since [225]. Here, $\langle\sigma_\chi^{\rm ann} v\rangle$ represents the thermally averaged cross-section for annihilation.

The relic density is determined by solving the Boltzmann equation for the number density n_χ, coupled with the law of entropy conservation [225]:

$$\frac{dn_\chi}{dt} = -3Hn_\chi - \langle\sigma_\chi^{\rm ann} v\rangle \left[n_\chi^2 - (n_\chi^{\rm eq})^2\right], \tag{12.3}$$

$$\frac{ds}{dt} = -3Hs, \tag{12.4}$$

where $n_\chi^{\rm eq}$ is the equilibrium number density, which depends on the temperature T, and is given by:

$$n_\chi^{\rm eq} = g_\chi \left(\frac{m_\chi T}{2\pi}\right)^{3/2} e^{-m_\chi/T}. \tag{12.5}$$

The Hubble parameter H is related to the temperature T as [215]:

$$H(T) = 2\pi\sqrt{\frac{\pi g_*}{45}}\frac{T^2}{M_{Pl}}, \tag{12.6}$$

where M_{Pl} is the Planck mass, and g_* is the number of relativistic degrees of freedom at the time. For the MSSM, $g_* \simeq 228.75$.

To simplify the Boltzmann equation, it is convenient to introduce the dimensionless variable $x = m_\chi/T$ and the comoving number density $Y = n_\chi/s$, where s is the entropy density. The equation then becomes:

$$\frac{dY}{dx} = \frac{1}{3H}\frac{ds}{dx}\langle\sigma_\chi^{\text{ann}} v\rangle \left(Y^2 - Y_{\text{eq}}^2\right). \tag{12.7}$$

During the radiation-dominated era, the entropy density is approximately:

$$s(x) = \frac{2\pi^2}{45} g_{*_s}(x)\, m_\chi^3\, x^{-3}, \tag{12.8}$$

with its derivative given by:

$$\frac{ds}{dx} = -\frac{3s}{x}. \tag{12.9}$$

Assuming $g_* \simeq g_{*_s}$, the Boltzmann equation simplifies to:

$$\frac{dY}{dx} = -\sqrt{\frac{\pi g_*}{45}} M_{Pl}\, m_\chi \frac{\langle\sigma_\chi^{\text{ann}} v\rangle}{x^2} \left(Y^2 - Y_{\text{eq}}^2\right). \tag{12.10}$$

If the DM candidate annihilates predominantly via s-wave and p-wave processes, the thermally averaged cross-section can be approximated as [226]:

$$\langle\sigma_\chi^{\text{ann}} v\rangle \simeq a_\chi + \frac{6b_\chi}{x}, \tag{12.11}$$

where a_χ and b_χ correspond to the s-wave and p-wave contributions, respectively. The present-day relic abundance of DM is given by:

$$\Omega h^2 = \frac{m_\chi s_0 Y_\chi(\infty)}{\rho_c/h^2}, \tag{12.12}$$

where $Y_\chi(\infty)$ is the asymptotic value of the comoving number density, s_0 is the current entropy density, and ρ_c is the critical density of the universe. Solving the Boltzmann equation yields:

$$Y_\chi(\infty) = \frac{1}{\lambda_\chi}\left(\frac{a_\chi}{x_f} + \frac{3b_\chi}{x_f^2}\right)^{-1}, \tag{12.13}$$

where $x_f = m_\chi/T_f$ is the freeze-out parameter and $\lambda_\chi = s(m_\chi)/H(m_\chi)$. The value of x_f is obtained from the transcendental equation:

$$x_f = \ln\left[\frac{\alpha_\chi \lambda_\chi c(c+2)}{\sqrt{x_f}}\left(a_\chi + \frac{6b_\chi}{x_f}\right)\right], \tag{12.14}$$

with $\alpha_\chi = \frac{45}{2\pi^4}\sqrt{\frac{\pi}{8}}\frac{g_\chi}{g_{*_s}(T_f)}$, and $c \approx 0.5$ is a fitting parameter that yields sufficient accuracy for most cosmological applications.

12.2.4 Lightest Neutralino as DM in SUSY Theories

The lightest neutralino (χ) in supersymmetric (SUSY) theories is among the most compelling dark matter (DM) candidates. In the Minimal Supersymmetric Standard Model (MSSM), it naturally arises as a linear combination of the bino ($\tilde{B}$), wino ($\tilde{W}^0$), and the neutral higgsinos ($\tilde{H}_1^0, \tilde{H}_2^0$) [33]. The mass eigenstate is given by:

$$\chi = N_{11}\tilde{B} + N_{12}\tilde{W}^0 + N_{13}\tilde{H}_1^0 + N_{14}\tilde{H}_2^0, \tag{12.15}$$

where N_{ij} are elements of the neutralino mixing matrix. The specific composition of the lightest neutralino determines its couplings and annihilation mechanisms.

Depending on its nature, the neutralino can annihilate into various final states such as fermion-antifermion pairs ($f\bar{f}$), W^+W^-, ZZ, Zh, ZH, ZA, H^+H^-, and others involving Higgs bosons. For a bino-like LSP (i.e., $N_{11} \simeq 1$, $N_{1i} \simeq 0$ for $i = 2, 3, 4$), the dominant annihilation channels are into fermion-antifermion pairs, as shown in Fig. 12.4, while all other channels are suppressed.

In this scenario, annihilation via Z-boson exchange is typically negligible due to the small $Z\chi\chi$ coupling, which is proportional to $N_{13}^2 - N_{14}^2$, unless the process occurs near the Z-resonance ($m_\chi \approx m_Z/2$). However, such a mass range is excluded by current experimental bounds.

The dominant annihilation proceeds through t-channel exchange of sleptons, particularly into lepton pairs via $\tilde{l}_L, \tilde{l}_R$ (with $l = e, \mu, \tau$), as shown in the first diagram of Fig. 12.4. Annihilation via squark exchange is generally suppressed due to their heavier masses.

In Fig. 12.5, we illustrate the constraints from the observed relic abundance Ωh^2 in the $(m_0, m_{1/2})$ plane for two values of the trilinear coupling, $A_0 = 0$ and 2000 GeV, and for $\tan\beta = 10$ and 50, with $\mu > 0$ [227]. The relic abundance of the lightest neutralino was computed using the micrOMEGAs package [62], which includes all relevant

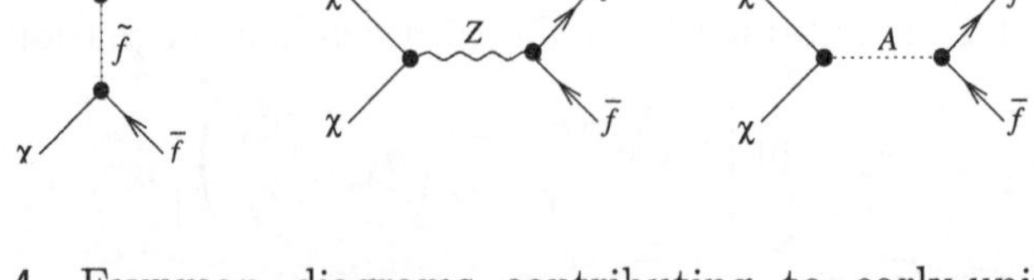

Figure 12.4 Feynman diagrams contributing to early-universe neutralino χ annihilation into fermions through sfermions, Z-gauge boson, and Higgs.

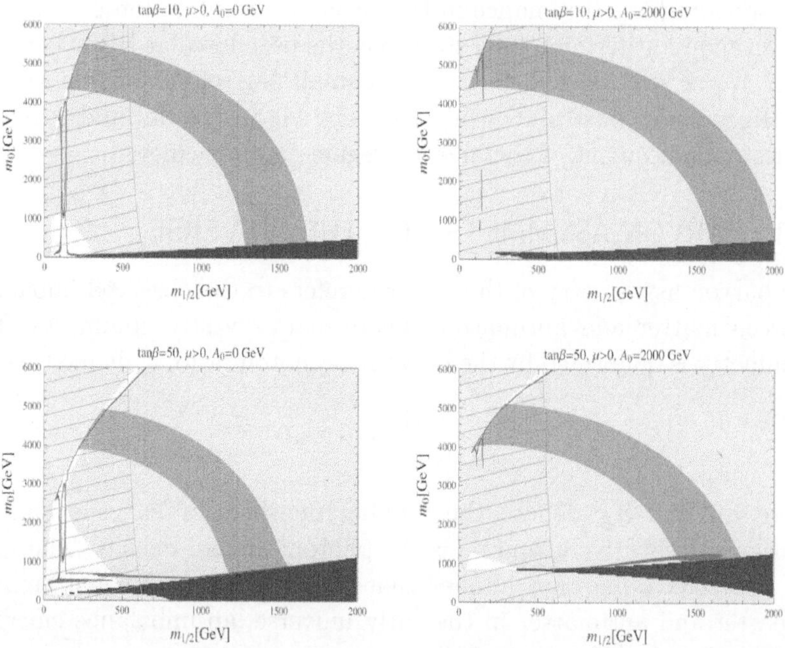

Figure 12.5 LSP relic–abundance constraints shown as differently shaded regions in the $(m_0 - m_{1/2})$ plane. The LUX constraint corresponds to the *light-shaded* region. The parameter points yielding the correct neutralino relic abundance, $0.09 < \Omega h^2 < 0.14$, appear as the *medium-shaded curved band.* The *dark-shaded* and hatched areas denote parameter points excluded by experimental or theoretical constraints.

annihilation and coannihilation channels. In particular, coannihilation with the next-to-lightest supersymmetric particle (NLSP), typically the lightest stau, $\tilde{\tau}_1$, is fully accounted for [228].

It is important to note that these coannihilation effects are not captured by the simplified analytic expressions in Eqs. (12.11)–(12.14). In the figure, the *medium-shaded region* corresponds to points where the computed relic abundance falls within the experimentally allowed range,

$$0.09 < \Omega h^2 < 0.14. \tag{12.16}$$

The LUX constraint is satisfied in the *light-shaded region.*

It is noticeable that for low values of $\tan\beta$ (around 10), the allowed region corresponds to light $m_{1/2}$ values (below 500 GeV), where efficient coannihilation between the LSP and the lightest stau occurs. However, this region is now excluded due to current constraints on the Higgs and gluino masses [230]. At large $\tan\beta$, an additional viable region appears, corresponding to a resonance in the s-channel annihilation of DM pairs into fermion-antifermion final states via the pseudoscalar Higgs boson A, when $M_A \simeq 2m_\chi$. For $A_0 = 0$, only a small part of this region satisfies the Higgs mass constraint, while for larger A_0 values (around 2 TeV), a modest enhancement of the allowed region can be achieved.

12.3 BARYON ASYMMETRY OF THE UNIVERSE

The baryon asymmetry of the universe refers to the observed imbalance between matter and antimatter, where matter vastly dominates. This asymmetry is quantified by the baryon-to-photon ratio, η, defined as [31]:

$$\eta \equiv \frac{n_B - n_{\bar{B}}}{n_\gamma} \approx 6 \times 10^{-10}, \tag{12.17}$$

where n_B and $n_{\bar{B}}$ denote the number densities of baryons and antibaryons, respectively, and n_γ is the photon number density. The small but non-zero value of η implies that, despite the near-equal creation of matter and antimatter in the early universe, an imbalance emerged, leaving a surplus of matter [231].

In the hot, dense conditions of the early universe, matter and antimatter were produced in almost equal amounts and subsequently annihilated into photons as the universe cooled. In the absence of any asymmetry, this process should have left a universe devoid of matter. However, the presence of a substantial matter component today, with negligible amounts of primordial antimatter, indicates that a fundamental asymmetry developed during the early stages of cosmic evolution. This leads to one of the most profound open questions in modern physics: Why does the universe exhibit a preference for matter over antimatter? Understanding the origin of this asymmetry is essential to comprehending the conditions and mechanisms that shaped the early universe.

12.3.1 Sakharov Conditions

In 1967, Andrei Sakharov identified three essential conditions required to generate the baryon asymmetry of the universe [40]. These conditions, known as the Sakharov conditions, provide the minimal ingredients for a matter-dominated universe:

1. Baryon Number Violation: Interactions that violate baryon number (B) must exist to allow a net excess of baryons over antibaryons.

2. C and CP Violation: Charge conjugation (C) and charge-parity (CP) symmetries must be violated so that baryons and antibaryons behave differently, enabling an imbalance in their production rates.

3. Departure from Thermal Equilibrium: The universe must have undergone periods out of thermal equilibrium; otherwise, any generated asymmetry would be washed out by inverse processes.

Although the SM includes mechanisms that partially satisfy these conditions, they fall short of explaining the observed asymmetry. This limitation motivates extensions of the SM and ongoing research into the early universe.

12.3.2 Leptogenesis

Leptogenesis is one of the most compelling mechanisms proposed to explain the observed baryon asymmetry of the universe. It provides a natural connection between the matter-antimatter imbalance and neutrino physics, relying on processes beyond the SM. The core idea is that a lepton asymmetry is first generated and subsequently converted into a baryon asymmetry via electroweak sphaleron transitions [232]. This elegantly links the baryon asymmetry to the properties of neutrinos, which are already known to possess non-zero masses and mixing angles not accounted for in the SM.

In the SM extended by three right-handed neutrinos ν_{R_i}, the relevant Lagrangian involving these fields is given by

$$\begin{aligned}\mathcal{L} = &\sum_i M_i\,(\bar{\nu}^c_{Ri}\nu_{Ri} + \bar{\nu}_{Ri}\nu^c_{Ri}) + \sum_{ij}\lambda^*_{ji}\,\bar{\nu}_{Ri}\phi L_j + \sum_{ij}\lambda_{ji}\,\bar{L}_j\bar{\phi}\nu_{Ri}\\ &+\sum_{ij}\lambda^*_{ji}\bar{L}^c_j\phi\nu^c_{Ri} + \sum_{ij}\lambda_{ji}\bar{\nu}^c_{Ri}\bar{\phi}L^c_j,\end{aligned} \tag{12.18}$$

where L denotes the lepton doublet, ϕ the Higgs doublet, and $\psi^c = C\gamma^0\psi^* = i\gamma^2\psi^* = C\bar{\psi}^T$ is the charge-conjugated spinor.

Asymmetry ε_i arises from the CP-violating decays of the heavy Majorana right-handed neutrinos N_i into the Higgs doublet ϕ and lepton doublets L_α, where $\alpha = e, \mu, \tau$, i.e., $N_i \to \phi + L_\alpha$. In most scenarios, the resulting lepton asymmetry is dominated by the decay of the lightest right-handed neutrino N_1. The N_i are Majorana fermions, representing the mass eigenstates of the right-handed neutrinos, and are defined as:

$$N_i = \nu_{Ri} + \nu^c_{Ri}, \qquad N^c = N. \tag{12.19}$$

The chiral components relate to N via the projection operators:

$$\nu_R = P_R N, \qquad \nu_R^c = P_L N, \tag{12.20}$$

where $P_L = \frac{1}{2}(1-\gamma^5)$ and $P_R = \frac{1}{2}(1+\gamma^5)$ are the standard left- and right-handed helicity projectors.

The CP asymmetry ε_1 generated by the decay of the lightest right-handed neutrino N_1 is defined as

$$\varepsilon_1 = \frac{\sum_\alpha \left(|A(N_1 \to \phi L_\alpha)|^2 - \left|A(N_1 \to \bar{\phi} \bar{L}_\alpha)\right|^2 \right)}{\sum_\alpha \left(|A(N_1 \to \phi L_\alpha)|^2 + \left|A(N_1 \to \bar{\phi} \bar{L}_\alpha)\right|^2 \right)}, \tag{12.21}$$

where $A(N_1 \to \phi L_\alpha)$ denotes the total decay amplitude, including both tree-level and one-loop contributions:

$$A(N_1 \to \phi L_\alpha) = A_0(N_1 \to \phi L_\alpha) + A_1(N_1 \to \phi L_\alpha). \tag{12.22}$$

The CP asymmetry arises from the interference between the tree-level amplitude and the loop corrections. In the SM extended by right-handed neutrinos, the decay of N_1 proceeds via the tree-level process, as well as through one-loop vertex and self-energy diagrams, as illustrated in Fig. 12.6.

The tree-level decay amplitude of ν_{R_1} into the Higgs doublet ϕ and lepton doublet L_α is given by:

$$A_0(N_1 \to \phi L_\alpha) = -(\lambda_\nu)_{\alpha 1} \left(\bar{u}(p) P_R u^c(q)\right), \tag{12.23}$$

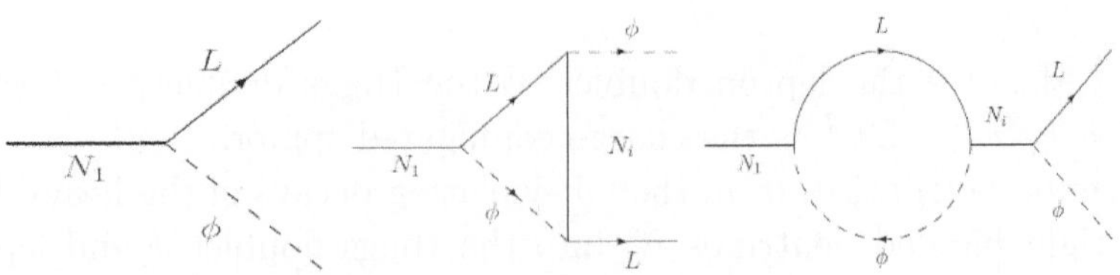

Figure 12.6 Feynman diagrams in SM with right-handed neutrinos that contribute to the decay $N_1 \to \phi L_\alpha$.

where $\bar{u}(p)$ is the Dirac spinor of the outgoing lepton l_α, $u^c(q)$ is the spinor of the incoming Majorana neutrino N_1, and P_R is the right-handed projection operator.

The one-loop vertex correction contributes:

$$A_{1V}(N_1 \to \phi L_\alpha) = \frac{1}{16\pi} \sum_i (\lambda_\nu)_{\alpha i} (\lambda_\nu^\dagger \lambda_\nu)_{1i} \left(\bar{u}(p) P_R u^c(q)\right) F_V \left(\frac{M_i^2}{M_1^2}\right), \tag{12.24}$$

where the loop function $F_V(x)$ has the imaginary and real parts:

$$\text{Im } F_V(x) = \sqrt{x} \left[1 - (1+x) \ln\left(\frac{1+x}{x}\right)\right], \tag{12.25}$$

$$\text{Re } F_V(x) = -\sqrt{x} \left[1 + \ln(x) \left(1 - (1+x) \ln\left(\frac{1+x}{x}\right)\right)\right]. \tag{12.26}$$

The self-energy contribution to the decay amplitude is:

$$A_{1S}(N_1 \to \phi L_\alpha) = \frac{1}{16\pi} \sum_i (\lambda_\nu)_{\alpha i} (\lambda_\nu^\dagger \lambda_\nu)_{1i} \left(\bar{u}(p) P_R u^c(q)\right) F_S \left(\frac{M_i^2}{M_1^2}\right), \tag{12.27}$$

where $F_S(x) = \text{Re } F_S(x) + i \, \text{Im } F_S(x)$, and in the limit $|M_i - M_1| \gg |\Gamma_i - \Gamma_1|$, the imaginary part simplifies to:

$$\text{Im } F_S(x) = \frac{\sqrt{x}}{1-x}. \tag{12.28}$$

The CP asymmetry in $N_1 \to \phi L_\alpha$ decays arises from complex phases in the Yukawa couplings λ_ν, which violate CP symmetry. These phases change sign under CP conjugation, leading to differences in the decay amplitudes of particles and antiparticles. In contrast, the phases of the loop functions $F_V(x)$ and $F_S(x)$ are CP-conserving, i.e., they remain unchanged under CP.

To analyze the asymmetry, the decay amplitudes can be factorized into CP-violating and CP-conserving components. Specifically, we write:

$$A_0(N_1 \to \phi L_\alpha) = A_{\text{tree}}, \quad \bar{A}_0(N_1 \to \bar{\phi} \bar{L}_\alpha) = A_{\text{tree}}^*, \tag{12.29}$$

$$A_1(N_1 \to \phi L_\alpha) = A_{\text{loop}} \cdot F, \quad \bar{A}_1(N_1 \to \bar{\phi} \bar{L}_\alpha) = A_{\text{loop}}^* \cdot F. \tag{12.30}$$

Here, F represents the loop function (from vertex and self-energy corrections), which is CP-even, while A_{tree} and A_{loop} carry the CP-odd phases from the Yukawa couplings.

The CP asymmetry then follows from the interference between tree and loop amplitudes:

$$\varepsilon = \frac{\mathrm{Im}\left[A_{\mathrm{tree}}A^*_{\mathrm{loop}}\right] \cdot \mathrm{Im}[F]}{|A_{\mathrm{tree}}|^2 + |A_{\mathrm{loop}}F|^2 + 2\,\mathrm{Re}[F]\,\mathrm{Re}\left[A_{\mathrm{tree}}A^*_{\mathrm{loop}}\right]}. \tag{12.31}$$

Since the tree-level amplitude typically dominates, this simplifies to:

$$\varepsilon_1 \simeq \frac{1}{|A_{\mathrm{tree}}|^2}\,\mathrm{Im}\left[A_{\mathrm{tree}}A^*_{\mathrm{loop}}\right] \cdot \mathrm{Im}[F]. \tag{12.32}$$

In the SM with right-handed neutrinos, the resulting CP asymmetry becomes:

$$\varepsilon_1 \simeq \frac{1}{8\pi}\frac{1}{(\lambda_\nu^\dagger\lambda_\nu)_{11}}\sum_{i=2,3}\mathrm{Im}\left[(\lambda_\nu^\dagger\lambda_\nu)^2_{1i}\right]\left[\mathrm{Im}F_V\left(\frac{M^2_{R_i}}{M^2_{R_1}}\right) + \mathrm{Im}F_S\left(\frac{M^2_{R_i}}{M^2_{R_1}}\right)\right]. \tag{12.33}$$

Several key points follow from this expression: (*i*) A non-zero lepton asymmetry requires $\mathrm{Im}[(\lambda_\nu^\dagger\lambda_\nu)_{1i}] \neq 0$, which translates to

$$\mathrm{Im}\left(\sqrt{M_R}\,R\,m_\nu^{\mathrm{diag}}\,R^\dagger\sqrt{M_R}\right) \neq 0, \quad \text{for } i = 2, 3.$$

(*ii*) Leptogenesis is independent of the phases in the PMNS matrix due to its unitarity. (*iii*) If both R and M_R are real, then $\varepsilon_1 = 0$, and leptogenesis cannot proceed. (*iv*) In the case of quasi-degenerate right-handed neutrino masses, the CP asymmetry can be resonantly enhanced due to the near-zero denominator in the loop function $g(x)$, a scenario known as resonant leptogenesis [233]. This becomes particularly relevant for models with suppressed Yukawa couplings.

Once a lepton asymmetry is generated, it can be partially converted into a baryon asymmetry via electroweak sphaleron processes. Sphalerons are non-perturbative configurations in the Standard Model that violate baryon number (B) and lepton number (L) individually, but preserve their difference, $B-L$. Active during the electroweak phase transition in the early universe, these processes convert part of the lepton asymmetry into a baryon asymmetry.

The final baryon asymmetry η_B is related to the initial lepton asymmetry $\eta_L \sim \kappa\,\varepsilon$, where ε is the CP asymmetry from heavy neutrino decays, and κ is an efficiency factor accounting for washout processes and the thermal history. In typical thermal leptogenesis scenarios, κ ranges from 10^{-3} to 10^{-1}, depending on the model parameters [209]. The conversion between lepton and baryon asymmetries can be approximated as:

$$\eta_B \simeq -\frac{1}{3}\eta_L,$$

with the factor $-1/3$ arising from the equilibrium conditions of sphaleron interactions and the Standard Model particle content.

12.4 INFLATION AND BSM

The SM of cosmology, grounded in the cosmological principle, has successfully predicted key phenomena such as the universe's expansion and the existence of the CMB. However, it originally faced several challenges related to initial conditions. To resolve these, the theory of inflation was introduced [234,235], providing a mechanism that explains the observed homogeneity, isotropy, and large-scale structure of the universe. Inflation thereby addresses many of the initial value problems inherent in the classical Big Bang framework.

The theoretical foundation of cosmology rests on General Relativity (GR), formulated by Einstein in 1915 [236], with field equations:

$$R_{\mu\nu} - \frac{1}{2}g_{\mu\nu}R = 8\pi G T_{\mu\nu}. \tag{12.34}$$

In the 1920s, Friedmann used these equations to derive a solution for a homogeneous and isotropic universe [237,238], leading to the Friedmann-Lemaitre-Robertson-Walker (FLRW) metric:

$$ds^2 = dt^2 - a^2(t)\left[\frac{dr^2}{1-\kappa r^2} + r^2\left(d\theta^2 + \sin^2\theta\, d\phi^2\right)\right], \tag{12.35}$$

where $\kappa = -1, 0, 1$ corresponds to open, flat, and closed geometries, respectively. The evolution of the scale factor $a(t)$ is governed by the Friedmann equations:

$$\ddot{a} = -\frac{4\pi G}{3}(\rho + 3p)a, \tag{12.36}$$

$$H^2 + \frac{\kappa}{a^2} = \frac{8\pi G}{3}\rho, \tag{12.37}$$

with the Hubble parameter defined as $H = \dot{a}/a$. For a spatially flat universe ($\kappa = 0$), two distinct eras emerge: Radiation domination, with

$a(t) \propto t^{1/2}, \quad \rho \propto a^{-4}$, and Matter (dust) domination, with $a(t) \propto t^{2/3}, \quad \rho \propto a^{-3}$. In both cases, if $\rho + 3p > 0$, the expansion is decelerating ($\ddot{a} < 0$).

The Friedmann equation can also be recast as:

$$\Omega - 1 = \frac{\kappa}{a^2 H^2}, \tag{12.38}$$

where $\Omega = \rho / \rho_c$ and the critical density is given by $\rho_c = \frac{3H^2}{8\pi G}$. Present observations, particularly those from the Planck satellite, indicate that the density parameter Ω is very close to unity ($\Omega \sim \mathcal{O}(1)$), implying that the universe is nearly flat [31]. This near-flatness today suggests that in the early universe, Ω must have been extraordinarily fine-tuned to be even closer to one. For instance, at the Planck time ($t \sim 10^{-43}$ s), the deviation from exact flatness would have had to satisfy:

$$|\Omega - 1| < \mathcal{O}(10^{-64}). \tag{12.39}$$

This extreme sensitivity to initial conditions is known as the flatness problem [239], and it poses a major challenge to the standard Big Bang model, which lacks a dynamical mechanism to explain why Ω should be so close to one.

Closely related is the horizon problem, which concerns the causal structure of the early universe. Observations of the CMB reveal that regions of the sky separated by large angular distances have nearly identical temperatures, as shown in Fig. 12.7, differing by less than one

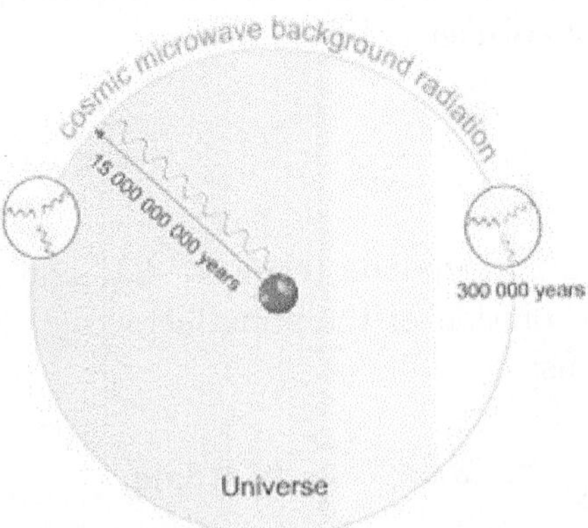

Figure 12.7 The CMB temperature map shows uniform temperatures across causally disconnected regions, highlighting the horizon problem and motivating cosmic inflation.

part in 10^5. However, under standard Big Bang evolution, these regions should have been causally disconnected, meaning they could not have exchanged information or energy due to the finite speed of light [240]. The observed uniformity suggests that these regions were once in thermal equilibrium, a state that cannot be explained without invoking a mechanism that allows causal contact over large distances.

Alan Guth proposed that at $t \sim 10^{-37}$ seconds, the universe underwent a brief period of exponential expansion, known as inflation, increasing in size by over 25 orders of magnitude within 10^{-34} seconds [234]. After inflation ended, the universe resumed a slower expansion over the next 13.7 billion years. Inflation requires accelerated expansion, $\ddot{a} > 0$, which from the Friedmann equations implies negative pressure:

$$p < -\frac{\rho}{3}. \tag{12.40}$$

This yields exponential growth of the scale factor:

$$a(t) \propto e^{Ht}, \quad t \in [t_i, t_f]. \tag{12.41}$$

A scalar field with potential $V(\phi)$ provides a natural mechanism for this behavior [241]. Its Lagrangian is:

$$\mathcal{L} = \frac{1}{2} g^{\mu\nu} \partial_\mu \phi \partial_\nu \phi - V(\phi), \tag{12.42}$$

with energy density and pressure:

$$\rho_\phi = \frac{1}{2}\dot{\phi}^2 + V(\phi), \quad p_\phi = \frac{1}{2}\dot{\phi}^2 - V(\phi). \tag{12.43}$$

When the potential dominates, $\dot{\phi}^2 \ll V(\phi)$, the condition for inflation is met [242]. This requires a flat potential for slow-roll and a minimum to end inflation [243].

The dynamics of a homogeneous scalar field in an expanding universe are governed by the following equations:

$$H^2 = \frac{8\pi G}{3} \left[V(\phi) + \frac{1}{2}\dot{\phi}^2 \right], \tag{12.44}$$

$$\ddot{\phi} + 3H\dot{\phi} = -V'(\phi), \tag{12.45}$$

which describe the evolution of the scalar field and the expansion during inflation, as shown in Fig. 12.8 [244]. Scalar fields, fundamental in many physical theories, are characterized by their potential forms. Notable examples include:

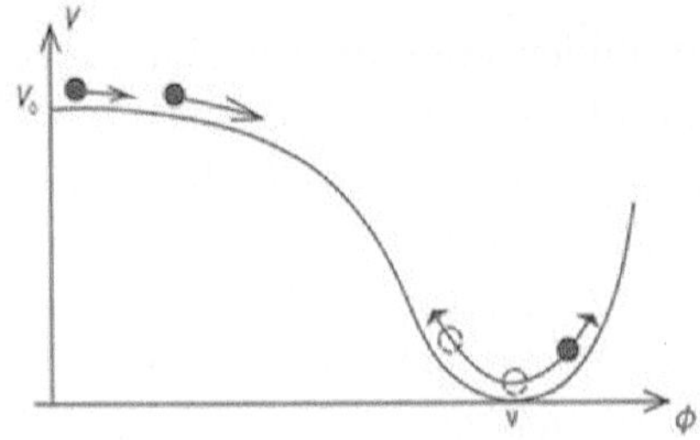

Figure 12.8 The flat region of the potential allows for slow-roll inflation, while the minimum provides a natural endpoint for the inflationary phase.

- The Higgs potential:

$$V(\phi) = \lambda \left(\phi^2 - M^2\right)^2, \tag{12.46}$$

central to the Higgs mechanism and spontaneous symmetry breaking [7],

- The massive scalar field potential:

$$V(\phi) = \frac{1}{2} m^2 \phi^2, \tag{12.47}$$

describing a scalar field with mass m [241],

- The self-interacting scalar field potential:

$$V(\phi) = \lambda \phi^4, \tag{12.48}$$

modeling a scalar field with quartic self-interactions [245].

In inflationary cosmology, the slow-roll approximation describes when a scalar field ϕ induces exponential expansion. This is quantified by two slow-roll parameters:

$$\epsilon = \frac{m_{\text{pl}}^2}{16\pi} \left(\frac{V'(\phi)}{V(\phi)}\right)^2, \quad \eta = \frac{m_{\text{pl}}^2}{8\pi} \frac{V''(\phi)}{V(\phi)}, \tag{12.49}$$

where $V(\phi)$ is the potential, $V'(\phi)$ and $V''(\phi)$ are its first and second derivatives, and m_{pl} is the Planck mass. The slow-roll condition holds when:

$$\epsilon \ll 1, \quad |\eta| \ll 1, \tag{12.50}$$

allowing for prolonged accelerated expansion [246]. Inflation ends when either ϵ or $|\eta|$ approaches unity [247].

The amount of inflation required to address the horizon and flatness problems of the Big Bang model is quantified by the number of e-foldings, denoted by N. The number of e-foldings is defined as:

$$N(\phi) = \ln\left(\frac{a_f}{a_i}\right) = \int_{t_i}^{t_f} H\,dt = -\frac{8\pi}{m_{\text{pl}}^2}\int_{\phi_i}^{\phi_f} \frac{V(\phi)}{V'(\phi)}\,d\phi, \tag{12.51}$$

where:

- a_i and a_f are the initial and final scale factors,
- H is the Hubble parameter,
- ϕ_i and ϕ_f are the initial and final values of the scalar field.

This formulation provides a detailed description of inflationary dynamics and the conditions necessary to resolve cosmological issues.

12.5 DARK ENERGY AND BSM

Dark energy, which accounts for approximately 68% of the total energy density of the universe [31], is one of the most profound mysteries in modern cosmology. It is responsible for the observed accelerated late expansion of the universe, first discovered through observations of Type Ia supernovae in the late 1990s. Despite its dominance, the nature of dark energy remains unknown, and its explanation often requires physics beyond the BSM. This section explores the dark energy problem, BSM explanations, and observational constraints.

12.5.1 Observational Evidence for Cosmic Acceleration

The accelerated expansion of the universe is strongly supported by multiple independent lines of observational evidence:

- Type Ia Supernovae (SN Ia): Observations of distant Type Ia supernovae provided the first direct evidence for cosmic acceleration, showing that the universe's expansion is speeding up over time [248, 249].

- CMB: Measurements of temperature anisotropies in the CMB, particularly by the Wilkinson Microwave Anisotropy Probe (WMAP), have confirmed the presence of dark energy and its influence on the universe's large-scale structure [31].

- Baryon Acoustic Oscillations (BAO): The detection of BAO in the distribution of galaxies has provided additional evidence for cosmic acceleration, serving as a "standard ruler" for measuring the expansion history of the universe.

12.5.2 The Nature of Dark Energy

Dark energy is a mysterious form of energy that plays a crucial role in the dynamics of the universe. It is believed to be the driving force behind the accelerated expansion, acting as a repulsive force that counteracts gravity on cosmic scales. Unlike matter, dark energy does not clump or cluster but is uniformly distributed throughout space, making it homogeneous. This uniformity ensures that its density remains constant across different regions of the universe, in contrast to visible matter and dark matter, which tend to cluster due to gravitational attraction.

The exact nature of dark energy remains one of the most profound questions in modern cosmology. Several theories have been proposed to explain its origin, including:

- Cosmological Constant (Λ): A constant energy density that fills space homogeneously, representing the simplest explanation for dark energy [203].
- Dynamical Fields (e.g., Quintessence): A time-varying scalar field that provides a dynamic explanation for dark energy [250], potentially addressing some of the theoretical challenges associated with the cosmological constant.

12.5.3 Cosmological Constant

The dark energy problem arises from the observed acceleration of the universe's expansion, as described by the Friedmann equation:

$$H^2 = \frac{8\pi G}{3}\left(\rho_m + \rho_r + \rho_\Lambda\right), \tag{12.52}$$

where H is the Hubble parameter, ρ_m and ρ_r represent the energy densities of matter and radiation, respectively, and ρ_Λ is the energy density associated with dark energy.

The concept of dark energy originates from Einstein's cosmological constant (Λ), introduced in 1916 to describe a static universe. He proposed Λ as a repulsive force to counteract gravity, ensuring a non-expanding universe. However, after Edwin Hubble's discovery of the expansion of the universe in the 1920s, Einstein discarded the cosmological constant, calling it his "biggest mistake."

Modern cosmology has revived the cosmological constant, driven by the observed acceleration of the universe's expansion, first confirmed in 1998 through Type Ia supernova measurements [248, 249]. In this context, the cosmological constant corresponds to a constant energy density:

$$\rho_\Lambda = \frac{\Lambda}{8\pi G}. \tag{12.53}$$

This energy density is uniform in space and time, providing a simple explanation for the accelerated expansion.

Despite its success, the cosmological constant faces two significant theoretical challenges.

- Cosmological Constant Problem: The observed value of ρ_Λ is exceedingly small ($\sim 10^{-47}$ GeV^4), and is very different from theoretical predictions from quantum field theory, suggesting a value many orders of magnitude larger. This discrepancy highlights the difficulty in reconciling the observed dark energy density with quantum physics [203].

- Coincidence Problem: The energy densities of dark energy and matter are comparable in the present epoch, even though they evolve differently over the universe's history. Matter density decreases with expansion, while dark energy remains constant. This raises the question of why the two densities are so similar today, challenging our understanding of the timing and nature of dark energy's dominance, as shown in Fig. 12.9 [251, 252].

These challenges highlight the elusive nature of dark energy and its deep implications for cosmology and fundamental physics. Addressing them may require new theoretical frameworks, such as modifications to general relativity, dynamical fields like quintessence, or more radical extensions beyond the SM. Observations from the CMB, large-scale structure surveys, and supernova data continue to offer essential insights, guiding the quest to understand dark energy and its role in the universe.

12.5.4 Quintessence

To address the limitations of the cosmological constant, several BSM explanations have been proposed. One prominent idea is that dark energy

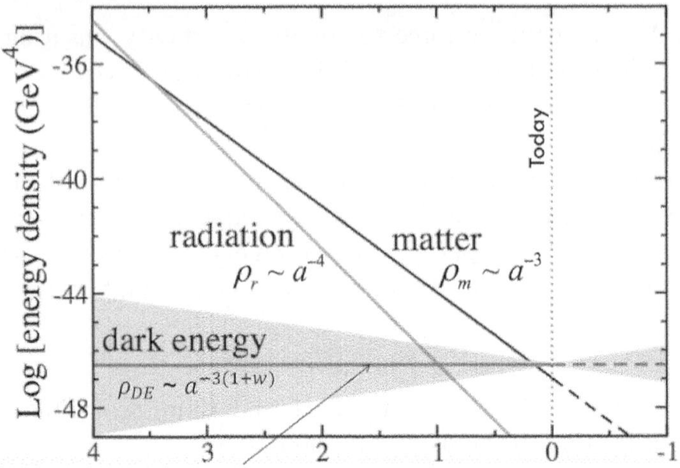

Figure 12.9 Energy density evolution in the universe as a function of scale factor a, showing logarithmic densities (in GeV^4) for radiation, matter, and dark energy. The vertical axis represents log density, and the horizontal axis the scale factor, with "Today" marking the present. Source: [253].

arises from a dynamical field, known as quintessence, which differs from the cosmological constant by varying over time and space. Unlike matter and radiation, which slow down the expansion through gravitational attraction, quintessence generates a repulsive force that accelerates the universe's expansion [250].

Quintessence is described by a dynamical scalar field ϕ that evolves over time, providing a time-varying dark energy density. The energy density ρ_ϕ and pressure p_ϕ of the quintessence field are:

$$\rho_\phi = \frac{1}{2}\dot{\phi}^2 + V(\phi), \tag{12.54}$$

$$p_\phi = \frac{1}{2}\dot{\phi}^2 - V(\phi), \tag{12.55}$$

where $V(\phi)$ is the scalar field potential [254].

A key feature of quintessence is the time-varying equation of state $w_\phi = p_\phi/\rho_\phi$, distinguishing it from the cosmological constant, which has a fixed equation of state $w_\Lambda = -1$. This dynamism may help resolve the coincidence problem, which questions why dark energy and matter densities are similar in the present epoch [255].

However, the specific form of the potential $V(\phi)$ remains unknown, and various models have been proposed, each with distinct implications for the evolution of the universe. Common choices for $V(\phi)$ include exponential, power-law, and tracker potentials, each leading to different cosmological behaviors [256]. Despite this diversity, none of these models have been conclusively supported by observational data.

A classic example of a quintessence potential is the exponential form:

$$V(\phi) = V_0 e^{-\lambda\kappa\phi}, \tag{12.56}$$

where V_0 is a constant with dimensions of energy density, λ is a dimensionless parameter controlling the slope of the potential, and $\kappa = \sqrt{8\pi G}$ is the gravitational coupling constant. The equation of state parameter w_ϕ for the quintessence field is given by:

$$w_\phi = \frac{p_\phi}{\rho_\phi} = \frac{\frac{1}{2}\dot{\phi}^2 - V(\phi)}{\frac{1}{2}\dot{\phi}^2 + V(\phi)}. \tag{12.57}$$

The evolution of the field depends on λ: for small λ, the field rolls slowly down the potential, leading to $w_\phi \approx -1$, mimicking a cosmological constant. For large λ, the field rolls faster, and w_ϕ deviates from -1, potentially resulting in a time-varying equation of state.

12.5.5 Modified Gravity

Modified gravity theories propose that dark energy is not a new form of energy but rather a manifestation of deviations from general relativity on cosmological scales [257, 258]. The action for $f(R)$ gravity, an arbitrary function of the Ricci scalar R, is given by:

$$S = \frac{1}{2\kappa^2}\int \mathrm{d}^4x\,\sqrt{-g}\,f(R) + \int \mathrm{d}^4x\,\mathcal{L}_M(g_{\mu\nu}, \Psi_M), \tag{12.58}$$

where $\kappa^2 = 8\pi G$ is the gravitational coupling constant, G is the Newtonian coupling constant, g is the determinant of the metric tensor $g_{\mu\nu}$, and $\mathcal{L}_M$ is the matter Lagrangian depending on the metric $g_{\mu\nu}$ and matter fields Ψ_M. Varying the action (12.58) with respect to the metric $g_{\mu\nu}$ leads to the field equations for $f(R)$ gravity [259]:

$$F(R)R_{\mu\nu}(g) - \frac{1}{2}f(R)g_{\mu\nu} - \nabla_\mu\nabla_\nu F(R) + g_{\mu\nu}\Box F(R) = \kappa^2 T_{\mu\nu}, \tag{12.59}$$

where $F(R) \equiv \partial f/\partial R$ is the derivative of $f(R)$ with respect to R, $R_{\mu\nu}$ is the Ricci tensor, and $T_{\mu\nu} = -\frac{2}{\sqrt{-g}}\frac{\delta\mathcal{L}_M}{\delta g^{\mu\nu}}$ is the energy-momentum tensor of matter [258].

In $f(R)$ gravity, the modified Friedmann equations for a flat Friedmann-Robertson-Walker (FRW) universe with the metric

$$ds^2 = -dt^2 + a(t)^2 dx^2,$$

are:

$$3FH^2 = \kappa^2(\rho_m + \rho_r) + \frac{FR - f}{2} - 3H\dot{F}, \tag{12.60}$$

$$-2F\dot{H} = \kappa^2\left(\rho_m + \frac{4}{3}\rho_r\right) + \ddot{F} - H\dot{F}. \tag{12.61}$$

Here, ρ_m and ρ_r are the energy densities of matter and radiation, respectively, and $H = \frac{\dot{a}}{a}$ is the Hubble parameter [257].

The total energy density ρ_{tot} and pressure p_{tot} in $f(R)$ gravity include contributions from matter, radiation, and the effective dark energy component arising from the modification of gravity. The conservation equation for the total energy-momentum tensor is:

$$\dot{\rho}_{\text{tot}} + 3H(\rho_{\text{tot}} + p_{\text{tot}}) = 0. \tag{12.62}$$

This equation can be split into separate conservation equations for matter and radiation:

$$\dot{\rho}_m + 3H\rho_m = 0, \quad \text{(matter)} \tag{12.63}$$

$$\dot{\rho}_r + 4H\rho_r = 0. \quad \text{(radiation)} \tag{12.64}$$

In $f(R)$ gravity, the dark energy component arises from the modification of the gravitational action. The effective dark energy density ρ_{DE} and pressure p_{DE} are defined as:

$$\rho_{\text{DE}} = \frac{1}{\kappa^2}\left(\frac{FR - f}{2} - 3H\dot{F}\right), \tag{12.65}$$

$$p_{\text{DE}} = \frac{1}{\kappa^2}\left(\frac{f - FR}{2} + \ddot{F} + 2H\dot{F}\right). \tag{12.66}$$

These definitions ensure that the Friedmann equations can be written in the standard form:

$$3H^2 = \kappa^2(\rho_m + \rho_r + \rho_{\text{DE}}), \tag{12.67}$$

$$-2\dot{H} = \kappa^2\left(\rho_m + \frac{4}{3}\rho_r + \rho_{\text{DE}} + p_{\text{DE}}\right). \tag{12.68}$$

The effective equation of state parameter w_{eff} is defined as the ratio of the total pressure to the total energy density of the universe:

$$w_{\text{eff}} = \frac{p_{\text{tot}}}{\rho_{\text{tot}}}. \tag{12.69}$$

From Friedmann equations, one can express $\dot{H}$ as follows:

$$\dot{H} = -\frac{3}{2}H^2 \left(1 + w_{\text{eff}}\right). \tag{12.70}$$

Solving for w_{eff}, we arrive at:

$$w_{\text{eff}} = -1 - \frac{2\dot{H}}{3H^2}. \tag{12.71}$$

The effective equation of state w_{eff} describes the overall behavior of the universe's expansion:

- If $w_{\text{eff}} < -\frac{1}{3}$, the universe undergoes accelerated expansion (dark energy domination).
- If $w_{\text{eff}} = -1$, the expansion mimics a cosmological constant (ΛCDM).
- If $w_{\text{eff}} > -\frac{1}{3}$, the expansion decelerates (matter or radiation domination).

It is worth mentioning that w_{eff} depends on the functional form of $f(R)$ [257, 259].

Bibliography

[1] Shaaban Khalil and Stefano Moretti. *Standard Model Phenomenology.* CRC Press, 6 2022.

[2] Sheldon L. Glashow. Partial-symmetries of weak interactions. *Nucl. Phys.*, 22:579–588, 1961.

[3] Steven Weinberg. A Model of Leptons. *Phys. Rev. Lett.*, 19:1264–1266, 1967.

[4] Abdus Salam. Weak and electromagnetic interactions. In Nils Svartholm, editor, *Elementary Particle Theory: Relativistic Groups and Analyticity (Nobel Symposium No. 8)*, pages 367–377, Stockholm, 1968. Almqvist & Wiksell.

[5] David J. Gross and Frank Wilczek. Ultraviolet Behavior of Non-Abelian Gauge Theories. *Phys. Rev. Lett.*, 30:1343–1346, 1973.

[6] H. David Politzer. Reliable Perturbative Results for Strong Interactions? *Phys. Rev. Lett.*, 30:1346–1349, 1973.

[7] Peter W. Higgs. Broken Symmetries and the Masses of Gauge Bosons. *Phys. Rev. Lett.*, 13:508–509, 1964.

[8] Francois Englert and Robert Brout. Broken Symmetry and the Mass of Gauge Vector Mesons. *Phys. Rev. Lett.*, 13:321–323, 1964.

[9] Gerald S. Guralnik, C. R. Hagen, and T. W. B. Kibble. Global Conservation Laws and Massless Particles. *Phys. Rev. Lett.*, 13:585–587, 1964.

[10] Nicola Cabibbo. Unitary Symmetry and Leptonic Decays. *Phys. Rev. Lett.*, 10:531–533, 1963.

[11] Makoto Kobayashi and Toshihide Maskawa. CP Violation in the Renormalizable Theory of Weak Interaction. *Prog. Theor. Phys.*, 49:652–657, 1973.

[12] Particle Data Group and R. L. et al. Workman. Review of Particle Physics. *PTEP*, 2022(8):083C01, 2022.

[13] Y. et al. [Super-Kamiokande Collaboration] Fukuda. Evidence for oscillation of atmospheric neutrinos. *Phys. Rev. Lett.*, 81:1562–1567, 1998.

[14] Q. R. et al. [SNO Collaboration] Ahmad. Direct evidence for neutrino flavor transformation from neutral current interactions in the Sudbury Neutrino Observatory. *Phys. Rev. Lett.*, 89:011301, 2002.

[15] B. et al. [Muon g-2 Collaboration] Abi. Measurement of the Positive Muon Anomalous Magnetic Moment to 0.46 ppm. *Phys. Rev. Lett.*, 126:141801, 2021.

[16] T. et al. Aoyama. The anomalous magnetic moment of the muon in the Standard Model. *Phys. Rept.*, 887:1–166, 2020.

[17] R. et al. [LHCb Collaboration] Aaij. Test of Lepton Universality Using $B^+ \to K^+\ell^+\ell^-$ Decays. *Phys. Rev. Lett.*, 113:151601, 2014.

[18] R. et al. [LHCb Collaboration] Aaij. Test of lepton universality with $B^0 \to K^{*0}\ell^+\ell^-$ decays. *JHEP*, 08:055, 2017.

[19] Tsutomu Yanagida. Horizontal gauge symmetry and masses of neutrinos. *Conf. Proc. C*, 7902131:95–99, 1979.

[20] Rabindra N. Mohapatra and Goran Senjanović. Neutrino mass and spontaneous parity nonconservation. *Phys. Rev. Lett.*, 44:912–915, 1980.

[21] M. Magg and C. Wetterich. Neutrino Mass Problem and Gauge Hierarchy. *Phys. Lett. B*, 94:61–64, 1980.

[22] Rabindra N. Mohapatra and Goran Senjanovic. Neutrino Masses and Mixings in Gauge Models with Spontaneous Parity Violation. *Phys. Rev. D*, 23:165, 1981.

[23] Robert Foot, H. Lew, X. G. He, and Girish C. Joshi. Seesaw Neutrino Masses Induced by a Triplet of Leptons. *Z. Phys. C*, 44:441, 1989.

[24] R. N. Mohapatra and J. W. F. Valle. Neutrino Mass and Baryon Number Nonconservation in Superstring Models. *Phys. Rev. D*, 34:1642, 1986.

[25] D. Wyler and L. Wolfenstein. Massless Neutrinos in Left-Right Symmetric Models. *Nucl. Phys. B*, 218:205–214, 1983.

[26] S. M. Barr. A Different seesaw formula for neutrino masses. *Phys. Rev. Lett.*, 92:101601, 2004.

[27] V. C. Rubin, N. Thonnard, and W. K. Ford, Jr. Rotational properties of 21 SC galaxies with a large range of luminosities and radii, from NGC 4605 /R = 4kpc/ to UGC 2885 /R = 122 kpc/. *Astrophys. J.*, 238:471, 1980.

[28] Edvige Corbelli and Paolo Salucci. The Extended Rotation Curve and the Dark Matter Halo of M33. *Mon. Not. Roy. Astron. Soc.*, 311:441–447, 2000.

[29] A. N. Taylor, S. Dye, Thomas J. Broadhurst, N. Benitez, and E. van Kampen. Gravitational lens magnification and the mass of Abell 1689. *Astrophys. J.*, 501:539, 1998.

[30] Douglas Clowe, Marusa Bradac, Anthony H. Gonzalez, Maxim Markevitch, Scott W. Randall, Christine Jones, and Dennis Zaritsky. A direct empirical proof of the existence of dark matter. *Astrophys. J. Lett.*, 648:L109–L113, 2006.

[31] N. Aghanim et al. Planck 2018 results. VI. Cosmological parameters. *Astron. Astrophys.*, 641:A6, 2020. [Erratum: Astron.Astrophys. 652, C4 (2021)].

[32] Gianfranco Bertone, Dan Hooper, and Joseph Silk. Particle dark matter: Evidence, candidates and constraints. *Phys. Rept.*, 405:279–390, 2005.

[33] Gerard Jungman, Marc Kamionkowski, and Kim Griest. Supersymmetric dark matter. *Phys. Rept.*, 267:195–373, 1996.

[34] Paolo Gondolo and Graciela Gelmini. Cosmic abundances of stable particles: Improved analysis. *Nucl. Phys. B*, 360:145–179, 1991.

[35] Leonard Susskind. Dynamics of Spontaneous Symmetry Breaking in the Weinberg-Salam Theory. *Phys. Rev. D*, 20:2619–2625, 1979.

[36] Steven Weinberg. Implications of Dynamical Symmetry Breaking. *Phys. Rev. D*, 13:974–996, 1976. [Addendum: Phys.Rev.D 19, 1277–1280 (1979)].

[37] Eldad Gildener. Gauge Symmetry Hierarchies. *Phys. Rev. D*, 14:1667, 1976.

[38] Riccardo Barbieri and G. F. Giudice. Upper Bounds on Supersymmetric Particle Masses. *Nucl. Phys. B*, 306:63–76, 1988.

[39] Steven Weinberg. Anthropic Bound on the Cosmological Constant. *Phys. Rev. Lett.*, 59:2607, 1987.

[40] A. D. Sakharov. Violation of CP Invariance, C asymmetry, and baryon asymmetry of the universe. *Pisma Zh. Eksp. Teor. Fiz.*, 5:32–35, 1967.

[41] N. Aghanim et al. Planck 2018 results. vi. cosmological parameters. *Astron. Astrophys.*, 641:A6, 2020.

[42] M. Fukugita and T. Yanagida. Baryogenesis Without Grand Unification. *Phys. Lett. B*, 174:45–47, 1986.

[43] T. D. Lee. A Theory of Spontaneous T Violation. *Phys. Rev. D*, 8:1226–1239, 1973.

[44] Sheldon L. Glashow and Steven Weinberg. Natural Conservation Laws for Neutral Currents. *Phys. Rev. D*, 15:1958, 1977.

[45] Junjie Cao, Peihua Wan, Lei Wu, and Jin Min Yang. Lepton-Specific Two-Higgs Doublet Model: Experimental Constraints and Implication on Higgs Phenomenology. *Phys. Rev. D*, 80:071701, 2009.

[46] Heather E. Logan and Deanna MacLennan. Charged Higgs phenomenology in the flipped two Higgs doublet model. *Phys. Rev. D*, 81:075016, 2010.

[47] T. P. Cheng and Marc Sher. Mass Matrix Ansatz and Flavor Nonconservation in Models with Multiple Higgs Doublets. *Phys. Rev. D*, 35:3484, 1987.

[48] G.C. Branco, P.M. Ferreira, L. Lavoura, M.N. Rebelo, M. Sher, and Joao P. Silva. Theory and phenomenology of two-Higgs-doublet models. *Phys. Rept.*, 516:1–102, 2012.

[49] John F. Gunion and Howard E. Haber. The CP conserving two Higgs doublet model: The Approach to the decoupling limit. *Phys. Rev. D*, 67:075019, 2003.

[50] Shinya Kanemura, Koji Tsumura, Kei Yagyu, and Hiroshi Yokoya. Fingerprinting nonminimal Higgs sectors. *Phys. Rev. D*, 90:075001, 2014.

[51] J.F. Gunion, H.E. Haber, G.L. Kane, and S. Dawson. *The Higgs Hunter's Guide*. Addison-Wesley, 1990.

[52] G. Passarino and M. J. G. Veltman. One Loop Corrections for e+ e- Annihilation Into mu+ mu- in the Weinberg Model. *Nucl. Phys. B*, 160:151–207, 1979.

[53] Vernon D. Barger, King-man Cheung, A. Djouadi, Bernd A. Kniehl, and P. M. Zerwas. Higgs bosons: Intermediate mass range at e+ e- colliders. *Phys. Rev. D*, 49:79–90, 1994.

[54] Tao Han and Danny Marfatia. h —> mu tau at hadron colliders. *Phys. Rev. Lett.*, 86:1442–1445, 2001.

[55] Albert M Sirunyan et al. Search for additional neutral MSSM Higgs bosons in the $\tau\tau$ final state in proton-proton collisions at $\sqrt{s} = 13$ TeV. *JHEP*, 09:007, 2018.

[56] Georges Aad et al. Search for heavy Higgs bosons decaying into two tau leptons with the ATLAS detector using pp collisions at $\sqrt{s} = 13$ TeV. *Phys. Rev. Lett.*, 125(5):051801, 2020.

[57] D. J. Miller, S. Moretti, D. P. Roy, and W. James Stirling. Detecting heavy charged Higgs bosons at the CERN LHC with four b quark tags. *Phys. Rev. D*, 61:055011, 2000.

[58] Athanasios Dedes and Howard E. Haber. Can the Higgs sector contribute significantly to the muon anomalous magnetic moment? *JHEP*, 05:006, 2001.

[59] Stephen M. Barr and A. Zee. Electric Dipole Moment of the Electron and of the Neutron. *Phys. Rev. Lett.*, 65:21–24, 1990. [Erratum: Phys.Rev.Lett. 65, 2920 (1990)].

[60] Darwin Chang, We-Fu Chang, Chung-Hsien Chou, and Wai-Yee Keung. Large two loop contributions to g-2 from a generic pseudoscalar boson. *Phys. Rev. D*, 63:091301, 2001.

[61] Riccardo Barbieri, Lawrence J. Hall, and Vyacheslav S. Rychkov. Improved naturalness with a heavy Higgs: An Alternative road to LHC physics. *Phys. Rev. D*, 74:015007, 2006.

[62] G. Belanger, F. Boudjema, A. Pukhov, and A. Semenov. micromegas_3: A program for calculating dark matter observables. *Comput. Phys. Commun.*, 185:960–985, 2014.

[63] Rabindra N. Mohapatra and R. E. Marshak. Local B-L Symmetry of Electroweak Interactions, Majorana Neutrinos and Neutron Oscillations. *Phys. Rev. Lett.*, 44:1316–1319, 1980. [Erratum: Phys.Rev.Lett. 44, 1643 (1980)].

[64] Shaaban Khalil. Low scale $B-L$ extension of the Standard Model at the LHC. *J. Phys. G*, 35:055001, 2008.

[65] W. Emam and S. Khalil. Higgs and Z-prime phenomenology in B-L extension of the standard model at LHC. *Eur. Phys. J. C*, 52:625–633, 2007.

[66] Lorenzo Basso, Alexander Belyaev, Stefano Moretti, and Claire H. Shepherd-Themistocleous. Phenomenology of the minimal B-L extension of the Standard Model: Z' and neutrinos. *Phys. Rev. D*, 80:055030, 2009.

[67] S. Khalil and S. Moretti. Heavy neutrinos, Z' and Higgs bosons at the LHC: new particles from an old symmetry. *J. Mod. Phys.*, 4(1):7–10, 2013.

[68] A. Leike. The Phenomenology of extra neutral gauge bosons. *Phys. Rept.*, 317:143–250, 1999.

[69] J. A. Casas and A. Ibarra. Oscillating neutrinos and $\mu \to e\gamma$. *Nucl. Phys. B*, 618:171–204, 2001.

[70] M. Carena, A. Daleo, B. A. Dobrescu, and T. M. P. Tait. Z' gauge bosons at the Tevatron. *Phys. Rev. D*, 70:093009, 2004.

[71] Lorenzo Basso, Alexander Belyaev, Stefano Moretti, and Claire H. Shepherd-Themistocleous. Phenomenology of the minimal B-L extension of the Standard model: Z' and neutrinos. *Phys. Rev. D*, 80:055030, 2009.

[72] Shaaban Khalil and Stefano Moretti. Heavy neutrinos, Z' and Higgs bosons at the LHC: new particles from an old symmetry. *J. Mod. Phys.*, 3:1302–1314, 2012.

[73] S. Khalil and A. Masiero. Radiative B-L symmetry breaking in supersymmetric models. *Phys. Lett. B*, 665:374–377, 2008.

[74] Jogesh C. Pati and Abdus Salam. Lepton Number as the Fourth Color. *Phys. Rev. D*, 10:275–289, 1974.

[75] Rabindra N. Mohapatra and Jogesh C. Pati. A Natural Left-Right Symmetry. *Phys. Rev. D*, 11:2558, 1975.

[76] N. G. Deshpande, J. F. Gunion, Boris Kayser, and Fredrick I. Olness. Left-right symmetric electroweak models with triplet Higgs. *Phys. Rev. D*, 44:837–858, 1991.

[77] P. S. Bhupal Dev, Rabindra N. Mohapatra, and Yongchao Zhang. Probing the higgs sector of the minimal left-right symmetric model at future hadron colliders. *Phys. Rev. D*, 93(11):115001, 2016.

[78] Paul Langacker. The physics of heavy z' gauge bosons. *Rev. Mod. Phys.*, 81:1199–1228, 2009.

[79] P. Langacker and S. Uma Sankar. Bounds on the mass of w_r and the $w_l - w_r$ mixing angle ξ in general $su(2)_l \times su(2)_r \times u(1)$ models. *Phys. Rev. D*, 40:1569, 1989.

[80] Joydeep Chakrabortty, Partha Konar, and Tirtha Sankar Ray. Constraining a class of $b-l$ extended models from vacuum stability and perturbativity. *JHEP*, 08:008, 2012.

[81] M. Ashry and Shaaban Khalil. Phenomenological aspects of a tev-scale alternative left-right model. *Phys. Rev. D*, 91:015009, 2015.

[82] Shaaban Khalil, Hye-Sung Lee, and Ernest Ma. Generalized lepton number and dark left-right gauge model. *Phys. Rev. D*, 79:041701, 2009.

[83] Ernest Ma. Variants of the dark left-right gauge model: Neutrinos and scotinos. *Phys. Rev. D*, 79:117701, 2009.

[84] Ernest Ma and Utpal Sarkar. Radiative left-right dirac neutrino mass. *Phys. Rev. D*, 96:095017, 2017.

[85] K. Ezzat, M. Ashry, and S. Khalil. Search for a heavy neutral higgs boson in a left-right model with an inverse seesaw mechanism at the lhc. *Phys. Rev. D*, 104(1):015016, 2021.

[86] M. T. Arun, T. Mandal, S. Mitra, A. Mukherjee, L. Priya, and A. Sampath. Testing left-right symmetry with an inverse seesaw mechanism at the lhc. *Phys. Rev. D*, 105(11):115007, 2022.

[87] W. Abdallah, A. Awad, S. Khalil, and H. Okada. Muon Anomalous Magnetic Moment and mu -> e gamma in B-L Model with Inverse Seesaw. *Eur. Phys. J. C*, 72:2108, 2012.

[88] Ziro Maki, Masami Nakagawa, and Shoichi Sakata. Remarks on the unified model of elementary particles. *Prog. Theor. Phys.*, 28:870–880, 1962.

[89] M. Ashry, K. Ezzat, and S. Khalil. Muon g-2 anomaly in a left-right model with an inverse seesaw mechanism. *Phys. Rev. D*, 107(5):055044, 2023.

[90] P. Osland, A. A. Pankov, and I. A. Serenkova. Bounds on the mass and mixing of Z' and W' bosons decaying into different pairings of W, Z, or Higgs bosons using CMS data at the LHC. 6 2022.

[91] S. L. Glashow, J. Iliopoulos, and L. Maiani. Weak Interactions with Lepton-Hadron Symmetry. *Phys. Rev. D*, 2:1285–1292, 1970.

[92] S. Antusch, C. Biggio, E. Fernandez-Martinez, M. B. Gavela, and J. Lopez-Pavon. Unitarity of the Leptonic Mixing Matrix. *JHEP*, 10:084, 2006.

[93] Michal Malinsky, Tommy Ohlsson, Zhi-zhong Xing, and He Zhang. Non-unitary neutrino mixing and CP violation in the minimal inverse seesaw model. *Phys. Lett. B*, 679:242–248, 2009.

[94] K. Ezzat, M. Ashry, and S. Khalil. Search for a heavy neutral Higgs boson in a left-right model with an inverse seesaw mechanism at the LHC. *Phys. Rev. D*, 104(1):015016, 2021.

[95] Angela Papa. Towards a new generation of Charged Lepton Flavour Violation searches at the Paul Scherrer Institut: The MEG upgrade and the Mu3e experiment. *EPJ Web Conf.*, 234:01011, 2020.

[96] Howard Georgi and Sheldon L. Glashow. Unity of All Elementary Particle Forces. *Phys. Rev. Lett.*, 32:438–441, 1974.

[97] Richard Slansky. Group Theory for Unified Model Building. *Phys. Rept.*, 79:1–128, 1981.

[98] Paul Langacker. Grand Unified Theories and Proton Decay. *Phys. Rept.*, 72:185, 1981.

[99] Pran Nath and Pavel Fileviez Perez. Proton stability in grand unified theories, in strings and in branes. *Phys. Rept.*, 441:191–317, 2007.

[100] John R. Ellis, Mary K. Gaillard, and Dimitri V. Nanopoulos. Fermion Masses and Higgs Representations in SU(5). *Phys. Lett. B*, 88:320–324, 1979.

[101] Graham G. Ross. Grand Unified Theories. 1985.

[102] Howard Georgi and Cecilia Jarlskog. A New Lepton - Quark Mass Relation in a Unified Theory. *Phys. Lett. B*, 86:297–300, 1979.

[103] S. Raby. Grand unified theories. *Rept. Prog. Phys.*, 67:755–811, 2004.

[104] S. Dimopoulos and F. Wilczek. Incomplete multiplets in supersymmetric unified models. Preprint NSF-ITP-82-07 (unpublished).

[105] R. N. Mohapatra. *Unification and Supersymmetry: The Frontiers of Quark-Lepton Physics*. Springer, 3rd edition, 2003.

[106] Particle Data Group and R.L. et al. Workman. Review of Particle Physics. *Prog. Theor. Exp. Phys.*, 2022:083C01, 2022.

[107] K. Abe et al. (Super-Kamiokande Collaboration). Search for proton decay via $p \to e^+\pi^0$ and $p \to \mu^+\pi^0$ in 0.31 megaton · years exposure of the super-kamiokande water cherenkov detector. *Phys. Rev. D*, 95(1):012004, 2017.

[108] H. Murayama and A. Pierce. Not even decoupling can save minimal supersymmetric su(5). *Phys. Rev. D*, 65:055009, 2002.

[109] J. Hisano. Proton decay in grand unified theories. *Lect. Notes Phys.*, 939:483–508, 2017.

[110] P. Eckert, J. M. Gerard, H. Ruegg, and Thomas Schucker. Minimization of the SU(5) Invariant Scalar Potential for the Fortyfive-dimensional Representation. *Phys. Lett. B*, 125:385–388, 1983.

[111] Ilja Dorsner and Pavel Fileviez Perez. Unification versus proton decay in SU(5). *Phys. Lett. B*, 642:248–252, 2006.

[112] Pavel Fileviez Perez. Renormalizable adjoint SU(5). *Phys. Lett. B*, 654:189–193, 2007.

[113] K. Abe et al. Search for proton decay via $p \to k^+\bar{\nu}$ using 260 kiloton·year data of super-kamiokande. *Phys. Rev. D*, 90(7):072005, 2014.

[114] S. Khalil and S. Salem. Enhancement of $H \to \gamma\gamma$ in $SU(5)$ model with 45_{H^1} plet. *Nucl. Phys. B*, 876:473–492, 2013.

[115] S. Khalil, S. Salem, and M. Allam. SU(5) Octet Scalar at the LHC. *Phys. Rev. D*, 89:095011, 2014. [Erratum: Phys.Rev.D 91, 119908 (2015)].

[116] Toru Goto, Satoshi Mishima, and Tetsuo Shindou. Flavor physics in SU(5) GUT with scalar fields in the 45 representation. *Phys. Rev. D*, 108(9):095012, 2023.

[117] Julius Wess and Jonathan Bagger. *Supersymmetry and Supergravity*. Princeton University Press, 2nd edition, 1992.

[118] Stefano Moretti and Shaaban Khalil. *Supersymmetry Beyond Minimality: From Theory to Experiment*. CRC Press, 2019.

[119] S. Dimopoulos, S. Raby, and Frank Wilczek. Supersymmetry and the Scale of Unification. *Phys. Rev. D*, 24:1681–1683, 1981.

[120] Ugo Amaldi, Wim de Boer, and Hermann Furstenau. Comparison of grand unified theories with electroweak and strong coupling constants measured at LEP. *Phys. Lett. B*, 260:447–455, 1991.

[121] Paul Langacker and Nir Polonsky. Uncertainties in coupling constant unification. *Phys. Rev. D*, 47:4028–4045, 1993.

[122] Yu. A. Golfand and E. P. Likhtman. Extension of the Algebra of Poincare Group Generators and Violation of p Invariance. *JETP Lett.*, 13:323–326, 1971.

[123] J. Wess and B. Zumino. Supergauge Transformations in Four-Dimensions. *Nucl. Phys. B*, 70:39–50, 1974.

[124] Sidney R. Coleman and J. Mandula. All Possible Symmetries of the S Matrix. *Phys. Rev.*, 159:1251–1256, 1967.

[125] Steven Weinberg. *The Quantum Theory of Fields. Vol. 3: Supersymmetry.* Cambridge University Press, 2000.

[126] Stephen P. Martin. A supersymmetry primer. 1997. [Adv. Ser. Direct. High Energy Phys. 21 (2010) 1-153].

[127] L. Girardello and M. T. Grisaru. Soft breaking of supersymmetry. *Nucl. Phys. B*, 194:65–76, 1982.

[128] Steven Weinberg. Supersymmetry at Ordinary Energies. 1. Masses and Conservation Laws. *Phys. Rev. D*, 26:287, 1982.

[129] Yasuhiro Okada, Masahiro Yamaguchi, and Tsutomu T. Yanagida. Upper bound of the lightest higgs boson mass in the minimal supersymmetric standard model. *Prog. Theor. Phys.*, 85:1–6, 1991.

[130] John R. Ellis, Giovanni Ridolfi, and Fabio Zwirner. Radiative corrections to the masses of supersymmetric higgs bosons. *Phys. Lett. B*, 257:83–91, 1991.

[131] Howard E. Haber and Ralf Hempfling. Can the mass of the lightest higgs boson of the minimal supersymmetric model be larger than m_z? *Phys. Rev. Lett.*, 66:1815–1818, 1991.

[132] ATLAS Collaboration. Search for supersymmetry in final states with missing transverse momentum and multiple b-jets in proton–proton collisions at $\sqrt{s} = 13$ tev with the atlas detector. *JHEP*, 04:084, 2019.

[133] CMS Collaboration. Search for new physics in final states with jets and missing transverse momentum in proton-proton collisions at 13 tev. *JHEP*, 10:005, 2017.

[134] ATLAS and CMS Collaborations. Searches for supersymmetry at the high-luminosity lhc: a summary of current status and prospects. *Nature Phys.*, 18:1326–1337, 2022.

[135] ATLAS Collaboration. Search for squarks and gluinos in final states with jets and missing transverse momentum using 139 fb^{-1} of $\sqrt{s} = 13$ tev pp collision data with the atlas detector. *Phys. Rev. D*, 101(5):052005, 2020.

[136] Fabiola Gianotti et al. Supersymmetry prospects at the hl-lhc and he-lhc. *Eur. Phys. J. C*, 78:110, 2018.

[137] Th. Kaluza. Zum Unitätsproblem der Physik. *Sitzungsber. Preuss. Akad. Wiss. Berlin (Math. Phys.)*, 1921:966–972, 1921.

[138] Oskar Klein. Quantum Theory and Five-Dimensional Theory of Relativity. (In German and English). *Z. Phys.*, 37:895–906, 1926.

[139] Nima Arkani-Hamed, Savas Dimopoulos, and Gia Dvali. The Hierarchy Problem and New Dimensions at a Millimeter. *Phys. Lett. B*, 429:263–272, 1998.

[140] Lisa Randall and Raman Sundrum. A Large Mass Hierarchy from a Small Extra Dimension. *Phys. Rev. Lett.*, 83:3370–3373, 1999.

[141] Lisa Randall and Raman Sundrum. An Alternative to Compactification. *Phys. Rev. Lett.*, 83:4690–4693, 1999.

[142] Joseph Polchinski. *String Theory: Volume 1, An Introduction to the Bosonic String*. Cambridge University Press, 1998.

[143] Edward Witten. String theory and m-theory at strong coupling. *Nuclear Physics B*, 443:85–126, 1995.

[144] Shamit Kachru, Renata Kallosh, Andrei D. Linde, and Sandip P. Trivedi. De sitter vacua in string theory. *Physical Review D*, 68:046005, 2003.

[145] G. R. Dvali and G. Gabadadze. Gravity on a brane in anti-de sitter space. *Physical Review D*, 63:065007, 2001.

[146] J. M. Overduin and P. S. Wesson. Kaluza-klein gravity. *Phys. Rept.*, 283:303–378, 1997.

[147] T. Appelquist, A. Chodos, and P. G. O. Freund. Modern kaluza-klein theories. *Front. Phys.*, 65, 1987. Originally published in 1983 by Benjamin/Cummings.

[148] Ignatios Antoniadis, Nima Arkani-Hamed, Savas Dimopoulos, and Gia Dvali. New dimensions at a millimeter to a Fermi and superstrings at a TeV. *Phys. Lett. B*, 436:257–263, 1998.

[149] C. D. Hoyle, U. Schmidt, B. R. Heckel, E. G. Adelberger, J. H. Gundlach, D. J. Kapner, and H. E. Swanson. Submillimeter tests of the gravitational inverse square law: a search for large extra dimensions. *Phys. Rev. Lett.*, 86:1418–1421, 2001.

[150] Petr Horava and Edward Witten. Heterotic and Type I string dynamics from eleven dimensions. *Nucl. Phys. B*, 460:506–524, 1996.

[151] Walter D. Goldberger and Mark B. Wise. Modulus stabilization with bulk fields. *Phys. Rev. Lett.*, 83:4922–4925, 1999.

[152] Y. S. Myung, Gungwon Kang, and H. W. Lee. Randall-Sundrum gauge in the brane world. *Phys. Lett. B*, 478:294–298, 2000.

[153] X. Cid Vidal et al. Beyond the Standard Model Physics at the HL-LHC and HE-LHC. *CERN Yellow Rep. Monogr.*, 7:585–865, 2019.

[154] Daniel Abercrombie et al. Dark Matter Benchmark Models for Early LHC Run-2 Searches: Report of the ATLAS/CMS Dark Matter Forum.

[155] O. Buchmueller, S.A. Malik, A. Saavedra, A. Weiler, and T. Weiler. Simplified models for displaced dark matter signatures. *Phys. Lett. B*, 771:100–106, 2017.

[156] Nathaniel Craig, Hou Keong Lou, Matthew McCullough, and Jesse Thaler. The Higgs Portal Above Threshold. *JHEP*, 02:127, 2016.

[157] Jonathan L. Feng. Dark matter candidates from particle physics and methods of detection. *Annual Review of Astronomy and Astrophysics*, 48:495–545, 2010.

[158] ATLAS Collaboration. Summary of the searches for squarks and gluinos using $\sqrt{s} = 8$ tev pp collisions with the atlas experiment at the lhc. *JHEP*, 10:054, 2015.

[159] CMS Collaboration. Search for long-lived particles using non-prompt jets and missing transverse momentum with proton-proton collisions at $\sqrt{s} = 13$ tev. *Phys. Lett. B*, 797:134876, 2019.

[160] Search for squarks and gluinos with the ATLAS detector in final states with jets and missing transverse momentum and 20.3 fb^{-1} of $\sqrt{s} = 8$ TeV proton-proton collision data. 5 2013.

[161] Serguei Chatrchyan et al. Inclusive search for squarks and gluinos in *pp* collisions at $\sqrt{s} = 7$ TeV. *Phys. Rev. D*, 85:012004, 2012.

[162] Georges Aad et al. Observation of a new particle in the search for the Standard Model Higgs boson with the ATLAS detector at the LHC. *Phys. Lett. B*, 716:1–29, 2012.

[163] Serguei Chatrchyan et al. Search for Top-Squark Pair Production in the Single-Lepton Final State in pp Collisions at $\sqrt{s} = 8$ TeV. *Eur. Phys. J. C*, 73(12):2677, 2013.

[164] Search for direct top squark pair production in final states with one isolated lepton, jets, and missing transverse momentum in $sqrts =$ 8,TeV *pp* collisions using 21 fb^{-1} of ATLAS data. 3 2013.

[165] Search for direct top squark pair production and dark matter production in final states with two leptons in $\sqrt{s} = 13$ TeV *pp* collisions using 13.3 fb^{-1} of ATLAS data. 8 2016.

[166] Search for direct top squark pair production in the single lepton final state at $\sqrt{s} = 13$ TeV. 2016.

[167] Search for electroweak production of charginos, neutralinos, and sleptons using leptonic final states in pp collisions at 8 TeV. 2013.

[168] Search for direct-slepton and direct-chargino production in final states with two opposite-sign leptons, missing transverse momentum and no jets in 20/fb of pp collisions at sqrt(s) = 8 TeV with the ATLAS detector. 5 2013.

[169] Search for electroweak SUSY production in multilepton final states in pp collisions at sqrt(s)=13 TeV with 12.9/fb. 2016.

[170] Morad Aaboud et al. Search for new high-mass phenomena in the dilepton final state using 36 fb^{-1} of proton-proton collision data at $\sqrt{s} = 13$ TeV with the ATLAS detector. *JHEP*, 10:182, 2017.

[171] Albert M Sirunyan et al. Search for high-mass resonances in dilepton final states in proton-proton collisions at $\sqrt{s} = 13$ TeV. *JHEP*, 06:120, 2018.

[172] Search for a narrow resonance in high-mass dilepton final states in proton-proton collisions using 140 fb^{-1} of data at $\sqrt{s} = 13$ TeV. 2019.

[173] Search for high-mass dilepton resonances using 139 fb^{-1} of pp collision data collected at $\sqrt{s} = 13\,\mathrm{TeV}$ with the ATLAS detector. 2 2019.

[174] Vardan Khachatryan et al. Search for heavy gauge W' boson in events with an energetic lepton and large missing transverse momentum at $\sqrt{s} = 13$ TeV. *Phys. Lett. B*, 770:278–301, 2017.

[175] Abdelhak Djouadi. The Anatomy of electro-weak symmetry breaking. I: The Higgs boson in the standard model. *Phys. Rept.*, 457:1–216, 2008.

[176] Aram Hayrapetyan et al. Dark sector searches with the CMS experiment. *Phys. Rept.*, 1115:448–569, 2025.

[177] Georges Aad et al. Search for new phenomena in events with an energetic jet and missing transverse momentum in pp collisions at $\sqrt{s}$ =13 TeV with the ATLAS detector. *Phys. Rev. D*, 103(11):112006, 2021.

[178] LHCb Collaboration. Updated constraints on rare decays and flavor observables. *LHCb-PAPER-2022-XXX*, 2022. CERN preprint, in preparation.

[179] The MEG Collaboration. Search for the lepton flavour violating decay $\mu^+ \to e^+\gamma$ with the full dataset of the meg experiment. *Eur. Phys. J. C*, 76:434, 2016.

[180] Super-Kamiokande Collaboration. Evidence for oscillation of atmospheric neutrinos. *Phys. Rev. Lett.*, 81:1562–1567, 1998.

[181] Gino Isidori, Yosef Nir, and Gilad Perez. Flavor physics constraints for physics beyond the standard model. *Ann. Rev. Nucl. Part. Sci.*, 60:355–389, 2010.

[182] Ferruccio Feruglio. Pieces of the flavor puzzle. *Eur. Phys. J. C*, 75:373, 2015.

[183] Guido Altarelli and Ferruccio Feruglio. Discrete flavor symmetries and models of neutrino mixing. *Rev. Mod. Phys.*, 82:2701–2729, 2010.

[184] Hiroshi Ishimori, Tatsuo Kobayashi, Hiroshi Ohki, Yusuke Shimizu, and Morimitsu Tanimoto. Non-abelian discrete symmetries in particle physics. *Prog. Theor. Phys. Suppl.*, 183:1–163, 2010.

[185] Stephen F. King and Christoph Luhn. Neutrino mass and mixing with discrete symmetry. *Rept. Prog. Phys.*, 76:056201, 2013.

[186] Guido Altarelli and Ferruccio Feruglio. Tri-bimaximal neutrino mixing from discrete symmetry in extra dimensions. *Nucl. Phys. B*, 720:64–88, 2005.

[187] C. D. Froggatt and H. B. Nielsen. Hierarchy of Quark Masses, Cabibbo Angles and CP Violation. *Nucl. Phys. B*, 147:277, 1979.

[188] M. Leurer, Y. Nir, and N. Seiberg. Mass matrix models: The sequel. *Nucl. Phys. B*, 420:468–504, 1994.

[189] A. Crivellin, C. Greub, and A. Kokulu. Explaining $B \to D\tau\nu$, $B \to D^*\tau\nu$ and $B \to \tau\nu$ in a 2HDM of type III. *Phys. Rev. D*, 86:054014, 2012.

[190] A. J. Buras and D. Guetta. The standard model and beyond: Implications for cp violation. *Phys. Lett. B*, 688(1):55–64, 2010.

[191] Martin Jung, Antonio Pich, and Paula Tuzon. Charged-Higgs phenomenology in the Aligned two-Higgs-doublet model. *JHEP*, 11:003, 2010.

[192] S. Inoue, M. J. Ramsey-Musolf, and Y. Zhang. CP-violating phenomenology of flavor conserving two Higgs doublet models. *Phys. Rev. D*, 89:115023, 2014.

[193] R. N. Mohapatra and J. C. Pati. A gauge theory of grand unification of electromagnetic, weak and strong interactions. *Phys. Rev. D*, 10(5):1507–1514, 1974.

[194] Goran Senjanović and Vladimir Tello. Right Handed Quark Mixing in Left-Right Symmetric Theory. *Phys. Rev. Lett.*, 114(7):071801, 2015.

[195] Ilja Dorsner and Pavel Fileviez Perez. Unification without supersymmetry: Neutrino mass, proton decay and light leptoquarks. *Nucl. Phys. B*, 723:53–76, 2005.

[196] A. Ismael and S. Khalil. Resolving R_D and R_{D^*} anomalies in adjoint SU(5). *JHEP*, 11:143, 2023.

[197] Minoru Tanaka and Ryoutaro Watanabe. New physics in the weak interaction of $b^- \to d^{(*)}\tau\nu^-$. *Phys. Rev. D*, 87(3):034028, 2013.

[198] Syuhei Iguro, Teppei Kitahara, Yuji Omura, Ryoutaro Watanabe, and Kei Yamamoto. d^* polarization vs. $r_{D^{(*)}}$ anomalies in the leptoquark models. *JHEP*, 02:194, 2019.

[199] Edward W. Kolb and Michael S. Turner. *The Early Universe*, volume 69. Taylor and Francis, 5 2019.

[200] Julien Lesgourgues and Sergio Pastor. Massive neutrinos and cosmology. *Phys. Rept.*, 429:307–379, 2006.

[201] Adam G. Riess et al. Observational evidence from supernovae for an accelerating universe and a cosmological constant. *Astron. J.*, 116:1009–1038, 1998.

[202] S. Perlmutter et al. Measurements of Ω and Λ from 42 High Redshift Supernovae. *Astrophys. J.*, 517:565–586, 1999.

[203] Steven Weinberg. The Cosmological Constant Problem. *Rev. Mod. Phys.*, 61:1–23, 1989.

[204] Jerome Martin. Everything You Always Wanted To Know About The Cosmological Constant Problem (But Were Afraid To Ask). *Comptes Rendus Physique*, 13:566–665, 2012.

[205] R. R. Caldwell, Rahul Dave, and Paul J. Steinhardt. Cosmological imprint of an energy component with general equation of state. *Phys. Rev. Lett.*, 80:1582–1585, 1998.

[206] Tim Clifton, Pedro G. Ferreira, Antonio Padilla, and Constantinos Skordis. Modified gravity and cosmology. *Physics Reports*, 513(1):1–189, 2012.

[207] Andrew G. Cohen, D. B. Kaplan, and A. E. Nelson. Progress in electroweak baryogenesis. *Ann. Rev. Nucl. Part. Sci.*, 43:27–70, 1993.

[208] Masanori Fukugita and Tsutomu Yanagida. Baryogenesis without grand unification. *Physical Letters B*, 174(1):45–47, 1986.

[209] W. Buchmüller, P. Di Bari, and M. Plumacher. Leptogenesis for pedestrians. *Annals of Physics*, 315(1):305–351, 2005.

[210] David E. Morrissey and Michael J. Ramsey-Musolf. Electroweak baryogenesis. *New J. Phys.*, 14:125003, 2012.

[211] Fritz Zwicky. Die rotverschiebung von extragalaktischen nebeln. *Helvetica Physica Acta*, 6:110–127, 1933.

[212] Vera C. Rubin and W. Kent Ford, Jr. Rotation of the Andromeda Nebula from a Spectroscopic Survey of Emission Regions. *Astrophys. J.*, 159:379–403, 1970.

[213] Mordehai Milgrom. A modification of the newtonian dynamics as a possible alternative to the hidden mass hypothesis. *The Astrophysical Journal*, 270:365–370, 1983.

[214] Masataka Fukugita and P. James E. Peebles. The Cosmic energy inventory. *Astrophys. J.*, 616:643–668, 2004.

[215] Giovanni Bertone, Dan Hooper, and Joao Silk. Particle dark matter: Evidence, candidates, and constraints. *Physics Reports*, 405(5):279–390, 2005.

[216] Keith A. Olive and P. Shaver. Big bang nucleosynthesis and the problem of dark matter. *Review of Modern Physics*, 86(4):1063–1099, 2014.

[217] Jonathan L. Feng. Dark matter candidates from particle physics and methods of detection. *Annual Review of Astronomy and Astrophysics*, 48:495–545, 2010.

[218] Mark Kamionkowski Gary Jungman and Kent A. Olive. Supersymmetric dark matter. *Physics Reports*, 267(5):195–373, 1996.

[219] Michael B. Wise John Preskill and Frank Wilczek. Cosmology of the axion. *Physical Review Letters*, 51:1231–1234, 1983.

[220] Scott D. H. Dodelson Peter Bode and Alex Kehayias. Warm dark matter. *Astrophysical Journal*, 556:93–105, 2001.

[221] Kevork Abazajian, George M. Fuller, and Mitesh Patel. Sterile neutrino hot, warm, and cold dark matter. *Phys. Rev. D*, 64:023501, 2001.

[222] Scott Dodelson and Lawrence M. Widrow. Sterile neutrinos as dark matter. *Phys. Rev. Lett.*, 72:17–20, 1994.

[223] M. Aaboud et al. Search for supersymmetry with the atlas detector at the lhc. *Nature*, 563:345–348, 2018.

[224] A. N. Taylor et al. Gravitational lens magnification and the mass of abell 1689. *The Astrophysical Journal*, 501(2):539–553, 1998.

[225] B. W. Lee and Steven Weinberg. Cosmological Lower Bound on Heavy-Neutrino Masses. *Phys. Rev. Lett.*, 39:165–168, 1977.

[226] J. Edsjö and P. Gondolo. Neutralino relic density including coannihilations. *Physics Review D*, 56:1879–1894, 1997.

[227] Jalal Abdallah et al. Simplified models for dark matter searches at the lhc. *Physics of the Dark Universe*, 9-10:8–23, 2015.

[228] Kim Griest and David Seckel. Three exceptions in the calculation of relic abundances. *Phys. Rev. D*, 43:3191–3203, 1991.

[229] P. A. R. Ade and others [Planck Collaboration]. Planck 2013 results. xvi. cosmological parameters. *Astronomy & Astrophysics*, 571:A16, 2014.

[230] S. Chakraborti, A. Ray, N. Chandra, S. S. S. V. Cheela, and R. S. Barniol Duran. Radio detection of a type ia supernova: Implications for the single degenerate scenario. *Astrophysical Journal*, 805:187, 2015.

[231] Gary Steigman. Primordial Nucleosynthesis in the Precision Cosmology Era. *Ann. Rev. Nucl. Part. Sci.*, 57:463–491, 2007.

[232] V. A. Kuzmin, V. A. Rubakov, and M. E. Shaposhnikov. On the Anomalous Electroweak Baryon Number Nonconservation in the Early Universe. *Phys. Lett. B*, 155:36, 1985.

[233] Apostolos Pilaftsis. Resonant cp violation induced by particle mixing in transition amplitudes. *Nucl. Phys. B*, 504:61–107, 1997.

[234] Alan H. Guth. Inflationary universe: A possible solution to the horizon and flatness problems. *Phys. Rev. D*, 23:347–356, 1981.

[235] A. A. Starobinsky. A new type of isotropic cosmological models without singularity. *Phys. Lett. B*, 91:99–102, 1980.

[236] Albert Einstein. The foundation of the general theory of relativity. *Annalen der Physik*, 49:769–822, 1916.

[237] A. Friedmann. Über die krümmung des raumes. *Zeitschrift für Physik*, 10:377–386, 1922.

[238] A. Friedmann. Über die möglichkeit einer welt mit konstanter negativer krümmung des raumes. *Zeitschrift für Physik*, 21:326–332, 1924.

[239] R. H. Dicke and P. J. E. Peebles. The big bang cosmology and the microwave background. *The Astrophysical Journal*, 192:E1–E5, 1979.

[240] P. J. E. Peebles. *Principles of Physical Cosmology.* Princeton University Press, Princeton, NJ, 1993.

[241] A. D. Linde. Chaotic inflation. *Physics Letters B*, 129:177–181, 1983.

[242] A. Albrecht and P. J. Steinhardt. Cosmology for grand unified theories with radiatively induced symmetry breaking. *Physical Review Letters*, 48:1220–1223, 1982.

[243] David H. Lyth and Antonio Riotto. Particle physics models of inflation and the cosmological density perturbation. *Physics Reports*, 314:1–146, 1999.

[244] Viatcheslav F. Mukhanov, H. A. Feldman, and Robert H. Brandenberger. Theory of cosmological perturbations. part 1. classical perturbations. part 2. quantum theory of perturbations. part 3. extensions. *Physics Reports*, 215:203–333, 1992.

[245] S. R. Coleman and E. J. Weinberg. Radiative corrections as the origin of spontaneous symmetry breaking. *Physical Review D*, 7:1888–1910, 1973.

[246] A. R. Liddle and D. H. Lyth. The cold dark matter density perturbation. *Physics Reports*, 231:1–105, 1993.

[247] Paul J. Steinhardt and Michael S. Turner. A prescription for successful new inflation. *Phys. Rev. D*, 29:2162, 1984.

[248] A. G. Riess et al. Observational evidence from supernovae for an accelerating universe and a cosmological constant. *The Astronomical Journal*, 116:1009–1038, 1998.

[249] S. Perlmutter et al. Discovery of a supernova explosion at half the age of the universe. *Nature*, 391:51–54, 1998.

[250] B. Ratra. Cosmological implications of a rolling scalar field. *Physical Review D*, 35:232–235, 1987.

[251] Ivaylo Zlatev, Li-Min Wang, and Paul J. Steinhardt. Quintessence, cosmic coincidence, and the cosmological constant. *Phys. Rev. Lett.*, 82:896–899, 1999.

[252] R. R. Caldwell, Romesh Dave, and Paul J. Steinhardt. Cosmological Imprint of an Energy Component with Generalized Equation of State. *Phys. Rev. Lett.*, 80:1582–1585, 2002.

[253] Galaxies Cosmology. History of the universe, 2015. Accessed: 2025-04-07.

[254] R. R. Caldwell, R. Dave, and P.J. Steinhardt. Cosmological imprint of an energy component with general equation of state. *Physical Review Letters*, 80:1582–1585, 2002.

[255] I. Zlatev, L. Wang, and P.J. Steinhardt. Quintessence, cosmic coincidence, and the cosmological constant. *Physical Review Letters*, 82:896–899, 1999.

[256] P.J. Steinhardt, L. Wang, and I. Zlatev. Cosmological consequences of a rolling homogeneous scalar field. *Physics Letters B*, 459:1–10, 1999.

[257] S. Capozziello et al. Curvature quintessence from modified gravity. *International Journal of Modern Physics D*, 12(6):1073–1088, 2003.

[258] S. Nojiri and S. D. Odintsov. Modified gravity with the help of the scalar field and the fate of the universe. *Phys. Lett. B*, 639:144–152, 2006.

[259] A. A. Starobinsky. Disappearing cosmological constant in f(R) gravity. *J. Exp. Theor. Phys. Lett.*, 86:157–163, 2007.

Index

Financing Energy Projects Deskbook

Financing Energy Projects Deskbook

Albert Thumann, PE, CEM
Fred E. Wainwright

Published by
THE FAIRMONT PRESS, INC.
700 Indian Trail
Lilburn, GA 30247

Library of Congress Cataloging-in-Publication Data

Thumann, Albert.
Financing energy projects deskbook / Albert Thumann, Fred Wainwright
p. cm.
Includes bibliographical references and index.
ISBN 0-88173-272-9
1. Energy conservation--Finance. 2. Industries--Energy conservation--Finanace. I. Wainwright, Fred. II. Title.
HD9502.A2T519 1997 658.2--dc21 97-5327
CIP

Published by The Fairmont Press, Inc.
700 Indian Trail
Lilburn, GA 30247

Printed in the United States of America

10 9 8 7 6 5 4 3 2 1

ISBN 0-88173-272-9 FP

ISBN 0-13-888520-6 PH

Distributed by Prentice Hall PTR
Prentice-Hall, Inc.
A Simon & Schuster Company
Upper Saddle River, NJ 07458

Prentice-Hall International (UK) Limited, London
Prentice-Hall of Australia Pty. Limited, Sydney
Prentice-Hall Canada Inc., Toronto
Prentice-Hall Hispanoamericana, S.A., Mexico
Prentice-Hall of India Private Limited, New Delhi
Prentice-Hall of Japan, Inc., Tokyo
Simon & Schuster Asia Pte. Ltd., Singapore
Editora Prentice-Hall do Brasil, Ltda., Rio de Janeiro

This book is dedicated to facility managers and energy engineers throughout the world. They are the true champions of energy conservation.

Table of Contents

Introduction

Energy projects are socially important because they help save energy and therefore save precious environmental resources, yet most organizations do not make their decisions for altruistic reasons. The reality is that management decides whether to approve or deny an energy project based on one factor and one factor only—money.

Every day that your company spends without installing the appropriate energy-efficiency measures in its facilities means less profits, decreased productivity, and lost opportunities. If the chief financial officer of your organization stood on the roof of your office building and threw a large basket of hundred-dollar bills into the wind, what would you say? Perhaps you can show him that he might as well be doing that if your energy project continues to be delayed.

More than $2 billion has been invested in energy conservation projects during the past 10 years. As utility-based rebates become scarce, alternative financing methods will gain favor. While the engineering that goes into developing energy project calculations is important, it is equally important to conduct a *financial engineering* study of the project. What are the real costs of the project? What about hidden costs like employee or customer inconveniences while the project is being constructed? What are the risks of the equipment not performing as promised? What if key manufacturing operations are shut down because the air conditioning fails? Can the money spent on the project be better applied in researching a new product or in a larger advertising campaign? Will the cash to pay for the project come from internal funds or from a bank loan? Are there other sources of cash, such as utility rebates? Since energy projects are ultimately financial decisions, you must become a financial engineer if you are going to successfully present your project to upper management.

Our goals as authors are threefold. First, we want to explore as many financing options as possible with you. Second, we want to show you the tools to make a comprehensive financial engineering analysis. Third, we want to broaden your horizons with new trends in the industry. If you have gained sufficient confidence after reading this book to make that convincing energy project presentation to the chief financial officer of your organization, then we are all winners.

Chapter 1

Financing Energy Projects

The main job of the chief financial officer of your organization is to reduce risks. The risk analysis of an energy project is not complete unless all technical and financial options are explored and understood.

For the purposes of this chapter, it is assumed that an energy services company (ESCO) will be responsible for providing the energy conservation measures (ECMs) and managing your project. These ECMs may include audits, design, construction, installation, monitoring, and maintenance.

It is also assumed that a separate party, such as a bank or investment company, is providing the capital to purchase the equipment. Sometimes ESCOs market themselves as being financiers, but most often they have a financing source such as a corporate parent or bank in the background. For simplicity's sake, the ultimate source of capital for the project will be considered a separate lender with its own guidelines and its own decision-making process.

Some of the most common financing alternatives are:

COMMERCIAL BANK LOAN

The obvious first alternative is approaching your local business bank and applying for a loan. The bank will review your company's credit history and financial statements in order to make a decision. Interest rates are usually based on the prime rate (the rate offered by the largest U.S. banks to their best customers) plus a margin. For example, as of this writing, a company with established, above-average credit can expect to pay 1% over prime. Most banks prefer to lend to businesses in the form of credit lines which can be used by the company as needed, paid back, and reused again without time-consuming reapplication procedures.

There is a well-known "Catch-22 Effect" in banking which can be summarized as follows: if you are sufficiently well-qualified to be approved for a loan, you probably don't need one. If your company's financial statements show a $1,000,000 net worth and positive net income for the past few years, chances are that you will be approved for a loan for a $200,000 energy project. On the other hand, if you have a similar net worth and income and you ask your banker to help you with a $5,000,000 project, your banker will help you—find the door. Commercial banks understand financial statements but they don't understand kilowatts.

GENERAL OBLIGATION BOND

A bond is a security instrument representing an obligation to pay by the issuer (the "borrowing entity") to the buyers (the public or investment companies). Bonds can be secured by certain assets or by the good faith and credit of the issuer. General Obligation Bonds (known as "GOBs" to bankers) are specialized bonds issued by local and state government entities in order to raise money for general business operations. The interest rate paid by these bonds is based on the current overall interest rate market as well as the credit quality of the state or local government issuer. The process of preparing and issuing a general obligation bond is long and complicated but the interest rate which the issuer will have to pay is relatively low. As of this writing, bonds are being issued with interest rates in the range of 5% to 7%. Cheap money is good as long as it's not too difficult or too time-consuming to get!

LEASE

Leasing has gained tremendous popularity in the 1980s and 1990s because of its flexibility and practicality. For example, many car buyers are choosing to use leases instead of bank loans because they can get a better quality car for the same payment. They also prefer to return their car to the dealership after a few years and get a newer model, instead of having to go through the hassle and uncertainty of selling the car themselves. As you can tell, the dealer is more than willing to take the car back because he can estimate the car's value fairly accurately and resell it at a profit.

For energy equipment, leases have the added benefit of offering 100% financing; that is, your company does not have to make a substantial down payment and can thus conserve its cash for daily operations. There are numerous types of leases. Operating, capital, and municipal leases deserve special attention.

A *capital lease* is one which appears on the balance sheet of the lessee (the borrower) as a debt in much the same way as a commercial bank loan. A typical commercial or industrial company such as a steel manufacturer is a good candidate for lease financing. The rates for such leases are usually 1% to 5% above prime.

Most leased energy equipment uses the capital lease structure, which may include any of the following: transfer of ownership of equipment at the end of the lease, a dollar buyout clause at the end of the lease, the lease term equals 75% or more of the economic life of the equipment, the net present value of the lease payments is equal to or greater than 90% of value of equipment. (For an explanation of financial terms such as *net present value* or *present worth*, see Chapters 5 and 7.)

An *operating lease* is designed to fulfill certain requirements of the Financial Accounting Standards Board (FASB) and the Internal Revenue Service. A well-structured operating lease does not have to be reported on your company's balance sheet as debt. That is, it does not appear as a liability like most bank loans or credit lines but is merely a regular expense obligation on the income statement such as office rent. Oftentimes this can be very important for a company or institution that does not want to affect its borrowing capacity or its cost of capital by adding more debt to its financial structure.

Operating leases are difficult to structure for energy equipment because it is not easy to estimate the useful life and the fair market value of the equipment several years after installation. What is the value of an energy management system or compact fluorescent lights five years after they have been installed? There are stories of energy equipment leases that were structured as operating leases but which, upon closer scrutiny by auditors, had to be converted to capital leases. This could cause a substantial effect to a company's financial structure or, worse, trigger default clauses under bank credit lines which are vital to the company for day-to-day business. (For energy project financing *success stories*, please turn to Chapter 10.)

Another kind of lease is the *municipal lease.* Any entity that raises revenue through taxation and qualifies according to the IRS as tax-exempt under Section 103(c) of the Internal Revenue Code of 1986 can have access to a municipal lease. This form of lease is often used by state and local governments because of its low rates (below prime and comparable to GOB rates).

It is important to note that in all of the categories listed above, your company will be obligated to make the loan or lease payments regardless of the performance of the energy-saving equipment or the ESCO. You may have recourse for repairs under a separate energy services contract with the ESCO or through an equipment manufacturer's warranty, but the loan or lease payments have to be made on time every month. The interest rate and terms of the loan or lease are almost exclusively based on the credit quality of your organization. Figure 1-1 shows the structure of a traditional commercial bank loan or equipment lease.

PROJECT FINANCING/SHARED SAVINGS/ PERFORMANCE CONTRACTING

The United States government spends over $8 billion a year on its utility bills. As you know, due to severe budget constraints, the government is facing tremendous pressure to cut costs. In addition, the Energy Policy Act of 1992 (EPACT) and subsequent executive orders have mandated the reduction of energy consumption by 30% from a 1985 baseline by the year 2005. Yet the government continues to suffer from decaying

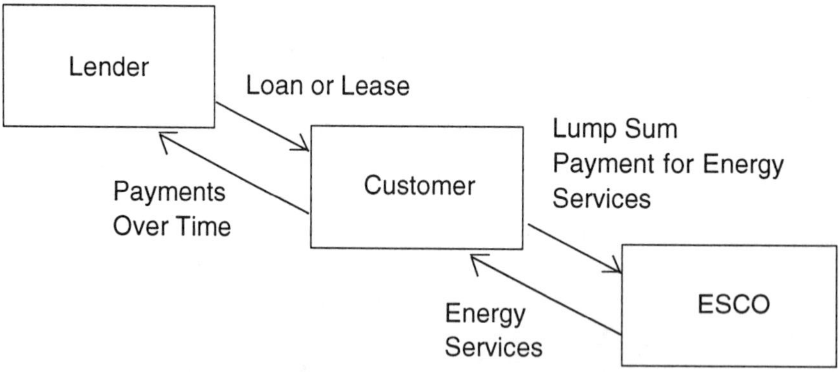

Figure 1-1. Commercial Bank Loan or Lease

and outdated energy systems in many of its facilities which cannot be repaired without the issuance of additional government debt. This problem is not isolated to just the government, as many institutions and corporations find themselves facing budget cutbacks due to downsizing or increased competition.

The solution to this problem is known as Energy Savings Performance Contracting (ESPC). In essence, the government (or any other organization) can contract with an ESCO for energy efficiency projects throughout its facilities but it will be the ESCO that will have to incur the cost of implementing the energy-savings measures. In exchange for this, both the ESCO and the government will derive the benefits from the savings generated by the more efficient equipment. For example, for every $10,000 saved, the ESCO will get paid $8,000 and the government keeps the other $2,000. This is a win-win situation since the ESCO will be very motivated to keep the equipment in peak operating condition in order to generate the greatest amount of revenue over time. Another benefit of this methodology is that the government or your organization will consider the ESCO payments an operating expense rather than a debt obligation. Your chief financial officer will appreciate that.

With ESPC, the ESCO, not the customer, has to find the financing. For small to mid-size energy projects, the ESCO may choose to use commercial banks or leasing firms, which will lend money based on the ESCO's creditworthiness. The federal projects mentioned above are so large (in the tens and hundreds of millions of dollars), that few if any ESCOs will qualify for traditional bank loans.

Bankers and specialized investment companies are increasingly becoming involved in ESPC. They are offering an innovative loan program known as Full-Recourse Project Financing, which is shown in Figure 1-2. They will review the ESCO's financial statements, but they will also review the customer's financial statements and the project specifics prior to making a lending decision. Chapter 6 will cover the lender's decision-making process in further detail. The term "Full-Recourse" is derived from the right which the lender will have to take back any asset of the borrower should the loan not be paid on schedule.

As some lenders become more comfortable with energy projects, they may be willing to structure what is called "Non-Recourse Project Financing." In this case, the loan is made to a single-purpose entity (SPE) which then owns the equipment of the project and contracts with the ESCO to perform the energy services. Figure 1-3 will show how Non-

Recourse Project Financing is different than a full recourse arrangement. If the project fails to perform as expected, the lender only has recourse to the project equipment in the SPE. The lender cannot reach the assets of either the customer or the ESCO. The single purpose entity concept is also used in the Asset Sale financing programs described below.

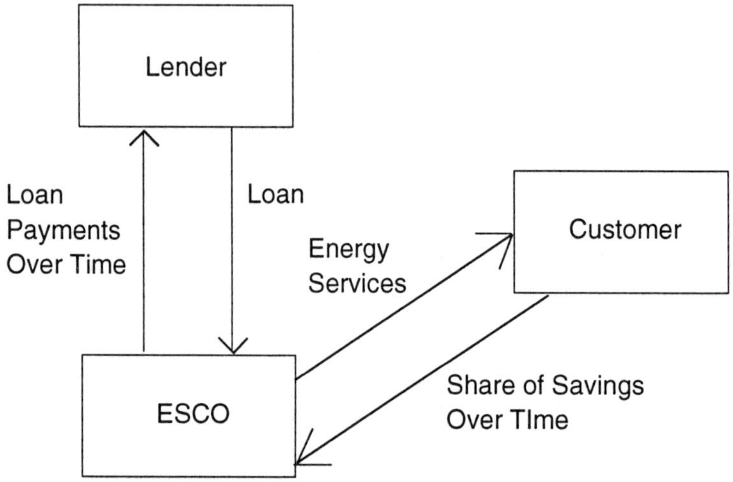

Figure 1-2. Full-Recourse Project Financing

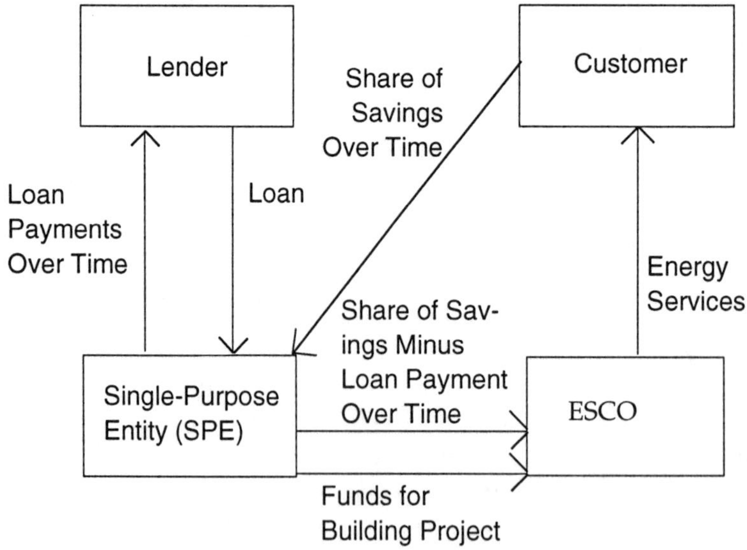

Figure 1-3. Non-Recourse Project Financing

ASSET SALE/BUYBACK

As of this writing, the first steps for the deregulation of the utility industry are being taken. As the definition of an energy provider becomes broader, so does the concept of an energy service company. The ESCO of the future will not only help customers with lighting installations but will also purchase energy on behalf of its customers. In effect, the pre-deregulation concepts of "supply-side" and "demand-side" will blend. Some of the issues and challenges of today's utilities are discussed in Chapter 9.

Large energy users, such as industrials or hospitals, often find themselves devoting too much personnel and resources to their energy management needs. If these companies want to focus on their core competencies; making paper and pulp products or healing patients; for example, they may choose to sell their physical energy plant to an ESCO, which will then sell back the raw materials, such as chilled water or emergency electricity, to the customers. This concept is known as *chauffage.* From the perspective of a chief financial officer of a company, this method allows the company to earn a better return on its assets (ROA) because the physical plant assets no longer have to appear on the financial statements.

UTILITY REBATES

As mentioned earlier, deregulation is becoming a fact of life for utilities. In order to remain competitive, many utilities are not only expanding internationally and into other service markets but they are also merging with other similar companies. An unfortunate additional effect of deregulation is the reduction or elimination of demand-side management rebates.

Some rebates are paid by the utility up front, upon project installation, while others are based on the measured savings of the project over time. For example, as of this writing, Public Service Electric and Gas in New Jersey and New York continues to operate its performance-based Standard Offer DSM Program. Contact your local utility for any available rebate, before it disappears (either the rebate or the utility).

Chapter 2

Fundamentals of Performance Contracting

Shirley Hansen, Ph.D.

Since energy and the great need for energy efficiency hit the center of our radar screen in the 1970s, there has been growing recognition that using energy more efficiently is good for the economy and the environment. It's just good business. Yet survey after survey in the late 70s and into the 80s revealed that many organizations were putting off energy-efficiency work for one major reason: the lack of money. Within this context it was not surprising that an industry which would guarantee funds to do energy work out of *future* energy savings would have great appeal.

What emerged was performance contracting, a financial mechanism which uses future energy costs to finance the rebuilding of decrepit heating and cooling systems. It is based on the same concept that stands behind commercial paper: the fact that a future obligation to pay already has value today. The future obligation in performance contracting is not derived from servicing a debt, but rather from the known, plannable and unavoidable cost of heating, cooling and illuminating buildings or fueling industrial processes.

Administrators around the world are burdened with obsolete, money-hungry schools, hospitals, etc., built in the low-energy-cost era of the 1960s. But they cannot afford to re-equip these structures with the immensely improved energy-use technologies that have emerged in recent years. Unfortunately, they are compelled to continue to throw good money after bad to keep buildings in service.

By a kind of financial judo, performance contracting turns this grim prospect into an asset. Future utility bills can be discounted, like commercial debt, and the resulting cash can be used today for retrofits.

PERFORMANCE CONTRACTING IN RETROSPECT

In the late 1970s, Scallop Thermal, a division of Royal Dutch Shell, introduced the concept of using third-party financing to improve energy efficiency and cut operating costs in North America. Scallop offered a hospital in Philadelphia conditioned space for 90 percent of its current utility bill. By upgrading the mechanical system and implementing energy-efficient practices, Scallop was able to bring consumption well below the 90 percent level. Scallop paid for its services and made a profit from this difference.

These early energy financing agreements generally turned on each party receiving a percentage of the energy cost savings. The energy service company (ESCO) received a share to cover its costs and make a profit. The owner also received a share (as well as capital improvements) as an incentive to participate. Since each party received a share of the energy cost savings, this procedure became known as "shared savings."

During the life of the contract, the ESCO expected its percentage of the cost savings to cover all the costs it had incurred, plus a profit. This concept worked quite well as long as energy prices stayed the same or increased.

In the mid-1980s, energy prices began to drop. With lower prices, it took longer than predicted for the firm to recover its costs. Some firms could not meet their payments to their suppliers or financial backers. Companies closed their doors and, in the process, defaulted on their commitments to their shared-savings customers. Some suppliers tried to recover costs from the building owners. Lawsuits were filed and "shared savings" was in trouble.

From this tenuous thread, the "shared savings" industry survived, but its character changed dramatically. Those supplying the financial backing and/or equipment recognized the risks of basing contracts on future energy prices. Higher risks meant higher interest rates, if the money were even available. Insurance, which had been available to ESCOs for "shared savings," almost disappeared. By 1983, shared *cost* savings agreements had shrunk to approximately 5 percent of the market.

From the end user's point of view, it soon became apparent that the "energy payments" to the ESCO were unpredictable. All too often, the customers found themselves paying far more than expected for the opportunity. A customer paying 70 percent of a predicted $1 million annual savings for five years assumed the total payment would be about $3.5 million for the acquired equipment and services. If the savings were greater than expected or the price of energy went up, however, the costs could easily become $5-$7 million for the same equipment. Payment procedures became confrontational. In the final analysis, the only real benefit of the shared savings transactions were their "off balance" feature. This was attractive to customers who did not wish to incur more debt, for whatever reason.

Out of the "shared savings" confusion of the 1970s-1980s evolved a new financing mechanism with the guarantee centered on the reduction in the *amount* of energy consumed and the dollar savings calculated at current billing rates. Typically, the projected dollar savings were guaranteed to cover any of the associated debt service obligations of the owner. To avoid the risks associated with falling energy prices, ESCOs began setting an energy floor price below which money guarantees would not apply. A new term, performance contracting, emerged that embraced all guaranteed energy efficiency financing schemes, including shared savings and guaranteed savings.

The initial attraction to performance contracting was the financing help. Through the years, the ESCO expertise has become equally attractive. For example, a survey of school administrators in the United States as recently as 1988 found that the majority still thought work on the building envelope (e.g., added insulation, double glazing of windows, etc., was the most cost-effective energy saving measures to take). As shown in Figure 2-1, a U.S. Department of Energy analysis at the time showed that building envelope measures average the longest payback (over eight years) of those measures studied. In contrast, control measures paid back in roughly two years. In attempts to cut energy costs, school administrators were not investing money in the most effective place; they were waiting eight years to get the return on investment that they should have achieved in two years.

ESCOs have learned, sometimes at great expense, what works—and what doesn't!! The earlier temptation to invest in elaborate control systems, for example, no longer exists. Today, experienced ESCOs know exactly how sophisticated a system should be in a given facility—right down to the number of sensing points and control points that are really needed. That expertise can keep owners from investing valuable money in the wrong measures.

Today, performance contracting has matured into a viable and reli-

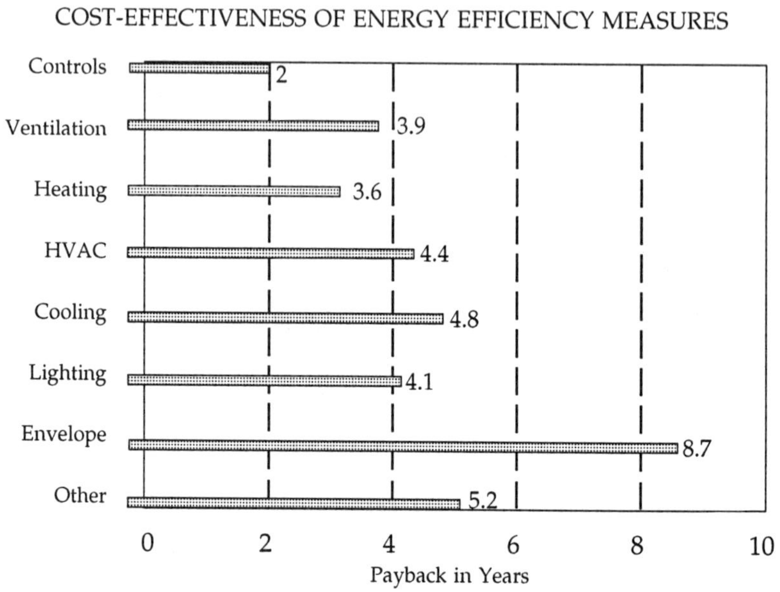

U.S. Department of Energy, 1989

Figure 2-1. Cost-Effectiveness Analysis for Buildings

able way of doing business in North America and many countries throughout the world. In the United States and Canada, federal law allows, even encourages, federal agencies to use performance contracting to cut operating costs. Over 4,000 U.S. school systems have some type of performance contract, and these school systems have as many as 800 buildings. Major manufacturers of energy-related equipment have ESCO services and divisions, utilities have developed or acquired ESCOs, and many regional ESCOs have grown from engineering firms, distributorships and mechanical contractors. It is conservative to say that in the United States there are at least ten major national ESCOs and another 40 operating on broad, regional bases. Local firms offering ESCO services undoubtedly exceed 100. In fact, the concept has become so popular there is great fear in the industry that "wannabe ESCOs" will begin offering services before they fully understand the complexity of the process and create credibility difficulties for the industry.

In 1994, the International Energy Agency went on record at the Conference on Energy Efficiency in Latin America to support and encourage the ESCO industry. Some version of guaranteed energy savings is now being practiced in France, Germany, Sweden, Spain, Italy, etc.

Some ESCO activity is reported in Poland and the Czech Republic has three projects underway. The U.S. Agency for International Development has been very active in fostering ESCO industries in Central and Eastern Europe as well as India. The multinational banks are increasingly supportive of the energy service company industry. The World Bank conducted a round table in September, 1994, to explore the energy-efficiency opportunities and assess the viability of creating criteria to foster an ESCO industry in a country as a condition of receiving a World Bank loan. The European Union's PHARE has begun funding performance contracting in countries which formerly had command-and-control economies.

TYPICAL ESCO SERVICES

Through the years, ESCO services have become more varied. It has become a customer-driven industry and the customer typically has a selection of ESCO services from which to choose. Services offered by an ESCO usually include:

- an investment-grade energy audit to identify energy and operational savings opportunities, assess risks and calculate cost-effectiveness of proposed measures over time;
- financing from its own resources or through arrangements with banks or other financing sources;
- the purchase, installation and maintenance of the installed energy-efficient equipment; possibly maintenance on all energy-consuming equipment;
- new equipment training of operations and maintenance (O&M) personnel;
- training of O&M personnel in energy-efficient practices;
- monitoring of the operations and energy savings, so reduced energy consumption and operating costs persist;
- measurement and savings verification; and

- a guarantee of the energy savings to be achieved.

The popularity of performance contracting rests on the many benefits it delivers. The highlights of the benefits it offers the customer and businesses, who offer ESCO services are addressed below.

CUSTOMER ADVANTAGES

The recipient of ESCO services can achieve many business benefits, including:

- an immediate upgrade of facilities and reduced operating costs—without any initial owner investment;
- access to the ESCO's energy-efficiency expertise. Through savings guarantees, ESCOs have learned from experience just how effective specific energy-efficiency measures are under specific conditions and how soon they can pay for themselves;
- positive cash flow—most projects generate savings that exceed the guarantee;
- the opportunity to use the money that would have been used for required upgrades to meet other needs;
- improved and more energy-efficient operations and maintenance;
- several normal business risks are assumed by the ESCO, including the guaranteed performance of the new equipment for the life of the contract (not just through a warranty period);
- a more comfortable, productive environment; and
- services paid for with the money the customer would have paid the utilities for wasted energy.

Performance contracting is probably the best and quickest way to be sure an organization is operating as efficiently as possible. It offers the customer a risk-shedding opportunity; however, risks do still exist.

In 1995-96, we have had the privilege of working with the City and County of San Francisco in its "Kilowatts to Megawatts" research and development project funded by the Urban Consortium.[1] A key component of this effort, ably led by Mr. John Deakin and Ms. Christine Vance, was to develop a decision model to aid other cities and counties in determining when to use their limited financial resources to do energy-efficiency work, when to outsource the work and which of the many outsourcing options available would best meet the municipalities' needs. A particularly challenging aspect of this work was the identification of specific risks associated with a broad range of energy-efficiency financing opportunities and the extent of those risks. The table on the following pages offers the general magnitude of those risks by option. Clearly, the specific magnitude of these risks will vary with the municipality's local conditions.

Risk-shedding always carries a price tag. It can be presumed, therefore, that the customer will typically pay more for the services provided through options on the right side of the table. It should be noted, however, that as the customer moves to the more comprehensive option, he or she also receives more services, has less administrative burden, and will probably achieve greater savings persistence. The costs associated with the "right-hand side of this menu" can also be mitigated by controlling the contributing factors listed. The greatest factor, however, could be the relative speed with which projects can be implemented, thus avoiding valuable dollars going up the smokestack while the customer tries to get more of the energy-efficiency work out of his or her organization's limited resources (options shown on the left in the table.)

In performance contracting the risks shed by the owner are largely assumed by the ESCO. Effective performance contracting then becomes a matter of risk assessment and management by the ESCO. The more accurately the ESCO assesses the risks and the more effectively it manages them, the greater the project benefits to all parties. Since performance contracting risks are primarily managed through the project's financial structure, effective risk management presents a major point of differentiation among ESCOs.

---

[1]Information on Urban Consortium projects, including "From Kilowatts to Megawatts," can be obtained from PTI (202) 626-2400.

Risk Assessment Matrix

Risk Factors	Relative Risk Level Associated with Financing Options — Option #1	#2	#3	#4	#5	#6	Major Variables	Mitigating Strategies
Equipment performance -warranty	3	3	3	3	3	1	Quality of specs Selection process Contract conditions	In-house/consultants' expertise Legal ability available
Technical expertise	4	4	4	3	2	1	Consultant qualifications	Selection process Negotiation
Audit quality; accuracy	3	3	3	3	3	2	Auditor ability Review capability Vender bias	Selection process Review a sample audit 3rd party validation Establish procedural criteria & scope of audit
Project management inadequate	3	3	3	3	3	2	Ability of PM or CM provided in-house and through project delivery option	Careful planning with DPW and clear definition of roles Service provider selection process
Construction/ installation	5	4	3	2	2	1	Vendor/subcontractor qualifications Contract provisions	Selection process Performance/payment bonds Legal ability
Maintenance & operations	5	5	5	5	3	1	Manpower In-house staff qualifications Training quality	Outsourcing Training; experience Selection process
Savings persistence	5	4	4	4	3	1	Varies by measure Administrative commitment O&M attitudes, training, experience	Contractual obligations Vendor selection ESCO selection Guarantees offered

Legend: N/A not applicable; 1 low risk; 2 low-medium risk; 3 medium risk; 4 medium-high risk; 5 high risk

Risk Factors	Relative Risk Level Associated with Financing Options — Option #1	#2	#3	#4	#5	#6	Major Variables	Mitigating Strategies
Savings verification instruments approach	3	3	3	3	3	1	Needs vary by measure/guarantee Accuracy	Costs paid for accuracy 3rd party validation
Establishing base year; baseline adjustments	3	3	3	3	3	4	Availability of historical data Clarity of formulas in the contract	Organizations energy management procedures; Combined legal& tech. expertise
Provisions & cost for M&V	3	3	3	3	3	5	Procedures accuracy	Instruments & approach used Cost
Cost of delay/ schedule adherence	5	4	3	2	2	1	Audit delays Financing/ESA Design delays Construction delays	Recognize costs incurred by delay; Consider penalty clauses when completion delayed; City expedite reviews, etc. when possible; Check past practice of firms during selection process
Hidden project costs	1	1	2	4	3	5	Margins, mark-ups Profit	Option 4-6; Reserve right to bid equipment; Open book pricing
Facility control problems	1	1	1	1	1	5	Contract provisions	Specify acceptable parameters in contract
Quality of maintenance training	5	5	5	5	2	1	Specification for maintenance trainer's abilities; cost	Selection of service provider to provide training
Intrusion/ interruption	4	4	4	4	4	5	Contract language—stipulate hours of installation	Close coordination with maintenance staff; careful attention to operation needs

Legend: N/A not applicable; 1 low risk; 2 low-medium risk; 3 medium risk; 4 medium-high risk; 5 high risk

ESCO RISK ASSESSMENT AND MANAGEMENT

Effective, experienced ESCOs have found a series of critical risk assessment components and management techniques, which include:

1) Prequalify the customer. Prequalification criteria should be established; customer organizational, technical and financial data should be carefully collected and validated.

2) Conduct an investment-grade audit. The typical energy audit gives a snapshot of predicted costs and savings. An investment-grade audit assesses administrative risks, operation and maintenance risks and the impact these risks will have on the measures' savings over time. An investment-grade audit also considers the time value of money. Projected payback calculations, for example, discount the value of the dollars saved each year for the life of the project. A typical four-year payback, for example, can easily become 5+ years in the real value of dollars saved, thus changing the dynamics of the cost-effectiveness calculations.

3) Establish sound base-year data. Base-year data are more than the average energy consumption over the last two to three years. They also consider what was happening in the facility (or industrial process) during that period. Hours of occupancy, level of occupancy, run times, etc., all become critical issues that must be verified prior to project implementation and signed off by the customer.

4) Secure a solid contract fair to all parties. Sometimes ESCOs get a little too zealous in managing their risks. The following paragraph, nearly a deal stopper, was put into a draft contract *in caps* as follows:

   IN NO EVENT SHALL [THE ESCO] BE LIABLE FOR ANY SPECIAL, INCIDENTAL, INDIRECT, SPECULATIVE, REMOTE, OR CONSEQUENTIAL DAMAGES ARISING FROM, RELATING TO, OR CONNECTED WITH THE WORK, THE SUPPORT SERVICES, EQUIPMENT, MATERIALS, OR ANY GOODS OR SERVICES PROVIDED HEREUNDER.

   After months of discussion, the paragraph was finally dropped and the project moved ahead.

5) Implement quality measures. Customers too often insist on quantity instead of quality. One federal procurement officer reportedly

told an ESCO the military base would rather have "15 Tempos than 5 Volvos." The agency may have reasons for needing 15 sets of wheels; but when the focus is energy efficiency, then the ESCO must look at parts, repairs, maintenance and savings persistence over time.

6) Operations and maintenance (O&M). An evaluation of the schools and hospitals federal energy-grants program after eight cycles revealed that in effective energy management programs up to 80 percent of the savings could be attributed to energy efficiency O&M practices. If ESCOs are to guarantee the savings from a measure, they must look beyond the equipment to its "care and feeding" over the life of the contract. If the ESCO does not perform the maintenance, it must: a) train the owner's staff; b) police the practices; and c) discount the predicted level of savings.

7) Measurement and Savings Verification. Once the recommended measures are known and approved, the ESCO and owner should cooperatively establish the level of accuracy and the associated costs they are willing to have the project carry. The question is: How much verification are the parties willing to pay for? When M&V costs come out of the project, they diminish the amount of equipment and other services provided and cut into the savings benefits.

8) Project management. The "business end" of the project, securing the savings, doesn't begin until the equipment is installed. Weak ESCOs tend to think of the project as "complete" once it's in the ground. Customers often think the ESCO is in business to sell equipment. Some of the big vendor-associated ESCOs actually lose money during equipment installation, but make up for it through the services and savings over the life of the project. A good project manager is an incredibly valuable part of the ESCO's project delivery.

When ESCOs have completed a thorough risk assessment and determined the procedures which will be needed to manage those risks, they then discount the savings to adjust for risk exposure. In other words, they never guarantee 100 percent of the expected savings.

PERFORMANCE CONTRACTING: A FINANCIAL TRANSACTION

Risk is managed through the financial structure. The ESCO's investment-grade audit will provide the information necessary to establish how much can be guaranteed and how much must be held back as a risk cushion. The financial structure has essentially three components:

(1) the guarantee, which covers design, acquisition, installation and the cost of money;

(2) the ESCO fee for services performed; and

(3) the positive cash flow (the savings in excess of the guarantee and ESCO fee).

The size of the ESCO fee will vary with the services provided and the risks it perceives and assumes. Some ESCOs put their fee in the guarantee package, others take it off the top of any savings above the guarantee and still others do some of each. Owners should realize that an ESCO that offers the customer "all of the positive cash flow" has all its fees in the guarantee—and the customer will pay finance charges on the ESCO fees for the life of the project.

Performance contracting has had its problems. Many of them in recent years, however, can be traced to the assumption, by fledgling ESCOs and customers alike, that performance contracting is a technical procedure. It is not. Technical issues are important; they are the foundation upon which the project is based.

On the other hand, lawyers would have you believe it's all managed through the contract. It is not. A good contract is the basis of a good project. But make no mistake: **Performance contracting is primarily a financial transaction.** The ultimate performance contract decisions cannot be made in the boiler room; they must be made in the business office. Engineers and lawyers must also be in that business office when the decisions are made.

Performance contracting is a simple idea, but a complex process. If we are to get the maximum benefit out of performance contracting, it helps if financial officers have some understanding of engineering and legal issues. But it is absolutely crucial that lawyers and engineers understand and appreciate the fundamental role finance plays in performance contracting.

Chapter 3

The Energy Audit

In financing any energy project for facilities modernization it is important to establish a benchmark. It is important to know how the facility presently uses energy and what energy-conservation opportunities can be utilized in a cost-effective manner. The purpose of this chapter is to review the fundamentals of auditing as they apply to project financing.

The energy audit serves to identify all of the energy streams in a facility and to quantify energy use according to discrete functions.

An energy audit may be considered as similar to the monthly closing statement of an accounting system. One series of entries consists of amounts of energy which were consumed during the month in the form of electricity, gas, fuel oil, steam, and the second series lists how the energy was used; how much for lighting, in air conditioning, in heating, in process, etc. The energy audit process must be carried out accurately enough to identify and qualify the energy and cost savings that are likely to be realized through investment in an energy-saving measure.

TYPES OF ENERGY AUDITS

The simplest definition for an energy audit is as follows: An energy audit serves the purpose of identifying where a building or plant facility uses energy and identifies energy conservation opportunities.

There is a direct relationship between the cost of the audit (amount of data collected and analyzed) and the number of energy conservation opportunities to be found. Thus, a first distinction is made between cost of the audit which determines the type of audit to be performed.

The second distinction is made between the type of facility. For example, a building audit may emphasize the building envelope, lighting, heating, and ventilation requirements. On the other hand, an audit of an industrial plant emphasizes the process requirements.

Most energy audits fall into three categories or types: Walk-Through, Mini-Audit, or Maxi-Audit.

Walk-Through

This type of audit is the least costly and identifies preliminary energy savings. A visual inspection of the facility is made to determine maintenance and operation energy-saving opportunities plus collection of information to determine the need for a more detailed analysis.

Mini-Audit

This type of audit requires tests and measurements to quantify energy uses and losses and determine the economics for changes.

Maxi-Audit

This type of audit goes one step further than the mini-audit. It contains an evaluation of how much energy is used for each function such as lighting, process, etc. It also requires a model analysis, such as a computer simulation, to determine energy-use patterns and predictions on a year-round basis, taking into account such variables as weather data.

To accomplish an energy-audit survey it is necessary to clarify energy uses and losses. This chapter illustrates various types of instruments which can aid in the energy-audit survey.

MEASURING BUILDING LOSSES

Infrared energy is an invisible part of the electromagnetic spectrum. It exists naturally and can be measured by remote, heat-sensing equipment. Within the last four years lightweight, portable infrared systems became available to help determine energy losses. Differences in the infrared emissions from the surfaces of objects cause color variations to appear on the scanner. The hotter the object, the more infrared radiated. With the aid of an isotherm circuit the intensity of these radiation levels can be accurately measured and quantified. In essence the infrared scanning device is a diagnostic tool which can be used to determine building heat losses.

MEASURING ELECTRICAL SYSTEM PERFORMANCE

The ammeter, voltmeter, wattmeter, power factor meter and footcandle meter are usually required to do an electrical survey.

Footcandle meters measure illumination in units of footcandles through a light-sensitive barrier layer of cells contained within them. They are usually pocket-size and portable and are meant to be used as field instruments to survey levels of illumination. Footcandle meters differ from conventional photographic lightmeters in that they are color- and cosine-corrected.

MEASURING COMBUSTION SYSTEMS

To maximize combustion efficiency it is necessary to know the composition of the flue gas. By obtaining a good air-fuel ratio substantial energy will be saved.

The combustion electronic analyzer permits fast, close adjustments. The unit contains digital displays. A standard sampler assembly with probe allows for stack measurements through a single stack or breaching hole.

MEASURING HEATING, VENTILATION AND AIR-CONDITIONING (HVAC) SYSTEM PERFORMANCE

- Air Velocity Measurement
- Temperature Measurement
- Pressure Measurement (absolute and differential)
- Humidity Measurement

The type of instrumentation required will depend on what energy using systems you have and how much data are desired. A typical instrumentation list may include:

- Lightmeter, digital-type with hold button
- Tachometer, stroboscope-type without need for reflecting tape
- Multimeter, digital

- Clamp-on kW meter
- Stopwatch, digital
- Pocket thermometer, digital
- Conductivity or total-dissolved-solids meter
- Clamp-on flowmeter
- Infrared thermometer
- Electric power/demand analyzer
- Air temperature, relative humidity, dew point meter
- Air balance, air velocity, temperature, pressure
- Vibration pen
- Auto-focus camera with flash override
- Portable gas analyzers
- Data loggers
- Hand-held calculator
- Miscellaneous tools and equipment:
  - Flashlight, dentist mirror, measuring tape (50 or 100 feet), extension cord, toothbrush, tissue, alcohol cleaner, screwdrivers, pliers, wrenches and other tools

ENERGY-AUDIT FORMS

Typical forms to perform an audit are illustrated in Figures 3-1 through 3-4. Figure 3-1 is used to summarize annual energy usage. Figure 3-2 is used to find the annual Btu usage. Figure 3-3 is used to determine the building occupancy schedule and Figure 3-4 is used for a CFM audit. There are many other forms that can be developed for performing lighting audits and audits on building systems such as motors.

BUILDING ____________________ ____________________ YEAR

MONTH*	HEATING DEGREE DAYS	COOLING DEGREE DAYS	ELECTRICITY				OIL				NATURAL GAS				COAL ☐ WOOD ☐	PURCHASED STEAM ☐ OTHER____			TOTAL ENERGY COST
			QUANTITY KWH	COST (DOLLARS)			QUANTITY GALLONS	COST (DOLLARS)			QUANTITY MCF	COST (DOLLARS)			QUANTITY UNIT	COST (DOLLARS)			
				TOTAL $	$/KWH	$/MMBTU		$/GAL.	TOTAL $	$/MMBTU		TOTAL $	$/MCF	$/MMBTU		TOTAL $	$/UNIT	$/MMBTU	
1	2	3	4	5	6	7	8	9	10	11	12	13	14	15	16	17	18	19	20
JANUARY																			
FEBRUARY																			
MARCH																			
APRIL																			
MAY																			
JUNE																			
JULY																			
AUGUST																			
SEPTEMBER																			
OCTOBER																			
NOVEMBER																			
DECEMBER																			
ANNUAL TOTALS																			
ANNUAL AVERAGES																			

* Or comparable time period

ELECTRICITY = 3412 Btu/Kwh
GAS = 1030 Btu/CF

OIL: #2 = .139 MMBTU/gal
#4 = .150 MMBTU/gal
#5 = .152 MMBTU/gal
#6 = .153 MMBTU/gal

MCF = 1000 cubic feet of gas
MMBTU = one million Btu

Figure 3-1. Energy Management Form

1. *Gross Annual Fuel and Energy Consumption*

Line No.	*A*	*B* Conversion Factor		*C* Thousands of Btus/yr
1. Oil—Gallons		× 138 (1)	=	
		× 146 (2)	=	
		× 1.0 (3)	=	
2. Gas—Cubic Feet		× 0.8 (4)	=	
3. Coal—Short Tons		× 26,000	=	
4. Steam-Pounds × 10^3		× 900	=	
5. Propane Gas—lbs		× 21.5	=	
6. Electricity—kW.Hrs		× 3.413	=	
7. Total Btus × 10^3/yr				
8. Btus × 10^3/Yr/Per Square Foot of Floor Area (Line 7 + Figure 4, Line 7)				

Use for (1) No. 2 Oil; (2) No. 6 Oil; (3) Natural Gas; (4) Mfg. Gas

2. *Annual Fuel and Energy Consumption for Heating*

Line No.	*A*	*B* Conversion Factor		*C* Thousands of Btus/yr
		× 138 (1)	=	
9. Oil—Gallons		× 146 (2)	=	
		× 1.0 (3)	=	
10. Gas—Cubic Feet		× 0.8 (4)	=	
11. Coal—Short Tons		× 26,000	=	
12. Steam—Pounds × 10^3		× 900	=	
13. Propane Gas—lbs		× 21.5	=	
14. Electricity—kW.Hrs		× 3.413	=	
15. Total Btus ×			=	
16. Btus × 10^3/Yr Per Square Foot of Floor Area (Line 15 Line 7)				

3. *Annual Fuel and Energy Consumption for Domestic Hot Water*

Line No.	*A*	*B* Conversion Factor		*C* Thousands of Btus/yr
17. Oil—Gallons		× 138 (1)	=	
		× 146 (2)	=	
18. Gas—Cubic Feet		× 1.0 (3)	=	
		× 0.8 (4)	=	

Figure 3-2. Energy Use Audit Form

(Source: Guidelines for Saving Energy in Existing Buildings—*Building Owners and Operators Manual*, ECM-1)

Line No.	*A*	*B Conversion Factor*		*C Thousands of Btus/yr*
19. Coal—Short Tons	______	× 26,000	=	______
20. Steam—Pounds × 10^3	______	× 900	=	______
21. Propane Gas—lbs	______	× 21.5	=	______
22. Electricity kW.Hrs	______	× 3.413	=	______
23. Total Btus/Yr × 10^3				
24. Btus × 10^3Yr/Per Square Foot of Floor Area (Line 23 + Figure 4, Line 7)				______

4. *Annual Fuel and/or Energy Consumption for Cooling (Compressors & Chillers)*

Line No.	*A*	*B Conversion Factor*		*C Thousands of Btus/yr*
a) if absorption cooling				
		× 138 (1)	=	______
25. Oil—Gallons	______	× 146 (2)	=	______
	______	× 1.0 (3)	=	______
26. Gas—Cubic Feet	______	× 0.8 (4)	=	______
27. Coal—Short Tons	______	× 26,000	=	______
28. Steam—Pounds × 10^3	______	× 900	=	______
29. Propane Gas—lbs	______	× 21.5	=	______
30. Total Btus/yr × 10^3				
31. Btus × 10^3/Yr Per Square Foot of Floor Area (Line 30 + Figure 4, Line 7)				______
b) if electric cooling				
32. Electricity—kWh	______	× 3.413	=	______
33. Btus × 10^3/Yr Per Square Foot of Floor Area (Line 32 + Figure 4, Line 7)				______

5. *Estimated Annual Energy Consumption for Interior Lighting*

Line No.	*A*	*B Conversion Factor*		*C Thousands of Btus/yr*
34. kWh Fig. 10, Line 3 × Fig. 10, Line 33 (1)	______	× 3.413	=	______
35. Btus × 10^3/Yr/Per Square Foot of Floor Area (Fig. 10, Line 35 Col. C + Fig. 4, Line 7)				______

6. *Estimated Annual Electrical Energy Consumption for all Motors and Machines if Building and Hot Water are Not Electrically Heated: (1)*

36. Total kW Hrs ________ Less kW Hrs Lighting ______ = ______ kW Hrs
(Line 22 Col. A)

37. kW Hrs/Yr/Sq Ft floor area = ______________________________ (1)
(Line 37, Col. C + Fig. 4, Line 7)

38. Btus × 10^3/Yr/Sq Ft floor area = (Line 37) × 3.431 ____________________ (2)

(1) and (2) If building heat and hot water are electrically heated, deduct the kW Hrs/Yr per sq ft and Btus/Yr per sq ft for heating and hot water. (Lines 37 and 38)

Figure 3-2. Energy Use Audit Form (concluded)

DATE ______________ PREPARED BY: ______________________________ PROJECT NO. __________

JOB NAME:			WEEKDAYS			SATURDAY			SUNDAYS/HOLIDAYS		
AREA SERVED	GROSS SQ. FEET	SYSTEM SERVING	NO. OF OCCU-PANTS	OCCUPANCY TIME A.M.	P.M.	NO. OF OCCU-PANTS	OCCUPANCY TIME A.M.	P.M.	NO. OF OCCU-PANTS	OCCUPANCY TIME A.M.	P.M.

Figure 3-3. Building Occupancy Schedule
(Source: Certified Test & Balance Company, Inc., Chicago, Illinois)

JOB NAME ______________________ DATE ______________________

PREPARED BY ______________________ PROJECT NO. ______________________

SYSTEM NUMBER	DESIGN		ACTUAL		REMARKS:
	TOTAL CFM	OUTSIDE AIR	TOTAL CFM	OUTSIDE AIR	

NOTE: These readings should be obtained by traverse and O.S. setting kept on minimum position ONLY.

Figure 3-4. CFM Audit
(Source: Certified Test & Balance Company, Inc., Chicago, Illinois)

Chapter 4

Measurement and Verification of Savings from Energy Conservation Projects

Martin A. Mozzo, Jr., PE, CEM, CLEP

INTRODUCTION

Engineers traditionally have had a tough time selling energy conservation projects to their management and project financiers. They typically understand principles and concepts of science and engineering, and intuitively know how projects will save energy dollars. However, in order to grant financing for projects, the lender must be satisfied that the project will save money and, once the project is installed, it is indeed achieving, at the very least, the minimum energy savings. Engineering theory is meaningless, actual dollars saved is the bottom line.

In order to insure the success of energy conservation projects, engineers should always consider the use of metering techniques to measure and verify savings achieved from the project and in, circumstances where financing is used, it is almost absolutely mandatory to meter. It is recognized that metering systems add costs to the project, and it is also recognized that these costs do not result in additional savings; however, it should also be recognized that measurement and verification, if properly done, will support energy-savings claims. If it can't be measured, savings can't be verified.

This chapter describes what verification means. It also contains case histories of measurement and verification techniques which were actually used.

WHAT IS VERIFICATION?

First of all, verification of energy projects is NOT master metering. It does not include reading of a main electrical or gas meter before and after the installation of a project to determine what energy savings may have been achieved. It also does not include engineering calculations or models of the project. Such calculations or models are theoretical in nature and hard to verify in actual life.

What verification of energy projects is about is the engineering and installation of metering systems and the development of monitoring protocols which will prove energy savings resulting from the project. Additionally, if the meters and protocol are allowed to remain in place and operational over the life of the project, then the savings over the life of the project will be verified.

Metering equipment, and the cost to monitor the project, is not cheap. Additionally, metering does not provide additional savings to the project. However, what should be strongly considered is the fact that a well-thought-out monitoring protocol coupled with a reasonably priced metering system will bring excellent returns to the engineer in proving his project. In developing the protocol and the system, the engineer should keep costs within reason, typically 3 to 5 percent of total costs, yet provide sufficient data to prove his savings. It has been this author's experience that all parties who have a vested interest in the project should actively participate in the engineering and development of the metering system and monitoring protocol. By this action, all parties have agreed how an energy project should be measured and savings verified.

The remainder of this chapter is devoted to case studies of successful measurement and verification activities for energy conservation projects.

CASE NUMBER ONE: LIGHTING PROJECT, MAJOR MEDICAL CENTER

Kilowatt (kW) savings for lighting projects are easy to determine. There have been numerous publications by manufacturers, utilities and standards labs which detail the wattage of commonly found fixtures. These are usually published in tables. For example, all utilities in New Jersey use a Luminaire Wattage Table for their lighting retrofit programs.

These tables contain wattages that standard-type fixtures draw when fully lit. For example, a four-lamp, 4' T-12 fixture with standard 40 watt lamps and two standard magnetic ballasts draws 188 watts. In contrast, a four-lamp, 4' T-8 fixture with 32 watt lamps and one electronic ballast draws 110 watts. These values have been determined through extensive testing by manufacturers and labs and the values published in these tables are accurate.

In determining the kW savings of a lighting project, the first step in the measurement and verification process is to conduct a fixture audit. This consists of an accurate count of all fixtures in the project, building by building, room by room. Quantities and types of fixtures are identified, and wattages for the existing fixtures are taken from standard tables. Using the audit data, a baseline kW can be calculated for the project.

Once the project has been constructed, a Post-Implementation audit is conducted. This audit will determine the number of fixtures retrofitted, the retrofit that was installed in the project, the kW associated with the retrofit and finally the kW savings achieved through the implementation of the project.

This project is a lighting retrofit of approximately 5,500 fixtures in a large, active hospital. kW savings were determined from the Luminaire Wattage Tables and an accepted audit of the quantity and content of the project. The next step was to determine kWh savings that were being achieved with an acceptable level of confidence and within reasonable costs. The lighting system throughout the hospital was 120 volts. Needless to say, the electrical system was mixed, that is, it was common to find coffee machines, electric heaters, and even copiers on lighting circuits. Lighting could not be segregated from other uses.

What was accepted in this case was a digital metering box. This box is capable of identifying event changes such as lights in an on or off condition. The measuring device is a photocell or a current transformer which is capable of sensing either light or current, but not a specific quantity of either. Thus, the device reads either only an on or off condition. The box has the capacity for measuring 32 distinct points. Additional boxes can be installed if more points are necessary.

To determine how to meter this project, areas were identified by use such as offices, hallways, laboratories, etc. The next step was to determine the expected hours of operation for each area. The objective of this classification is to develop areas and hours of operation categories which accurately portray the operation of the building; however, it must be recognized that each additional measurement point represents additional cost. The engineer should weigh the benefit of a large number of use categories which increases accuracy versus the cost of many points.

Once use and hour of operations classifications were identified, groupings were established. For example, Group A is Hallways, 24 hours/day, 365 days/year; Group B is Hallways, 12 hours/day, 6 days/week; Group C is Offices, 8 hours/day, 5 days/week. Once all the groupings were made, representative points or fixtures were selected which best represented each group, and a Monitoring Protocol was written and approved by all participating parties. After the Plan was approved, metering equipment was purchased and installed and monthly verification made. Appendix A shows a typical report for this type of an installation.

CASE NUMBER TWO— LIGHTING PROJECT, RESEARCH BUILDING

This project involved the retrofitting of approximately 15,000 lighting fixtures in four buildings. The buildings were all multistoried research facilities and consisted of offices, hallways and laboratories. Existing lighting consisted of fluorescent fixtures with 34-watt, energy-saving lamps and a mixture of standard and energy-saving magnetic ballasts. New lighting technologies were used in the retrofit, namely T-8 lamps and electronic ballasts. Through a detailed energy audit prior to the retrofit, a baseline kW consumption was established and agreed to by all participants. After the retrofit, a second audit was accomplished, and a post-retrofit kW was established and a kW savings agreed upon.

Determining hours of operation was difficult in this project. As previously stated, this project is in a research facility with many laboratories. As is typical in this type of operation, determining hours of operation is difficult. Many individuals working in this facility were busy on research projects, projects which could consume many hours. As a result, it was common to see a lab being in operation virtually around the clock during key phases of an experiment, followed by long periods of non-occupancy during lull periods. In order to accurately portray the hours of operation for such a facility with its hundreds of labs, many 32-point digital meters would have to have been purchased and installed. This was not economically justifiable, and another monitoring protocol had to be not only developed, but accepted by all parties.

During review of the lighting system of the research facility, it was discovered that each building had its own electrical substation, and each building had dedicated 277-volt lighting risers. Further investigation confirmed indeed that each riser fed power for lighting only; there were no parasitic loads on any of the risers. A data logger was installed on the lighting riser for each of the buildings early on in the project to evaluate electrical consumption patterns. These loggers measure current consumed versus time. After a couple of weeks, each logger was downloaded to a laptop computer and the data were analyzed.

Each building in this project showed the same consistent pattern. Figure 4-1 shows a typical weekly electrical consumption in one of the buildings. In the figure, there are five peak periods which correspond to Monday to Friday daytime lighting needs. Each weekday, the lighting system is turned on to a consistent level of consumption and basically remains at that level throughout the day. As night approaches, lighting is turned off, first to a cleaning level, then to a security level. Saturday and Sunday show a lighting consumption above the security level, though not at the weekday peak.

Figure 4-2 shows a typical daily weekday pattern. Lighting consumption increases to a high level beginning from 7:00-8:00 AM. It remains at this level until about 5:00-6:00 PM, when it decreases as workers depart. The early evening hours, until about 11:00 PM, show a reduced load, probably for a cleaning crew, and sometime after midnight, lights are turned down to security levels. Because all the buildings showed the same, consistent patterns, it was mutually decided to develop a monitoring protocol using this relatively inexpensive metering system.

Loggers were installed on the lighting risers for each building, and current is recorded every thirty (30) minutes. The following steps are used to determine hours of operation. A sample of a few days' worth of data is included in Appendix B. First, current readings from each leg of the lighting riser are placed into a table and summed by time period. Second, a maximum current draw for the riser during the month is determined. An assumption is made here that at the time of maximum current consumption, all the lights are on. Third, for each time interval, a percent of maximum is calculated. This will determine the time equivalents that the lights are on. For example, the percent of maximum for Thursday, September 2nd at midnight is 33.36%. From this we can state that all of

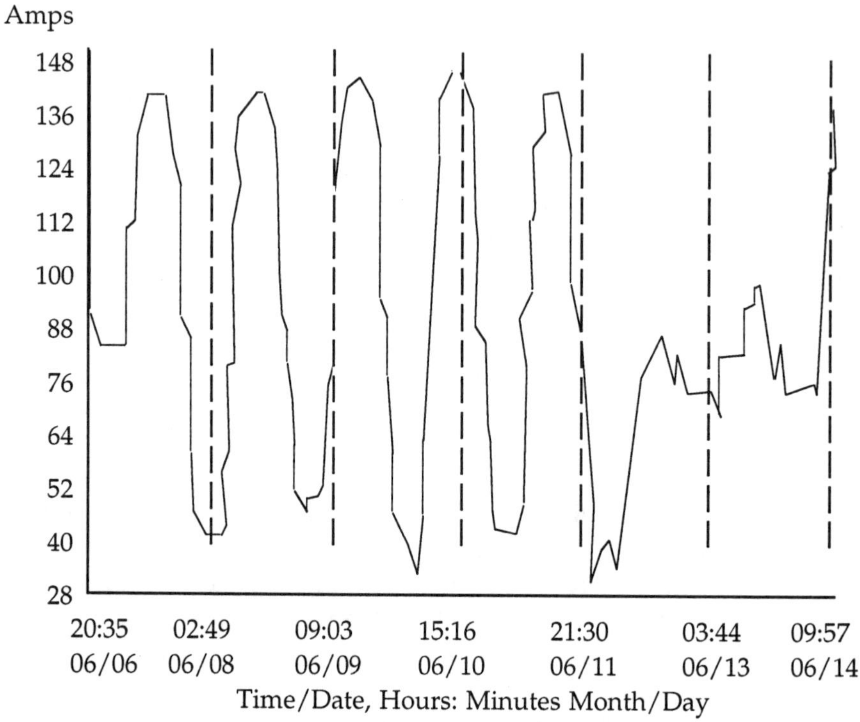

Figure 4-1

the lights were on 33.36% of that 30-minute interval, or approximately 10 minutes. This calculation is performed for each half-hour interval throughout the month. Kilowatt-hour savings are then obtained by multiplying the kW saved by the equivalent full-time hours the lights were on in the month. This methodology has proven to be accurate in several projects and has been accepted by all parties responsible for the project.

CASE NUMBER THREE—PROCESS PROJECT

In this project, process changes were to be made. Essentially, the project consists of a black box within which certain process changes were to be made. Electrical power is input to the black box at 480 volts. During baseline measurements, it was determined that the power consumption remained the same and production rates had no impact on electrical consumption. The process changes, however, did change electrical con-

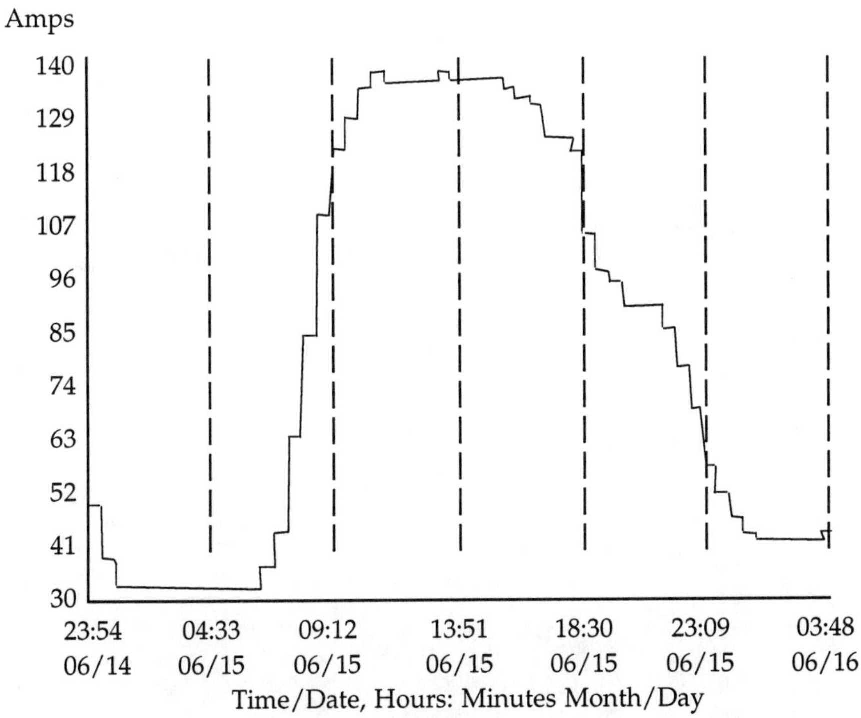

Figure 4-2

sumption.

After study of the process contained within the black box, it was determined that metering equipment which would measure input power to the black box would suffice. Meters that met utility standards were selected and installed. Initial measurements from the meter were used to establish baseline conditions. Electrical consumption essentially remained at a constant level as long as there was power to the black box; that is, the process was either in a production state or a ready-to-produce state. Figure 4-3 shows a typical baseline power consumption curve for this process.

After installation of the energy project, electrical power consumption was measured on a continual basis, and post-implementation curves were developed (see Figure 4-4). Kilowatt savings for the project is the difference between the baseline curves and the post-implementation curve. Kilowatt-hour savings is the sum of the kW savings for each hour in the time period.

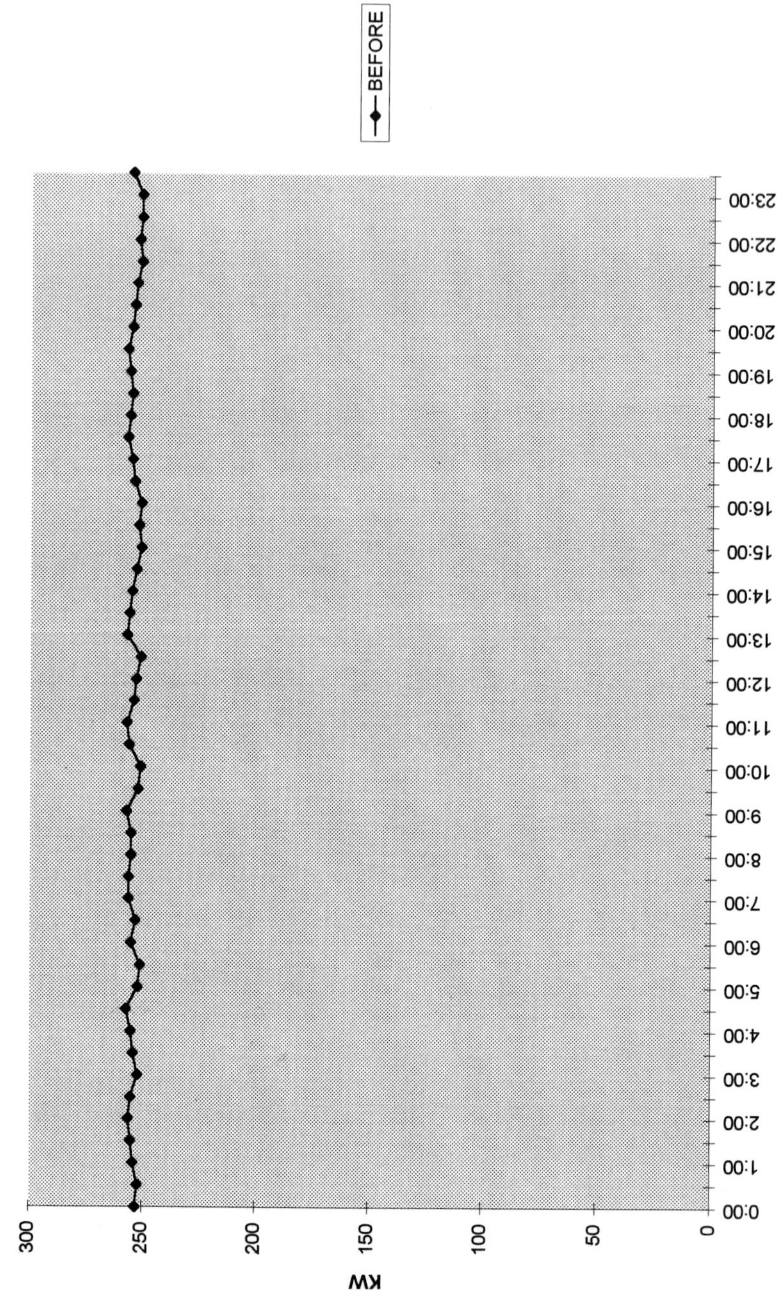
BEFORE
KW
300
250
200
150
100
50
0
0:00
1:00
2:00
3:00
4:00
5:00
6:00
7:00
8:00
9:00
10:00
11:00
12:00
13:00
14:00
15:00
16:00
17:00
18:00
19:00
20:00
21:00
22:00
23:00

Figure 4-3

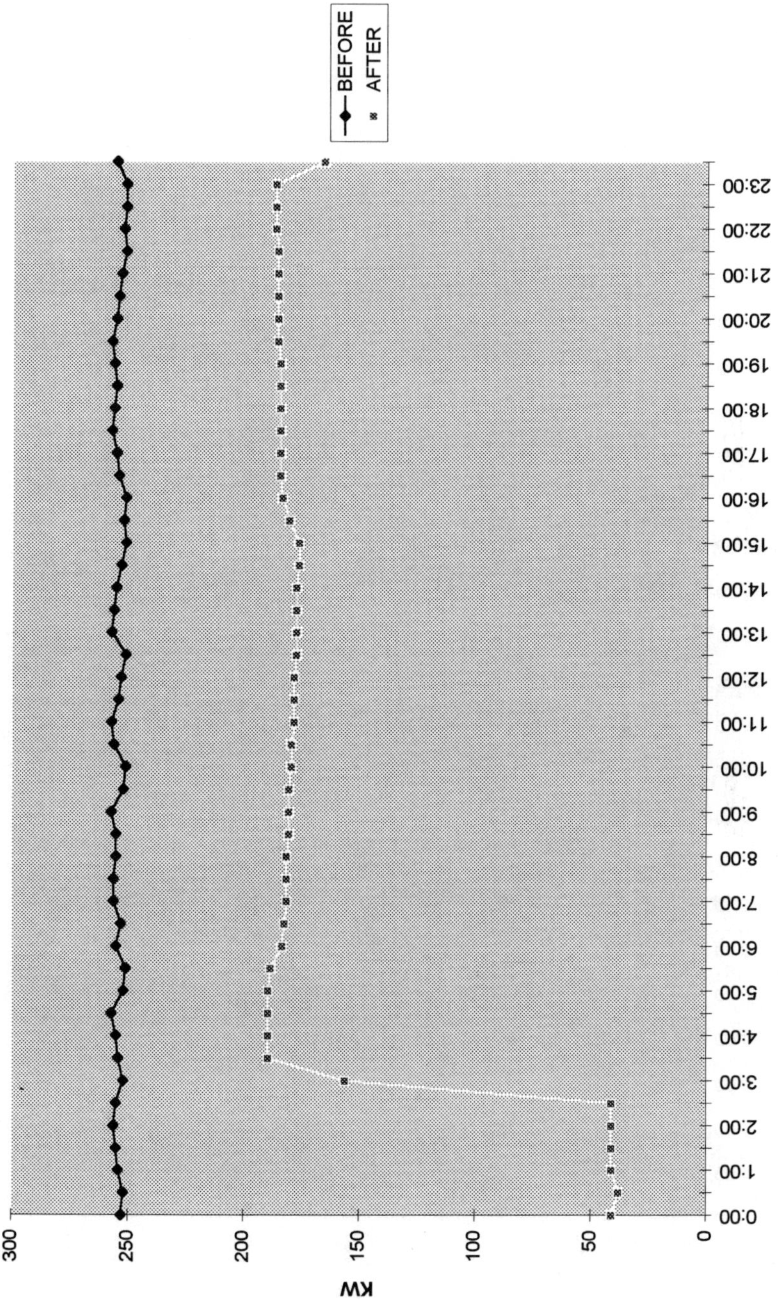
BEFORE
AFTER
KW
300
250
200
150
100
50
0
0:00
1:00
2:00
3:00
4:00
5:00
6:00
7:00
8:00
9:00
10:00
11:00
12:00
13:00
14:00
15:00
16:00
17:00
18:00
19:00
20:00
21:00
22:00
23:00

Figure 4-4

CASE NUMBER FOUR: VARIABLE FREQUENCY DRIVES

In this project, HVAC fan motors are run at a constant load with the flow controlled by outlet dampers. Baseline consumption can be a one-time measurement of kW consumed. The energy-saving retrofit project is to install variable frequency drives on the fans using static pressure in the ductwork to control fan speed. Essentially, as static pressure exceeds the set point, the VFD will slow the fan speed, thus dropping the fan output. As the speed of the fan, and thus the motor, drops, energy savings are achieved. Fan laws specify that power varies as a function of the fan speed cubed. Thus, if the fan speed is reduced by one half, energy savings of as much as 7/8 can be realized.

This project has an Energy Management System (EMS) installed for control of the HVAC system. The EMS also included metering devices to measure kW. Continuous kW readings over time, as compared to the baseline kW consumption, will provide the kWh savings achieved through the installation of the VFDs.

SUMMARY

In order to assure the success of energy projects and their financing strategies, all parties must have confidence in the savings levels achieved. The only sure way to do this is through measurement and verification. Metering equipment does not come cheap; however, in the design of energy projects, engineers should design metering systems which are cost-effective and provide a comfort level of savings achieved during the operation of the project. The success of energy conservation projects is dependent upon how well savings verification is achieved and maintained throughout the life of the project.

APPENDIX A

			ABC HOSPITAL KWH SAVED MAIN BUILDING Apr-96 TOTAL HOURS: 720 ON PEAK HOURS: 280 OFF PEAK HOURS: 440						
CIRCUIT NO.	BUILDING	FLOOR	LOCATION	KW SAVED	ON PEAK HOURS	ON PEAK KWH SAVED	OFF PEAK HOURS	OFF PEAK KWH SAVED	TOTAL KWH SAVED
1	1	G	ELEVATOR LOBBY	24.092	280	6,746	438	10,552	17,298
2	1	B	BIO-MED ENG'G OFFICE	8.145	223	1,816	83	676	2,492
3	1	G	HEALTH SERVICES OFFICE	15.284	165	2,522	29	443	2,965
4	1	3	HALL ROOM 338-PORTER	5.830	280	1,632	183	1,067	2,699
5	1	5	HALL ROOM 521-528	3.446	279	961	278	958	1,919
6	1	6	ELEVATOR LOBBY	2.616	241	630	386	1,010	1,640
7	2	G	HALLWAY	9.965	280	2,790	438	4,365	7,155
8	2	G	PHYSICAL THERAPY	16.615	272	4,519	219	3,639	8,158
9	2	1	DENTAL WIATING ROOM	13.166	264	3,476	39	513	3,989
10	2	B	HALLWAY RADIATION	1.587	269	427	145	230	657
11	2	5	ELEVATOR LOBBY	7.953	271	2,155	180	1,432	3,587
12	2	2	HALLWAY	2.235	214	478	193	431	910
13	2	5	STAFF LOUNGE	10.408	258	2,685	421	4,382	7,067
14	1	B	PHARMACY	13.094	280	3,666	438	5,735	9,401
15	3	G	CHEMISTRY	17.910	280	5,015	438	7,845	12,859
16	3	1	LIBRARY	19.537	263	5,138	45	879	6,017
17	3	2	NURSES STATION	3.368	280	943	438	1,475	2,418
18	3	2	HALL RMS 221-229	11.248	278	3,127	324	3,644	6,771
19	3	1	HALLWAY	1.386	280	388	438	607	995
20	3	5	RENAL DIALYSIS	14.918	196	2,924	28	418	3,342
21	4	1	RADIOLOGY RECEPTION	15.074	280	4,221	416	6,271	10,492
22	4	3	NURSURY	5.227	123	643	346	1,809	2,451
23	5	G	CENTRAL SUPPLY	4.276	280	1,197	438	1,873	3,070
24	5	G	MEDICAL RECORDS	15.219	280	4,261	365	5,555	9,816
25	5	2	OPERATING ROOM	10.101	246	2,485	113	1,141	3,626
26	5	4	ADMIN OFFICE	7.237	204	1,476	28	203	1,679
27	6	B	MAIN LAUNDRY ROOM	5.265	228	1,200	132	695	1,895
28	7	B	BOILER ROOM	2.253	280	631	438	987	1,618
29	5	G	MEDICAL RECORDS HALL	1.415	280	396	438	620	1,016
30	4	2	DAR	10.887	280	3,048	145	1,579	4,627
31	5	1	ADMITTING	11.242	238	2,676	383	4,306	6,981
32	ALL	ALL	EXIT SIGNS	8.736	280	2,446	440	3,844	6,290
			TOTALS	299.735		76,722		79,182	155,903

APPENDIX B

ABC CORPORATION
BUILDING 1, MAIN LIGHTING RISER
SEPTEMBER 1993 DATA

TIME INTERVAL: 30 MINUTES
UNITS: Amps
MAX AMPS: 491.05

TIME	DATE	DAY OF WEEK	READING #1	READING #2	READING #3	TOTAL	PERCENT OF MAX
0:00	2-Sep	THURSDAY	47.60	52.86	63.37	163.83	33.36
0:30	2-Sep	THURSDAY	44.10	47.60	63.37	155.07	31.58
1:00	2-Sep	THURSDAY	44.10	47.60	63.37	155.07	31.58
1:30	2-Sep	THURSDAY	44.10	47.60	63.37	155.07	31.58
2:00	2-Sep	THURSDAY	44.10	47.60	63.37	155.07	31.58
2:30	2-Sep	THURSDAY	44.10	47.60	63.37	155.07	31.58
3:00	2-Sep	THURSDAY	44.10	47.60	63.37	155.07	31.58
3:30	2-Sep	THURSDAY	44.10	47.60	63.37	155.07	31.58
4:00	2-Sep	THURSDAY	44.10	47.60	63.37	155.07	31.58
4:30	2-Sep	THURSDAY	44.10	47.60	63.37	155.07	31.58
5:00	2-Sep	THURSDAY	44.10	47.60	61.61	153.32	31.22
5:30	2-Sep	THURSDAY	52.86	58.11	65.12	176.09	35.86
6:00	2-Sep	THURSDAY	75.63	84.38	89.64	249.64	50.84
6:30	2-Sep	THURSDAY	91.39	94.89	106.77	293.05	59.68
7:00	2-Sep	THURSDAY	106.77	108.40	116.53	331.70	67.55
7:30	2-Sep	THURSDAY	116.53	137.67	132.79	386.99	78.81
8:00	2-Sep	THURSDAY	124.66	155.55	149.05	429.26	87.42
8:30	2-Sep	THURSDAY	134.42	162.06	157.18	453.65	92.38
9:00	2-Sep	THURSDAY	132.79	163.68	163.68	460.16	93.71
9:30	2-Sep	THURSDAY	132.79	165.31	163.68	461.78	94.04
10:00	2-Sep	THURSDAY	131.16	166.94	165.31	463.41	94.37
10:30	2-Sep	THURSDAY	131.16	166.94	165.31	463.41	94.37
11:00	2-Sep	THURSDAY	131.16	166.94	165.31	463.41	94.37
11:30	2-Sep	THURSDAY	131.16	166.94	163.68	461.78	94.04
12:00	2-Sep	THURSDAY	131.16	166.94	163.68	461.78	94.04
12:30	2-Sep	THURSDAY	131.16	168.56	165.31	465.03	94.70
13:00	2-Sep	THURSDAY	136.04	168.56	165.31	469.91	95.70
13:30	2-Sep	THURSDAY	140.92	168.56	165.31	474.79	96.69
14:00	2-Sep	THURSDAY	139.29	170.19	165.31	474.79	96.69
14:30	2-Sep	THURSDAY	131.16	171.81	165.31	468.29	95.36
15:00	2-Sep	THURSDAY	131.16	173.44	163.68	468.29	95.36
15:30	2-Sep	THURSDAY	137.67	171.81	157.18	466.66	95.03
16:00	2-Sep	THURSDAY	137.67	168.56	155.55	461.78	94.04
16:30	2-Sep	THURSDAY	126.29	162.06	149.05	437.39	89.07
17:00	2-Sep	THURSDAY	118.16	144.17	139.29	401.62	81.79
17:30	2-Sep	THURSDAY	108.40	124.66	126.29	359.34	73.18
18:00	2-Sep	THURSDAY	103.52	110.03	118.16	331.70	67.55
18:30	2-Sep	THURSDAY	65.12	72.12	84.38	221.62	45.13
19:00	2-Sep	THURSDAY	59.86	68.62	79.13	207.61	42.28
19:30	2-Sep	THURSDAY	72.12	82.63	93.14	247.89	50.48
20:00	2-Sep	THURSDAY	59.86	70.37	75.63	205.86	41.92
20:30	2-Sep	THURSDAY	56.36	68.62	72.12	197.10	40.14
21:00	2-Sep	THURSDAY	56.36	63.37	66.87	186.60	38.00
21:30	2-Sep	THURSDAY	56.36	59.86	66.87	183.09	37.29
22:00	2-Sep	THURSDAY	68.62	77.38	84.38	230.38	46.92
22:30	2-Sep	THURSDAY	61.61	68.62	77.38	207.61	42.28
23:00	2-Sep	THURSDAY	80.88	87.88	98.39	267.16	54.41
23:30	2-Sep	THURSDAY	68.62	73.87	84.38	226.88	46.20

Chapter 5

Understanding Rate-of-Return Analysis

An investment in an energy project must meet the desired rate of return of the lender or investor. This chapter reviews the basics of determining the true economics of the project.

USING THE PAYBACK PERIOD METHOD

The payback period is the time required to recover the capital investment out of the earnings or savings. This method ignores all savings beyond the payback years, thus penalizing projects that have long-life potentials for those that offer high savings for a relatively short period.

The payback period criterion is used when funds are limited and it is important to know how fast dollars will come back. The payback period is simply computed as:

$$\text{Payback period} = \frac{\text{initial investment}}{\text{after–tax savings}} \qquad \text{(Eq. 5-1)}$$

The energy manager who must justify energy equipment expenditures based on a payback period of one year or less has little chance for long-range success. Some companies have set higher payback periods for energy utilization methods. These longer payback periods are justified on the basis that:

- Fuel pricing will increase at a higher rate than the general inflation rate.
- The "risk analysis" for not implementing energy utilization measures may mean loss of production and losing a competitive edge.

USING LIFE-CYCLE COSTING

Life-cycle costing is an analysis of the total cost of a system, device, building, machine, etc., over its anticipated useful life. The name is new but the subject has, in the past, gone by such names as "engineering economic analysis" or "total owning and operating cost summaries."

Life-cycle costing has brought about a new emphasis on the comprehensive identification of all costs associated with a system. The most commonly included costs are initial in-place cost, operating costs, maintenance costs, and interest on the investment. Two factors enter into appraising the life of the system—namely, the expected physical life and the period of obsolescence. The lesser factor is governing time period. The effect of interest can then be calculated by using one of several formulas which take into account the time value of money.

When comparing alternative solutions to a particular problem, the system showing the lowest life-cycle cost will usually be the first choice (performance requirements are assessed as equal in value).

Life-cycle costing is a tool in value engineering. Other items, such as installation time, pollution effects, aesthetic considerations, delivery time, and owner preferences will temper the rule of always choosing the system with the lowest life-cycle cost. Good overall judgement is still required.

The life-cycle cost analysis still contains judgement factors pertaining to interest rates, useful life, and inflation rates. Even with the judgement element, life-cycle costing is the most important tool in value engineering, since the results are quantified in terms of dollars.

As the price for energy changes, and as governmental incentives are initiated, processes or alternatives which were not economically feasible will be considered. This chapter will concentrate on the principles of the life-cycle cost analysis as they apply to energy conservation decision making.

THE TIME VALUE OF MONEY

Most energy-saving proposals require the investment of capital to accomplish them. By investing today in energy conservation, yearly operating dollars over the life of the investment will be saved. A dollar in hand today is more valuable than one to be received at some time in the future. For this reason, a *time value* must be placed on all cash flows into and out of the company.

Money transactions are thought of as a cash flow to or from a company. Investment decisions also take into account alternate investment opportunities and the minimum return on the investment. In order to compute the rate of return on an investment, it is necessary to find the interest rate which equates payments outgoing and incoming, present and future. The method used to find the rate of return is referred to as *discounted cash flow.*

INVESTMENT DECISION-MAKING

To make investment decisions, the energy manager must follow one simple principle: Relate annual cash flows and lump-sum deposits to the same time base. The six factors used for investment decision making simply convert cash from one time base to another; since each company has various financial objectives, these factors can be used to solve *any* investment problem.

Single-Payment Compound Amount—F/P

Use the F/P factor to determine future amount F that present sum P will accumulate at i percent interest, in n years. If P (present worth) is known, and F (future worth) is to be determined, use Equation 5-2.

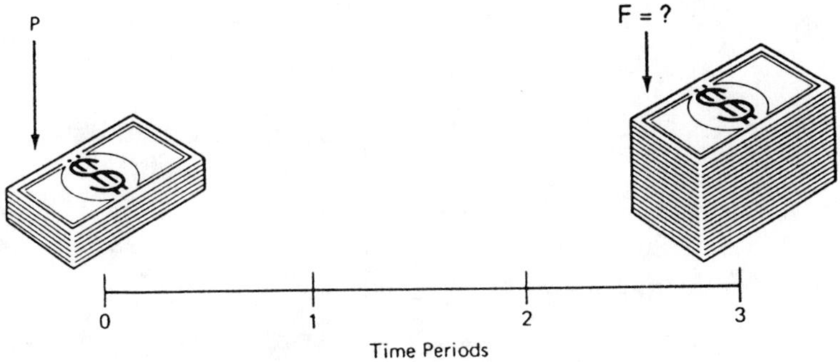

Figure 5-1. Single-Payment Compound Amount (F/P)

$$F = P \times (1 + i)^n \qquad \text{(Eq. 5-2)}$$

$$F/P = (1 + i)^n \qquad \text{(Eq. 5-3)}$$

Table 5-1. 10% Interest Factors

Period n	Single-payment compound-amount F/P	Single-payment present-worth P/F	Uniform-series compound-amount F/A	Sinking-fund payment A/F	Capital recovery A/P	Uniform-series present-worth P/A
	Future value of $1 $(1+i)^n$	Present value of $1 $\frac{1}{(1+i)^n}$	Future value of uniform series of $1 $\frac{(1+i)^n-1}{i}$	Uniform series whose future value is $1 $\frac{i}{(1+i)^n-1}$	Uniform series with present value of $1 $\frac{i(1+i)^n}{(1+i)^n-1}$	Present value of uniform series of $1 $\frac{(1+i)^n-1}{i(1+i)^n}$
1	1.100	0.9091	1.000	1.00000	1.10000	0.909
2	1.210	0.8264	2.100	0.47619	0.57619	1.736
3	1.331	0.7513	3.310	0.30211	0.40211	2.487
4	1.464	0.6830	4.641	0.21547	0.31147	3.170
5	1.611	0.6209	6.105	0.16380	0.26380	3.791
6	1.772	0.5645	7.716	0.12961	0.22961	4.355
7	1.949	0.5132	9.487	0.10541	0.20541	4.868
8	2.144	0.4665	11.436	0.08744	0.18744	5.335
9	2.358	0.4241	13.579	0.07364	0.17364	5.759
10	2.594	0.3855	15.937	0.06275	0.16275	6.144
11	2.853	0.3505	18.531	0.05396	0.15396	6.495
12	3.138	0.3186	21.384	0.04676	0.14676	6.814
13	3.452	0.2897	24.523	0.04078	0.14078	7.103
14	3.797	0.2633	27.975	0.03575	0.13575	7.367
15	4.177	0.2394	31.772	0.03147	0.13147	7.606
16	4.595	0.2176	35.950	0.02782	0.12782	7.824
17	5.054	0.1978	40.545	0.02466	0.12466	8.022
18	5.560	0.1799	45.599	0.02193	0.12193	8.201
19	6.116	0.1635	51.159	0.01955	0.11955	8.365
20	6.727	0.1486	57.275	0.01746	0.11746	8.514
21	7.400	0.1351	64.002	0.01562	0.11562	8.649
22	8.140	0.1228	71.403	0.01401	0.11401	8.772
23	8.954	0.1117	79.543	0.01257	0.11257	8.883
24	9.850	0.1015	88.497	0.01130	0.11130	8.985
25	10.835	0.0923	98.347	0.01017	0.11017	9.077
26	11.918	0.0839	109.182	0.00916	0.10916	9.161
27	13.110	0.0763	121.100	0.00826	0.10826	9.237
28	14.421	0.0693	134.210	0.00745	0.10745	9.307
29	15.863	0.0630	148.631	0.00673	0.10673	9.370
30	17.449	0.0673	164.494	0.00608	0.10608	9.427
35	28.102	0.0356	271.024	0.00369	0.10369	9.644
40	45.259	0.0221	442.593	0.00226	0.10226	9.779
45	72.890	0.0137	718.905	0.00139	0.10139	9.863
50	117.391	0.0085	1163.909	0.00086	0.10086	9.915
55	189.059	0.0053	1880.591	0.00053	0.10053	9.947
60	304.482	0.0033	3034.816	0.00033	0.10033	9.967
65	490.371	0.0020	4893.707	0.00020	0.10020	9.980
70	789.747	0.0013	7887.470	0.00013	0.10013	9.987
75	1271.895	0.0008	12708.954	0.00008	0.10008	9.992
80	2048.400	0.0005	20474.002	0.00005	0.10005	9.995
85	3298.969	0.0003	32979.690	0.00003	0.10003	9.997
90	5313.023	0.0002	53120.226	0.00002	0.10002	9.998
95	8556.676	0.0001	85556.760	0.00001	0.10001	9.999

Table 5-2. 12% Interest Factors

Period n	Single-payment compound-amount F/P	Single-payment present-worth P/F	Uniform-series compound-amount F/A	Sinking-fund payment A/F	Capital recovery A/P	Uniform-series present-worth P/A
	Future value of $1	Present value of $1	Future value of uniform series of $1	Uniform series whose future value is $1	Uniform series with present value of $1	Present value of uniform series of $1
	$(1 + i)^n$	$\frac{1}{(1+i)^n}$	$\frac{(1+i)^n-1}{i}$	$\frac{i}{(1+i)^n-1}$	$\frac{i(1+i)^n}{(1+i)^n-1}$	$\frac{(1+i)^n-1}{i(1+i)^n}$
1	1.120	0.8929	1.000	1.00000	1.12000	0.893
2	1.254	0.7972	2.120	0.47170	0.59170	1.690
3	1.405	0.7118	3.374	0.29635	0.41635	2.402
4	1.574	0.6355	4.779	0.20923	0.32923	3.037
5	1.762	0.5674	6.353	0.15741	0.27741	3.605
6	1.974	0.5066	8.115	0.12323	0.24323	4.111
7	2.211	0.4523	10.089	0.09912	0.21912	4.564
8	2.476	0.4039	12.300	0.08130	0.20130	4.968
9	2.773	0.3606	14.776	0.06768	0.18768	5.328
10	3.106	0.3220	17.549	0.05698	0.17698	5.650
11	3.479	0.2875	20.655	0.04842	0.16842	5.938
12	3.896	0.2567	24.133	0.04144	0.16144	6.194
13	4.363	0.2292	28.029	0.03568	0.15568	6.424
14	4.887	0.2046	32.393	0.03087	0.15087	6.628
15	5.474	0.1827	37.280	0.02682	0.14682	6.811
16	6.130	0.1631	42.753	0.02339	0.14339	6.974
17	6.866	0.1456	48.884	0.02046	0.14046	7.120
18	7.690	0.1300	55.750	0.01794	0.13794	7.250
19	8.613	0.1161	63.440	0.01576	0.13576	7.366
20	9.646	0.1037	72.052	0.01388	0.13388	7.469
21	10.804	0.0926	81.699	0.01224	0.13224	7.562
22	12.100	0.0826	92.503	0.01081	0.13081	7.645
23	13.552	0.0738	104.603	0.00956	0.12956	7.718
24	15.179	0.0659	118.155	0.00846	0.12846	7.784
25	17.000	0.0588	133.334	0.00750	0.12750	7.843
26	19.040	0.0525	150.334	0.00665	0.12665	7.896
27	21.325	0.0469	169.374	0.00590	0.12590	7.943
28	23.884	0.0419	190.699	0.00524	0.12524	7.984
29	26.750	0.0374	214.583	0.00466	0.12466	8.022
30	29.960	0.0334	241.333	0.00414	0.12414	8.055
35	52.800	0.0189	431.663	0.00232	0.12232	8.176
40	93.051	0.0107	767.091	0.00130	0.12130	8.244
45	163.988	0.0061	1358.230	0.00074	0.12074	8.283
50	289.002	0.0035	2400.018	0.00042	0.12042	8.304
55	509.321	0.0020	4236.005	0.00024	0.12024	8.317
60	897.597	0.0011	7471.641	0.00013	0.12013	8.324
65	1581.872	0.0006	13173.937	0.00008	0.12008	8.328
70	2787.800	0.0004	23223.332	0.00004	0.12004	8.330
75	4913.056	0.0002	40933.799	0.00002	0.12002	8.332
80	8658.483	0.0001	72145.692	0.00001	0.12001	8.332

Table 5-3. 15% Interest Factors

Period n	Single-payment compound-amount F/P	Single-payment present-worth P/F	Uniform-series compound-amount F/A	Sinking-fund payment A/F	Capital recovery A/P	Uniform-series present-worth P/A
	Future value of $1	Present value of $1	Future value of uniform series of $1	Uniform series whose future value is $1	Uniform series with present value of $1	Present value of uniform series of $1
	$(1+i)^n$	$\frac{1}{(1+i)^n}$	$\frac{(1+i)^n-1}{i}$	$\frac{i}{(1+i)^n-1}$	$\frac{i(1+i)^n}{(1+i)^n-1}$	$\frac{(1+i)^n-1}{i(1+i)^n}$
1	1.150	0.8696	1.000	1.00000	1.15000	0.870
2	1.322	0.7561	2.150	0.46512	0.61512	1.626
3	1.521	0.6575	3.472	0.28798	0.43798	2.283
4	1.749	0.5718	4.993	0.20027	0.35027	2.855
5	2.011	0.4972	6.742	0.14832	0.29832	3.352
6	2.313	0.4323	8.754	0.11424	0.26424	3.784
7	2.660	0.3759	11.067	0.09036	0.24036	4.160
8	3.059	0.3269	13.727	0.07285	0.22285	4.487
9	3.518	0.2843	16.786	0.05957	0.20957	4.772
10	4.046	0.2472	20.304	0.04925	0.19925	5.019
11	4.652	0.2149	24.349	0.04107	0.19107	5.234
12	5.350	0.1869	29.002	0.03448	0.18448	5.421
13	6.153	0.1625	34.352	0.02911	0.17911	5.583
14	7.076	0.1413	40.505	0.02469	0.17469	5.724
15	8.137	0.1229	47.580	0.02102	0.17102	5.847
16	9.358	0.1069	55.717	0.01795	0.16795	5.954
17	10.761	0.0929	65.075	0.01537	0.16537	6.047
18	12.375	0.0808	75.836	0.01319	0.16319	6.128
19	14.232	0.0703	88.212	0.01134	0.16134	6.198
20	16.367	0.0611	102.444	0.00976	0.15976	6.259
21	18.822	0.0531	118.810	0.00842	0.15842	6.312
22	21.645	0.0462	137.632	0.00727	0.15727	6.359
23	24.891	0.0402	159.276	0.00628	0.15628	6.399
24	28.625	0.0349	184.168	0.00543	0.15543	6.434
25	32.919	0.0304	212.793	0.00470	0.15470	6.464
26	37.857	0.0264	245.712	0.00407	0.15407	6.491
27	43.535	0.0230	283.569	0.00353	0.15353	6.514
28	50.066	0.0200	327.104	0.00306	0.15306	6.534
29	57.575	0.0174	377.170	0.00265	0.15265	6.551
30	66.212	0.0151	434.745	0.00230	0.15230	6.566
35	133.176	0.0075	881.170	0.00113	0.15113	6.617
40	267.864	0.0037	1779.090	0.00056	0.15056	6.642
45	538.769	0.0019	3585.128	0.00028	0.15028	6.654
50	1083.657	0.0009	7217.716	0.00014	0.15014	6.661
55	2179.622	0.0005	14524.148	0.00007	0.15007	6.664
60	4383.999	0.0002	29219.992	0.00003	0.15003	6.665
65	8817.787	0.0001	58778.583	0.00002	0.15002	6.666

Table 5-4. 20% Interest Factors

Period n	Single-payment compound-amount F/P	Single-payment present-worth P/F	Uniform-series compound-amount F/A	Sinking-fund payment A/F	Capital recovery A/P	Uniform-series present-worth P/A
	Future value of $1	Present value of $1	Future value of uniform series of $1	Uniform series whose future value is $1	Uniform series with present value of $1	Present value of uniform series of $1
	$(1+i)^n$	$\frac{1}{(1+i)^n}$	$\frac{(1+i)^n-1}{i}$	$\frac{i}{(1+i)^n-1}$	$\frac{i(1+i)^n}{(1+i)^n-1}$	$\frac{(1+i)^n-1}{i(1+i)^n}$
1	1.200	0.8333	1.000	1.00000	1.20000	0.833
2	1.440	0.6944	2.200	0.45455	0.65455	1.528
3	1.728	0.5787	3.640	0.27473	0.47473	2.106
4	2.074	0.4823	5.368	0.18629	0.38629	2.589
5	2.488	0.4019	7.442	0.13438	0.33438	2.991
6	2.986	0.3349	9.930	0.10071	0.30071	3.326
7	3.583	0.2791	12.916	0.07742	0.27742	3.605
8	4.300	0.2326	16.499	0.06061	0.26061	3.837
9	5.160	0.1938	20.799	0.04808	0.24808	4.031
10	6.192	0.1615	25.959	0.03852	0.23852	4.192
11	7.430	0.1346	32.150	0.03110	0.23110	4.327
12	8.916	0.1122	39.581	0.02526	0.22526	4.439
13	10.699	0.0935	48.497	0.02062	0.22062	4.533
14	12.839	0.0779	59.196	0.01689	0.21689	4.611
15	15.407	0.0649	72.035	0.01388	0.21388	4.675
16	18.488	0.0541	87.442	0.01144	0.21144	4.730
17	22.186	0.0451	105.931	0.00944	0.20944	4.775
18	26.623	0.0376	128.117	0.00781	0.20781	4.812
19	31.948	0.0313	154.740	0.00646	0.20646	4.843
20	38.338	0.0261	186.688	0.00536	0.20536	4.870
21	46.005	0.0217	225.026	0.00444	0.20444	4.891
22	55.206	0.0181	271.031	0.00369	0.20369	4.909
23	66.247	0.0151	326.237	0.00307	0.20307	4.925
24	79.497	0.0126	392.484	0.00255	0.20255	4.937
25	95.396	0.0105	471.981	0.00212	0.20212	4.948
26	114.475	0.0087	567.377	0.00176	0.20176	4.956
27	137.371	0.0073	681.853	0.00147	0.20147	4.964
28	164.845	0.0061	819.223	0.00122	0.20122	4.970
29	197.814	0.0051	984.068	0.00102	0.20102	4.975
30	237.376	0.0042	1181.882	0.00085	0.20085	4.979
35	590.668	0.0017	2948.341	0.00034	0.20034	4.992
40	1469.772	0.0007	7343.858	0.00014	0.20014	4.997
45	3657.262	0.0003	18281.310	0.00005	0.20005	4.999
50	9100.438	0.0001	45497.191	0.00002	0.20002	4.999

Table 5-5. 25% Interest Factors

Period n	Single-payment compound-amount F/P	Single-payment present-worth P/F	Uniform-series compound-amount F/A	Sinking-fund payment A/F	Capital recovery A/P	Uniform-series present-worth P/A
	Future value of $1	Present value of $1	Future value of uniform series of $1	Uniform series whose future value is $1	Uniform series with present value of $1	Present value of uniform series of $1
	$(1+i)^n$	$\frac{1}{(1+i)^n}$	$\frac{(1+i)^n-1}{i}$	$\frac{i}{(1+i)^n-1}$	$\frac{i(1+i)^n}{(1+i)^n-1}$	$\frac{(1+i)^n-1}{i(1+i)^n}$
1	1.250	0.8000	1.000	1.00000	1.25000	0.800
2	1.562	0.6400	2.250	0.44444	0.69444	1.440
3	1.953	0.5120	3.812	0.26230	0.51230	1.952
4	2.441	0.4096	5.766	0.17344	0.42344	2.362
5	3.052	0.3277	8.207	0.12185	0.37185	2.689
6	3.815	0.2621	11.259	0.08882	0.33882	2.951
7	4.768	0.2097	15.073	0.06634	0.31634	3.161
8	5.960	0.1678	19.842	0.05040	0.30040	3.329
9	7.451	0.1342	25.802	0.03876	0.28876	3.463
10	9.313	0.1074	33.253	0.03007	0.28007	3.571
11	11.642	0.0859	42.566	0.02349	0.27349	3.656
12	14.552	0.0687	54.208	0.01845	0.26845	3.725
13	18.190	0.0550	68.760	0.01454	0.26454	3.780
14	22.737	0.0440	86.949	0.01150	0.26150	3.824
15	28.422	0.0352	109.687	0.00912	0.25912	3.859
16	35.527	0.0281	138.109	0.00724	0.25724	3.887
17	44.409	0.0225	173.636	0.00576	0.25576	3.910
18	55.511	0.0180	218.045	0.00459	0.25459	3.928
19	69.389	0.0144	273.556	0.00366	0.25366	3.942
20	86.736	0.0115	342.945	0.00292	0.25292	3.954
21	108.420	0.0092	429.681	0.00233	0.25233	3.963
22	135.525	0.0074	538.101	0.00186	0.25186	3.970
23	169.407	0.0059	673.626	0.00148	0.25148	3.976
24	211.758	0.0047	843.033	0.00119	0.25119	3.981
25	264.698	0.0038	1054.791	0.00095	0.25095	3.985
26	330.872	0.0030	1319.489	0.00076	0.25076	3.988
27	413.590	0.0024	1650.361	0.00061	0.25061	3.990
28	516.988	0.0019	2063.952	0.00048	0.25048	3.992
29	646.235	0.0015	2580.939	0.00039	0.25039	3.994
30	807.794	0.0012	3227.174	0.00031	0.25031	3.995
35	2465.190	0.0004	9856.761	0.00010	0.25010	3.998
40	7523.164	0.0001	30088.655	0.00003	0.25003	3.999

Table 5-6. 30% Interest Factors

Period n	Single-payment compound-amount F/P	Single-payment present-worth P/F	Uniform-series compound-amount F/A	Sinking-fund payment A/F	Capital recovery A/P	Uniform-series present-worth P/A
	Future value of $1	Present value of $1	Future value of uniform series of $1	Uniform series whose future value is $1	Uniform series with present value of $1	Present value of uniform series of $1
	$(1 + i)^n$	$\frac{1}{(1+i)^n}$	$\frac{(1+i)^n - 1}{i}$	$\frac{i}{(1+i)^n - 1}$	$\frac{i(1+i)^n}{(1+i)^n - 1}$	$\frac{(1+i)^n - 1}{i(1+i)^n}$
1	1.300	0.7692	1.000	1.00000	1.30000	0.769
2	1.690	0.5917	2.300	0.43478	0.73478	1.361
3	2.197	0.4552	3.990	0.25063	0.55063	1.816
4	2.856	0.3501	6.187	0.16163	0.46163	2.166
5	3.713	0.2693	9.043	0.11058	0.41058	2.436
6	4.827	0.2072	12.756	0.07839	0.37839	2.643
7	6.275	0.1594	17.583	0.05687	0.35687	2.802
8	8.157	0.1226	23.858	0.04192	0.34192	2.925
9	10.604	0.0943	32.015	0.03124	0.33124	3.019
10	13.786	0.0725	42.619	0.02346	0.32346	3.092
11	17.922	0.0558	56.405	0.01773	0.31773	3.147
12	23.298	0.0429	74.327	0.01345	0.31345	3.190
11	30.288	0.0330	97.625	0.01024	0.31024	3.223
14	39.374	0.0254	127.913	0.00782	0.30782	3.249
15	51.186	0.0195	167.286	0.00598	0.30598	3.268
16	66.542	0.0150	218.472	0.00458	0.30458	3.283
17	86.504	0.0116	285.014	0.00351	0.30351	3.295
18	112.455	0.0089	371.518	0.00269	0.30269	3.304
19	146.192	0.0068	483.973	0.00207	0.30207	3.311
20	190.050	0.0053	630.165	0.00159	0.30159	3.316
21	247.065	0.0040	820.215	0.00122	0.30122	3.320
22	321.184	0.0031	1067.280	0.00094	0.30094	3.323
23	417.539	0.0024	1388.464	0.00072	0.30072	3.325
24	542.801	0.0018	1806.003	0.00055	0.30055	3.327
25	705.641	0.0014	2348.803	0.00043	0.30043	3.329
26	917.333	0.0011	3054.444	0.00033	0.30033	3.330
27	1192.533	0.0008	3971.778	0.00025	0.30025	3.331
28	1550.293	0.0006	5164.311	0.00019	0.30019	3.331
29	2015.381	0.0005	6714.604	0.00015	0.30015	3.332
30	2619.996	0.0004	8729.985	0.00011	0.30011	3.332
0	9727.8060	0.0001	32422.868	0.00003	0.30003	3.333

Table 5-7. 40% Interest Factors

Period n	Single-payment compound-amount F/P	Single-payment present-worth P/F	Uniform-series compound-amount F/A	Sinking-fund payment A/F	Capital recovery A/P	Uniform-series present-worth P/A
	Future value of $1	Present value of $1	Future value of uniform series of $1	Uniform series whose future value is $1	Uniform series with present value of $1	Present value of uniform series of $1
	$(1+i)^n$	$\frac{1}{(1+i)^n}$	$\frac{(1+i)^n-1}{i}$	$\frac{i}{(1+i)^n-1}$	$\frac{i(1+i)^n}{(1+i)^n-1}$	$\frac{(1+i)^n-1}{i(1+i)^n}$
1	1.400	0.7143	1.000	1.00000	1.40000	0.714
2	1.960	0.5102	2.400	0.41667	0.81667	1.224
3	2.744	0.3644	4.360	0.22936	0.62936	1.589
4	3.842	0.2603	7.104	0.14077	0.54077	1.849
5	5.378	0.1859	10.946	0.09136	0.49136	2.035
6	7.530	0.1328	16.324	0.06126	0.46126	2.168
7	10.541	0.0949	23.853	0.04192	0.44192	2.263
8	14.758	0.0678	34.395	0.02907	0.42907	2.331
9	20.661	0.0484	49.153	0.02034	0.42034	2.379
10	28.925	0.0346	69.814	0.01432	0.41432	2.414
11	40.496	0.0247	98.739	0.01013	0.41013	2.438
12	56.694	0.0176	139.235	0.00718	0.40718	2.456
13	79.371	0.0126	195.929	0.00510	0.40510	2.469
14	111.120	0.0090	275.300	0.00363	0.40363	2.478
15	155.568	0.0064	386.420	0.00259	0.40259	2.484
16	217.795	0.0046	541.988	0.00185	0.40185	2.489
17	304.913	0.0033	759.784	0.00132	0.40132	2.492
18	426.879	0.0023	1064.697	0.00094	0.40094	2.494
19	597.630	0.0017	1491.576	0.00067	0.40067	2.496
20	836.683	0.0012	2089.206	0.00048	0.40048	2.497
21	1171.356	0.0009	2925.889	0.00034	0.40034	2.498
22	1639.898	0.0006	4097.245	0.00024	0.40024	2.498
23	2295.857	0.0004	5737.142	0.00017	0.40017	2.499
24	3214.200	0.0003	8032.999	0.00012	0.40012	2.499
25	4499.880	0.0002	11247.199	0.00009	0.40009	2.499
26	6299.831	0.0002	15747.079	0.00006	0.40006	2.500
27	8819.764	0.0001	22046.910	0.00005	0.40005	2.500

Table 5-8. 50% Interest Factors

Period n	Single-payment compound-amount F/P	Single-payment present-worth P/F	Uniform-series compound-amount F/A	Sinking-fund payment A/F	Capital recovery A/P	Uniform-series present-worth P/A
	Future value of $1 $(1+i)^n$	Present value of $1 $\frac{1}{(1+i)^n}$	Future value of uniform series of $1 $\frac{(1+i)^n-1}{i}$	Uniform series whose future value is $1 $\frac{i}{(1+i)^n-1}$	Uniform series with present value of $1 $\frac{i(1+i)^n}{(1+i)^n-1}$	Present value of uniform series of $1 $\frac{(1+i)^n-1}{i(1+i)^n}$
1	1.500	0.6667	1.000	1.00000	1.50000	0.667
2	2.250	0.4444	2.500	0.40000	0.90000	1.111
3	3.375	0.2963	4.750	0.21053	0.71053	1.407
4	5.062	0.1975	8.125	0.12308	0.62308	1.605
5	7.594	0.1317	13.188	0.07583	0.57583	1.737
6	11.391	0.0878	20.781	0.04812	0.54812	1.824
7	17.086	0.0585	32.172	0.03108	0.53108	1.883
8	25.629	0.0390	49.258	0.02030	0.52030	1.922
9	38.443	0.0260	74.887	0.01335	0.51335	1.948
10	57.665	0.0173	113.330	0.00882	0.50882	1.965
11	86.498	0.0116	170.995	0.00585	0.50585	1.977
12	129.746	0.0077	257.493	0.00388	0.50388	1.985
13	194.620	0.0051	387.239	0.00258	0.50258	1.990
14	291.929	0.0034	581.859	0.00172	0.50172	1.993
15	437.894	0.0023	873.788	0.00114	0.50114	1.995
16	656.841	0.0015	1311.682	0.00076	0.50076	1.997
17	985.261	0.0010	1968.523	0.00051	0.50051	1.998
18	1477.892	0.0007	2953.784	0.00034	0.50034	1.999
19	2216.838	0.0005	4431.676	0.00023	0.50023	1.999
20	3325.257	0.0003	6648.513	0.00015	0.50015	1.999
21	4987.885	0.0002	9973.770	0.00010	0.50010	2.000
22	7481.828	0.0001	14961.655	0.00007	0.50007	2.000

Table 5-9. Five-Year Escalation Table

Present Worth of a Series of Escalating Payments Compounded Annually
Discount-Escalation Factors for n = 5 Years

Discount Rate	Annual Escalation Rate 0.10	0.12	0.14	0.16	0.18	0.20
0.10	5.000000	5.279234	5.572605	5.880105	6.202627	6.540569
0.11	4.866862	5.136200	5.420152	5.717603	6.029313	6.355882
0.12	4.738562	5.000000	5.274242	5.561868	5.863289	6.179066
0.13	4.615647	4.869164	5.133876	5.412404	5.704137	6.009541
0.14	4.497670	4.742953	5.000000	5.269208	5.551563	5.847029
0.15	4.384494	4.622149	4.871228	5.131703	5.404955	5.691165
0.16	4.275647	4.505953	4.747390	5.000000	5.264441	5.541511
0.17	4.171042	4.394428	4.628438	4.873699	5.129353	5.397964
0.18	4.070432	4.287089	4.513947	4.751566	5.000000	5.259749
0.19	3.973684	4.183921	4.403996	4.634350	4.875619	5.126925
0.20	3.880510	4.084577	4.298207	4.521178	4.755725	5.000000
0.21	3.790801	3.989001	4.196400	4.413341	4.640260	4.877689
0.22	3.704368	3.896891	4.098287	4.308947	4.529298	4.759649
0.23	3.621094	3.808179	4.003835	4.208479	4.422339	4.645864
0.24	3.540773	3.722628	3.912807	4.111612	4.319417	4.536517
0.25	3.463301	3.640161	3.825008	4.018249	4.220158	4.431144
0.26	3.388553	3.560586	3.740376	3.928286	4.124553	4.329514
0.27	3.316408	3.483803	3.658706	3.841442	4.032275	4.231583
0.28	3.246718	3.409649	3.579870	3.757639	3.943295	4.137057
0.29	3.179393	3.338051	3.503722	3.676771	3.857370	4.045902
0.30	3.114338	3.268861	3.430201	3.598653	3.774459	3.957921
0.31	3.051452	3.201978	3.359143	3.523171	3.694328	3.872901
0.32	2.990618	3.137327	3.290436	3.450224	3.616936	3.790808
0.33	2.931764	3.074780	3.224015	3.379722	3.542100	3.711472
0.34	2.874812	3.014281	3.159770	3.311524	3.469775	3.634758

Table 5-10. Ten-Year Escalation Table

Present Worth of a Series of Escalating Payments Compounded Annually
Discount-Escalation Factors for n = 10 Years

Discount Rate	Annual Escalation Rate 0.10	0.12	0.14	0.16	0.18	0.20
0.10	10.000000	11.056250	12.234870	13.548650	15.013550	16.646080
0.11	9.518405	10.508020	11.613440	12.844310	14.215140	15.741560
0.12	9.068870	10.000000	11.036530	12.190470	13.474590	14.903510
0.13	8.650280	9.526666	10.498990	11.582430	12.786980	14.125780
0.14	8.259741	9.084209	10.000000	11.017130	12.147890	13.403480
0.15	7.895187	8.672058	9.534301	10.490510	11.552670	12.731900
0.16	7.554141	8.286779	9.099380	10.000000	10.998720	12.106600
0.17	7.234974	7.926784	8.693151	9.542653	10.481740	11.524400
0.18	6.935890	7.589595	8.312960	9.113885	10.000000	10.980620
0.19	6.655455	7.273785	7.957330	8.713262	9.549790	10.472990
0.20	6.392080	6.977461	7.624072	8.338518	9.128122	10.000000
0.21	6.144593	6.699373	7.311519	7.987156	8.733109	9.557141
0.22	5.911755	6.437922	7.017915	7.657542	8.363208	9.141752
0.23	5.692557	6.192047	6.742093	7.348193	8.015993	8.752133
0.24	5.485921	5.960481	6.482632	7.057347	7.690163	8.387045
0.25	5.290990	5.742294	6.238276	6.783767	7.383800	8.044173
0.26	5.106956	5.536463	6.008083	6.526298	7.095769	7.721807
0.27	4.933045	5.342146	5.790929	6.283557	6.824442	7.418647
0.28	4.768518	5.158489	5.585917	6.054608	6.568835	7.133100
0.29	4.612762	4.984826	5.392166	5.838531	6.327682	6.864109
0.30	4.465205	4.820429	5.209000	5.634354	6.100129	6.610435
0.31	4.325286	4.664669	5.035615	5.441257	5.885058	6.370867
0.32	4.192478	4.517015	4.871346	5.258512	5.681746	6.144601
0.33	4.066339	4.376884	4.715648	5.085461	5.489304	5.930659
0.34	3.946452	4.243845	4.567942	4.921409	5.307107	5.728189

Table 5-11. Fifteen-Year Escalation Table

Present Worth of a Series of Escalating Payments Compounded Annually
Discount-Escalation Factors for n = 15 Years

Discount Rate	Annual Escalation Rate 0.10	0.12	0.14	0.16	0.18	0.20
0.10	15.000000	17.377880	20.199780	23.549540	27.529640	32.259620
0.11	13.964150	16.126230	18.690120	21.727370	25.328490	29.601330
0.12	13.026090	15.000000	17.332040	20.090360	23.355070	27.221890
0.13	12.177030	13.981710	16.105770	18.616160	21.581750	25.087260
0.14	11.406510	13.057790	15.000000	17.287320	19.985530	23.169060
0.15	10.706220	12.220570	13.998120	16.086500	18.545150	21.442230
0.16	10.068030	11.459170	13.088900	15.000000	17.244580	19.884420
0.17	9.485654	10.766180	12.262790	14.015480	16.066830	18.477610
0.18	8.953083	10.133630	11.510270	13.118840	15.000000	17.203010
0.19	8.465335	9.555676	10.824310	12.303300	14.030830	16.047480
0.20	8.017635	9.026333	10.197550	11.560150	13.148090	15.000000
0.21	7.606115	8.540965	9.623969	10.881130	12.343120	14.046400
0.22	7.227109	8.094845	9.097863	10.259820	11.608480	13.176250
0.23	6.877548	7.684317	8.614813	9.690559	10.936240	12.381480
0.24	6.554501	7.305762	8.170423	9.167798	10.320590	11.655310
0.25	6.255518	6.956243	7.760848	8.687104	9.755424	10.990130
0.26	5.978393	6.632936	7.382943	8.244519	9.236152	10.379760
0.27	5.721101	6.333429	7.033547	7.836080	8.757889	9.819020
0.28	5.481814	6.055485	6.710042	7.458700	8.316982	9.302823
0.29	5.258970	5.797236	6.410005	7.109541	7.909701	8.827153
0.30	5.051153	5.556882	6.131433	6.785917	7.533113	8.388091
0.31	4.857052	5.332839	5.872303	6.485500	7.184156	7.982019
0.32	4.675478	5.123753	5.630905	6.206250	6.860492	7.606122
0.33	4.505413	4.928297	5.405771	5.946343	6.559743	7.257569
0.34	4.345926	4.745399	5.195502	5.704048	6.280019	6.933897

Table 5-12. Twenty-Year Escalation Table

Present Worth of a Series of Escalating Payments Compounded Annually
Discount-Escalation Factors for n = 20 Years

Discount Rate	Annual Escalation Rate 0.10	0.12	0.14	0.16	0.18	0.20
0.10	20.000000	24.295450	29.722090	36.592170	45.308970	56.383330
0.11	18.213210	22.002090	26.776150	32.799710	40.417480	50.067940
0.12	16.642370	20.000000	24.210030	29.505400	36.181240	44.614710
0.13	15.259850	18.243100	21.964990	26.634490	32.502270	39.891400
0.14	14.038630	16.694830	20.000000	24.127100	29.298170	35.789680
0.15	12.957040	15.329770	18.271200	21.929940	26.498510	32.218060
0.16	11.995640	14.121040	16.746150	20.000000	24.047720	29.098950
0.17	11.138940	13.048560	15.397670	18.300390	21.894660	26.369210
0.18	10.373120	12.093400	14.201180	16.795710	20.000000	23.970940
0.19	9.686791	11.240870	13.137510	15.463070	18.326720	21.860120
0.20	9.069737	10.477430	12.188860	14.279470	16.844020	20.000000
0.21	8.513605	9.792256	11.340570	13.224610	15.527270	18.353210
0.22	8.010912	9.175267	10.579620	12.282120	14.355520	16.890730
0.23	7.555427	8.618459	9.895583	11.438060	13.309280	15.589300
0.24	7.141531	8.114476	9.278916	10.679810	12.373300	14.429370
0.25	6.764528	7.657278	8.721467	9.997057	11.533310	13.392180
0.26	6.420316	7.241402	8.216490	9.380883	10.778020	12.462340
0.27	6.105252	6.862203	7.757722	8.823063	10.096710	11.626890
0.28	5.816151	6.515563	7.339966	8.316995	9.480940	10.874120
0.29	5.550301	6.198027	6.958601	7.856833	8.922847	10.194520
0.30	5.305312	5.906440	6.609778	7.437339	8.416060	9.579437
0.31	5.079039	5.638064	6.289875	7.054007	7.954518	9.021190
0.32	4.869585	5.390575	5.995840	6.702967	7.533406	8.513612
0.33	4.675331	5.161809	5.725066	6.380829	7.148198	8.050965
0.34	4.494838	4.949990	5.475180	6.084525	6.795200	7.628322

The *F/P* can be computed by an interest formula, but usually its value is found by using the interest tables. Interest tables for interest rates of 10 to 50 percent are found in this chapter (Tables 5-1 through 5-8). In predicting future costs, there are many unknowns. For the accuracy of most calculations, interest rates are assumed to be compounded annually unless otherwise specified. Linear interpolation is commonly used to find values not listed in the interest tables.

Tables 5-9 through 5-12 can be used to determine the effect of fuel escalation on the life-cycle cost analysis.

Single-Payment Present Worth—P/F

The P/F factor is used to determine the present worth, *P*, that a future amount, *F*, will be at interest of *i*-percent, in *n* years. If *F* is known, and *P* is to be determined, then Equation 5-4 is used.

$$P = F \times 1/(1+i)^n \quad \text{(Eq. 5-4)}$$

$$P/F = \frac{1}{(1+i)^n} \quad \text{(Eq. 5-5)}$$

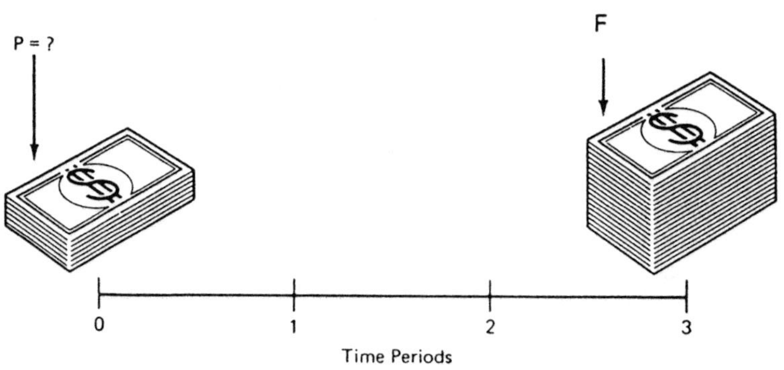

Figure 5-2. Single-Payment Present Worth (P/F).

Uniform-Series Compound Amount—F/A

The F/A factor is used to determine the amount *F* that an equal annual payment *A* will accumulate to in *n* years at *i* percent interest. If *A* (uniform annual payment) is known, and *F* (the future worth of these payments) is required, then Equation 5-6 is used.

$$F = A \times \frac{(1+i)^n - 1}{i} \qquad \text{(Eq. 5-6)}$$

$$F/A = \frac{(1+i)^n - 1}{i} \qquad \text{(Eq. 5-7)}$$

Uniform-Series Present Worth—(P/A)

The P/A factor is used to determine the present amount *P* that can be paid by equal payments of *A* (uniform annual payment) at *i* percent interest, for *n* years. If *A* is known, and *P* is required, then Equation 5-8 is used.

$$P = A \times \frac{(1+i)^n - 1}{i(1+i)^n} \qquad \text{(Eq. 5-8)}$$

$$P/A = \frac{(1+i)^n - 1}{i(1+i)^n} \qquad \text{(Eq. 5-9)}$$

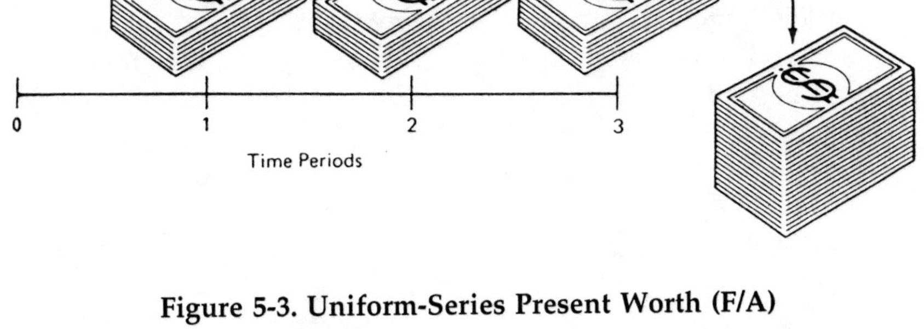

Figure 5-3. Uniform-Series Present Worth (F/A)

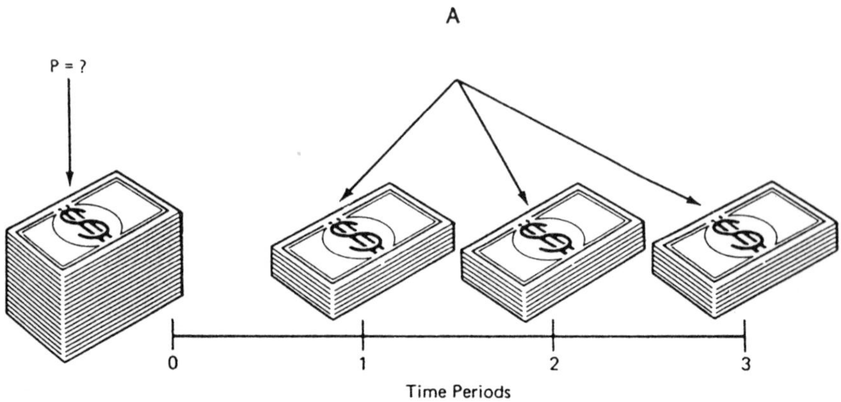

Figure 5-4. Uniform Series Present Worth (P/A)

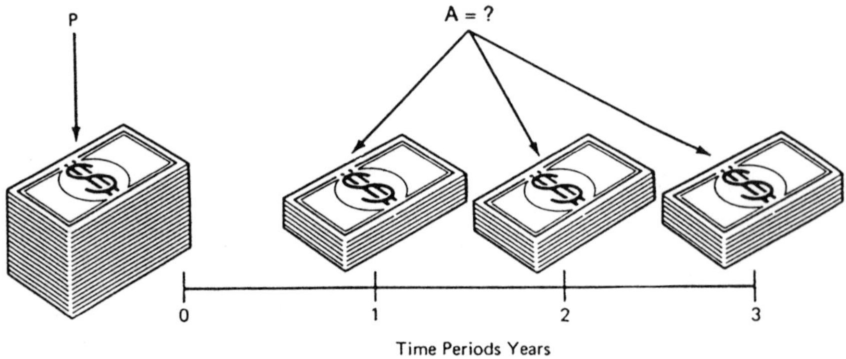

Figure 5-5. Capital Recovery (A/P)

Capital Recovery—A/P

The A/P factor is used to determine an annual payment A required to pay off a present amount P at i percent interest, for n years. If the present sum of money, P, spent today is known, and the uniform payment A needed to pay back P over a stated period of time is required, then Equation 5-10 is used.

$$A = P \times \frac{i(1+i)^n}{(1+i)^n - 1} \qquad \text{(Eq. 5-10)}$$

$$A/P = \frac{i(1+i)^n}{(1+i)^n - 1} \quad \text{(Eq. 5-11)}$$

Sinking-Fund Payment—A/F

The A/F factor is used to determine the equal annual amount *R* that must be invested for *n* years at *i* percent interest in order to accumulate a specified future amount. If *F* (the future worth of a series of annual payments) is known, and *A* (value of those annual payments) is required, then Equation 5-12 is used.

$$A = F \times \frac{i}{(1+i)^n - 1} \quad \text{(Eq. 5-12)}$$

$$A/F = \frac{i}{(1+i)^n - 1} \quad \text{(Eq. 5-13)}$$

Gradient Present Worth—GPW

The GPW factor is used to determine the present amount *P* that can be paid by annual amounts *A′* which escalate at *e* percent, at *i* percent interest, for *n* years. If *A′* is known, and *P* is required, then Equation 5-14 is used. The GPW factor is a relatively new term which has gained in importance due to the impact of inflation.

$$P = A' \times (\text{GPW})\, i_n \quad \text{(Eq. 5-14)}$$

$$P/A' = \text{GPW} = \frac{\frac{1+e}{1+i}\left[1 - \left(\frac{1+e}{1+i}\right)^n\right]}{1 - \frac{1+e}{1+i}} \quad \text{(Eq. 5-15)}$$

The three most commonly used methods in life-cycle costing are the annual cost, present worth and rate-of-return analysis.

In the present-worth method a minimum rate of return (i) is stipulated. All future expenditures are converted to present values using the interest factors. The alternative with lowest effective first cost is the most desirable.

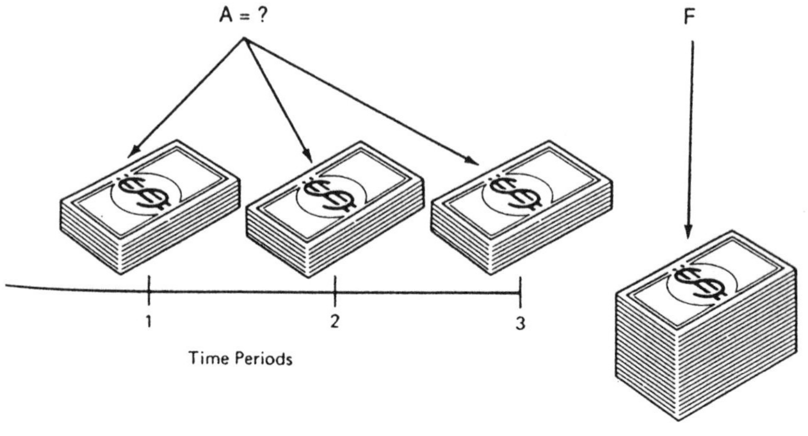

Figure 5-6. Sinking-Fund Payment (A/F)

A similar procedure is implemented in the annual cost method. The difference is that the first cost is converted to an annual expenditure. The alternative with lowest effective annual cost is the most desirable.

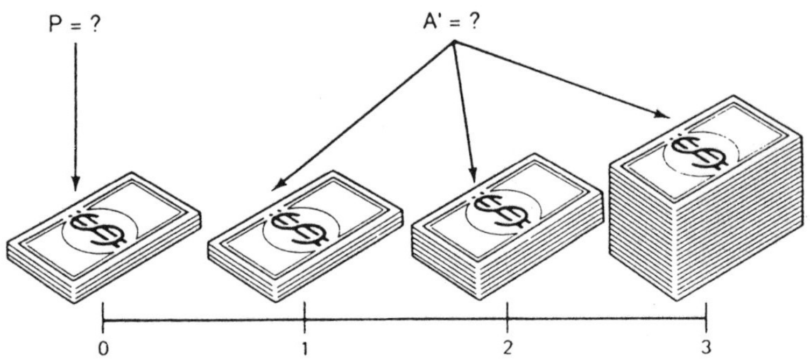

Figure 5-7. Gradient Present Worth

In the rate-of-return method, a trial-and-error procedure is usually required. Interpolation from the interest tables can determine what rate of return (i) will give an interest factor which will make the overall cash flow balance. The rate-of-return analysis gives a good indication of the overall ranking of independent alternates.

The effect of escalation in fuel costs can influence greatly the final decision. When an annual cost grows at a steady rate it may be treated as a gradient and the gradient present worth factor can be used.

Special thanks are given to Rudolph R. Yanuck and Dr. Robert Brown for the use of their specially designed interest and escalation tables used in this text.

When life-cycle costing is used to compare several alternatives the differences between costs are important. For example, if one alternate forces additional maintenance or an operating expense to occur, then these factors as well as energy costs need to be included. Remember, what was previously spent for the item to be replaced is irrelevant. The only factor to be considered is whether the new cost can be justified based on projected savings over its useful life.

Example 5-1

An evaluation needs to be made to replace all 40-watt fluorescent lamps with new lamps that save 12 percent, or 4.8 watts, and give the same output. The cost of each lamp is $2.80.

Assuming a rate of return before taxes of 25 percent is required, can the immediate replacement be justified? Hours of operation are 5,800 and the lamp life is two years. Electricity costs 7.0¢/kWh.

Answer

$$A = 5800 \times 4.8 \times 0.070/1000 = \$1.94$$

$$A/P = 1.94/2.80 = .69$$

From Table 5-5 a rate of return of 25 percent is obtained. When analyzing energy conservation measures, never look at what was previously spent or the life remaining. Just determine if the new expenditure will pay for itself.

Example 5-2

An electrical-energy audit indicates electrical motor consumption is 4×10^6 kWh per year. By upgrading the motor spares with high efficiency motors a 10% savings can be realized. The additional cost for these motors is estimated at $80,000. Assuming an 8¢ per kWh energy charge and 20-year life, is the expenditure justified based on a minimum rate of return of 20% before taxes? Solve the problem using the present worth, annual cost and rate-of-return methods.

Analysis

Present Worth Method

	Alternate 1 Present Method	*Alternate 2 Use High-Efficiency Motor Spares*
(1) First Cost (P)	—	\$80,000
(2) Annual Cost (A)	$4 \times 10^6 \times .08$	$.9 \times \$320,000$
	= \$320,000	= \$288,000
P/A (Table 5-4)	4.87	4.87
(2) A × 4.87 =	\$1,558,400	\$1,402,560
Present Worth	\$1,558,400	\$1,482,560
(1) + (3)		Choose Alternate with Lowest First Cost

Annual Cost Method

	Alternate 1	*Alternate 2*
(1) First Cost (P)	—	\$80,000
(2) Annual Cost (A)	\$320,000	\$288,000
A/P (Table 5-4)	.2	.2
(3) P × .2	—	\$16,000
Annual Cost	\$320,000	\$304,000
(2) + (3)		Choose Alternate with Lowest First Cost

Rate -of-Return Method

$$P = (\$320{,}000 - \$288{,}000)$$

$$P/A = \frac{80{,}000}{32{,}000} = 2.5$$

What value of i will make P/A = 2.5? i = 40% (Table 5-7).

Example 5-3

Show the effect of 10 percent escalation on the rate-of-return analysis given the

Energy equipment investment	= \$20,000
After-tax savings	= \$ 2,600
Equipment life (n)	= 15 years

Answer

Without escalation:

$$\frac{A}{P} = \frac{2{,}600}{20{,}000} = 0.13$$

From Table 5-1, the rate of return is 10 percent. With 10 percent escalation assumed:

$$\frac{P}{A} = \frac{20{,}000}{2{,}600} = 7.69$$

From Table 5-11, the rate of return is 21 percent.

Thus we see that taking into account a modest escalation rate can dramatically affect the justification of the project.

MAKING DECISIONS FOR ALTERNATE INVESTMENTS

There are several methods for determining which energy conservation alternative is the most economical. Probably the most familiar and trusted method is the annual cost method.

When evaluating replacement of processes or equipment *do not* consider what was previously spent. The decision will be based on whether the new process or equipment proves to save substantially enough in operating costs to justify the expenditure.

Equation 5-16 is used to convert the lump sum investment P into the annual cost. In the case where the asset has a value after the end of its useful life, the annual cost becomes:

$$AC = (P - L) * A/P + iL \qquad \text{(Eq. 5-16)}$$

where

AC is the annual cost

L is the net sum of money that can be realized for a piece of equipment, over and above its removal cost, when it is returned at the end of the service life. L is referred to as the salvage value.

As a practical point, the salvage value is usually small and can be neglected, considering the accuracy of future costs. The annual-cost technique can be implemented by using the following format:

	Alternate 1	Alternate 2
1. First cost (*P*)		
2. Estimated life (*n*)		
3. Estimated salvage value at end of life (*L*)		
4. Annual disbursements, including energy costs & maintenance (*E*)		
5. Minimum acceptable return *before* taxes (*i*)		
6. A/P *n, i*		
7. $(P - L)$ *A/P		
8. Li		
9. $AC = (P - L)$ * A/P + $Li + E$		

Choose alternate with lowest *AC*

The alternative with the lowest annual cost is the desired choice.

Example 5-4

A new water line must be constructed from an existing pumping station to a reservoir. Estimates of construction and pumping costs for each pipe size have been made.

Pipe Size	Estimated Construction Costs	Cost/Hour for Pumping
8"	$80,000	$4.00
10"	$100,000	$3.00
12"	$160,000	$1.50

The annual cost is based on a 16-year life and a desired return on investment, before taxes, of 10 percent. Which is the most economical pipe size for pumping 4,000 hours/year?

Answer

	8" Pipe	10" Pipe	12" Pipe
P	\$80,000	\$100,000	\$160,000
n	16	16	16
E	16,000	12,000	6,000
i	10%	10%	10%
A/P = 0.127	—	—	—
(*P* – *L*) A/P	10,160	12,700	20,320
Li	—	—	—
AC	\$26,160	\$24,700 *(Choice)*	\$26,320

DEPRECIATION, TAXES AND THE TAX CREDIT

Depreciation

Depreciation affects the "accounting procedure" for determining profits and losses and the income tax of a company. In other words, for tax purposes the expenditure for an asset such as a pump or motor cannot be fully expensed in its first year. The original investment must be charged off for tax purposes over the useful life of the asset. A company usually wishes to expense an item as quickly as possible.

The Internal Revenue Service allows several methods for determining the annual depreciation rate.

Straight-Line Depreciation. The simplest method is referred to as a straight-line depreciation and is defined as:

$$D = \frac{P - L}{n} \qquad \text{(Eq. 5-17)}$$

where

D is the annual depreciation rate

L is the value of equipment at the end of its useful life, commonly referred to as salvage value

n is the life of the equipment, which is determined by Internal Revenue Service guidelines

P is the initial expenditure.

Sum-of-Years Digits. Another method is referred to as the sum-of-years digits. In this method the depreciation rate is determined by finding the sum of digits using the following formula:

$$N = n\,\frac{(n+1)}{2} \qquad \text{(Eq. 5-18)}$$

where n *is* the life of equipment.

Each year's depreciation rate is determined as follows.

First year
$$D = \frac{n}{N}(P - L) \qquad \text{(Eq. 5-19)}$$

Second year
$$D = \frac{n-1}{N}(P - L) \qquad \text{(Eq. 5-20)}$$

n year
$$D = \frac{1}{N}(P - L) \qquad \text{(Eq. 5-21)}$$

Declining-Balance Depreciation. The declining-balance method allows for larger depreciation charges in the early years, which is sometimes referred to as fast write-off.

The rate is calculated by taking a constant percentage of the declining undepreciated balance. The most common method used to calculate the declining balance is to predetermine the depreciation rate. Under certain circumstances a rate equal to 200 percent of the straight-line depreciation rate may be used. Under other circumstances the rate is limited to 1-1/2 or 1/4 times as great as straight-line depreciation. In this method the salvage value or undepreciated book value is established once the depreciation rate is pre-established.

To calculate the undepreciated book value, Equation 5-22 used.

$$D = 1 - \left(\frac{L}{P}\right)^{1/N} \qquad \text{(Eq. 5-22)}$$

where

D is the annual depreciation rate
L is the salvage value
P is the first cost.

Tax Considerations

Tax-deductible expenses such as maintenance, energy, operating costs, insurance and property taxes reduce the income subject to taxes.

For the after-tax life-cycle analysis and payback analysis the actual incurred and annual savings is given as follows.

$$AS = (1 - \mathrm{I})E + ID \qquad \text{(Eq. 5-23)}$$

where

- AS is the annual after-tax savings (excluding effect of tax credit)
- E is the annual energy savings (difference between original expenses and expenses after modification)
- D is the annual depreciation rate in the income tax bracket.
- I is the income tax bracket

Equation 5-23 takes into account that the annual energy savings is partially offset by additional taxes which must be paid due to reduced operating expenses. On the other hand, the depreciation allowance reduces taxes directly.

After-Tax Analysis

To compute a rate of return which accounts for taxes, depreciation, escalation and tax credits, a cash-flow analysis is usually required. This method analyzes all transactions including first and operating costs. To determine the after-tax rate of return a trial and error or computer analysis is required.

All money is converted to the present assuming an interest rate. The summation of all present dollars should equal zero when the correct interest rate is selected, as illustrated in Figure 5-8.

This analysis can be made assuming a fuel escalation rate by using the gradient present worth interest of the present worth factor.

SUMMARY OF LIFE-CYCLE COSTING

Always draw a cash-flow diagram on a time-basis scale. Show cash influx as positive and cash outflow as negative.

In determining which interest formula to use, the following proce-

Year	1 Investment	2 Tax Credit	3 After-Tax Savings (*AS*)	4 Single-Payment Present-Worth Factor	(2+3) × 4 Present Worth
0	$-P$				$-P$
1		$+TC$	AS	P/F1	+P1
2			AS	P/F2	P2
3			AS	P/F3	P3
4			AS	P/F4	P4
Total					Σp

$AS = (1 - I)\,E + ID$
Trial-and-Error Solution:
Correct i when $\Sigma P = 0$

Figure 5-8. Cash-Flow Rate-of-Return Analysis

dure may be helpful. First, put the symbols in two rows, one above the other as below:

$$\frac{\text{PAF (unknown)}}{\text{PAF (known)}}$$

The top represents the unknown values, and the bottom line represents the known. From information you have and desire, simply circle one of each line, and you have the correct factor.

For example, if you want to determine the annual saving "A" required when the cost of the energy device "P" is known, circle P on the bottom and A on the top. The factor A/P or capital recovery is required for this example. Table 5-13 summarizes the cash analysis for interest formulas.

GIVEN	FIND	USE
P	F	F/P
F	P	P/F
A	F	F/A
F	A	A/F
P	A	A/P
A	P	P/A

Table 5-13. Cash Analysis for Interest Formulas

Chapter 6

How to Work with Lenders

Financing is an art, not a science. The best way to appreciate the process is to put yourself in the lender's shoes. If you were considering lending someone several million of your hard-earned dollars, what sort of questions would you ask? How concerned would you be about certain details of your borrower's financial history that were unclear? What weight, if any, would you give to the borrower's reputation?

Traditionally, lenders evaluate a possible loan from several perspectives. These are known as the "C's of Lending":

- *Capital*. Is there capital at risk on the part of the borrower? How motivated is the borrower to pay back the loan if its own capital is involved in the transaction?

- *Cash flow*. What are the revenues and expenses of the borrower and/or the project? Are there noticeable trends in past financial performance? What are expected trends in the energy industry or the industry of the borrower?

- *Collateral*. What is the security for the loan? If the loan is not paid back as scheduled, what items of value remain for the lender to own and/or sell? Is it the project equipment or another corporate asset, such as stock?

- *Conditions*. What are the desired terms of the loan?

- *Character*. How experienced is the management team? What is the reputation of the company?

Regardless of lending policies and rules, developing the trust of your lender will go a long way when you need to secure a loan approval. Many borrowers approach the lending process as a "torture chamber" of difficult questions and paperwork. This type of attitude is certainly not going to make a lot of friends in the lending community.

A lender relationship is something well worth cultivating. The best way to create a foundation for a long-term lending relationship is to offer accurate information about the project and the financial status of the parties involved in addition to presenting your references and project history.

Lenders have what is called a "due diligence period." This is a time when, even though a loan approval may have been issued, they are looking for ways to verify everything disclosed in the initial application package. They want to expose all the weaknesses of the project in order to properly evaluate the risks of the loan.

Never try to hide or manipulate information, as this will land you in court or bankruptcy proceedings. Lenders should have all the relevant information, both good and bad. You should make the effort to meet face-to-face with the bankers and, if possible, make a presentation to one or more members of the loan committee. This "relationship" aspect of the loan decision-making process is an opportunity for you to turn what could be a marginal loan application into an approval.

If you are preparing to send a package of information about your energy project to a lender, this is what you should include:

- Your company's financial statements (two years, audited)
- ESCO financial statements (two years, audited)
- Project pro formas (revenue and expense projections)
- Energy services agreement between your company and the ESCO (sample shown in Appendix B)
- Contract between your company (or the ESCO) and the utility (if a rebate program is available)
- Description of equipment to be installed

- Energy audit results summary
- Measurment and verification plan
- Construction schedule
- Management team summary
- References from suppliers and customers

Once you have submitted this information to a lender, it's your turn to start asking for information. Here are some important questions to ask a lender:

- Are you interested in a lease on the equipment, a loan to my company or a loan to the ESCO?
- Is construction financing being offered as well as a long-term loan? What factors will affect the rate and terms of the loan?
- How soon can you send me a proposal to finance the project?
- What is the timeline for loan approval, due diligence and funding?
- How often does the loan committee meet? Can I make a presentation to them?
- Do you require the ESCO to have a construction bond?
- What will be your course of action if the project fails to generate the expected savings?
- Is there a prepayment penalty on the loan? If so, under what conditions?
- What happens if utility rates go down and the dollar amount of savings is insufficient to pay for the loan?

- Can payments be made quarterly or annually instead of monthly?

Lastly, and most importantly, remember these words of wisdom: everything is negotiable. Don't be afraid to ask for a lower rate or better terms than initially offered by a lender. As long as you have clearly demonstrated to the loan committee that your project is well-planned and has a high likelihood of success, you will be surprised at how flexible lenders can be.

Chapter 7

The Financial Analysis Of a Project

When lenders are making a loan based on the full faith and credit of a company, they tend to focus on the financial statements and general business performance of the proposed borrower. With project financing, however, it is the project itself which receives much attention. Even though the ESCO is the borrower, the project must perform in order for the loan to be paid back on schedule.

The best way for the ESCO and lender to analyze a project is to view it as a stand-alone company. This "company" has revenues, expenses, assets and debt which will vary over time. The revenues are the estimated share of savings payments your organization agrees to make to the ESCO or to the single-purpose entity (see Chapter 1). The expenses may include equipment maintenance, measurement and verification ("M&V") or even the cost of raw materials to produce sold "goods" such as chilled water.

Lenders will be most interested in the debt service coverage analysis. How do the net revenues (net income after expenses) compare to the project loan payments? If the net revenues are $25,000 per month and the lease payments are $20,000 per month, the Debt Service Coverage Ratio is 1.25. Most lenders prefer debt service ratios above 1.10 in order to provide some protection against fluctuations in project revenues and expenses.

If the debt service coverage ratio is too low for your project, you may want to run some alternative scenarios. First, try to extend the term of the loan. This will have the effect of lowering the payment per month and increasing the coverage ratio. Second, if the expenses of the project are high, relative to other projects, you may want to consider restructuring the project. Eliminating a costly piece of equipment with high maintenance expenses may well lower the simple payback of the project and increase the chances of the project's overall success. A third option is for the ESCO to use its own cash to pay for a small portion of the project thus decreasing the loan amount.

In terms of a presentation to your ogranization's finance team, any project needs to be compared to other projects which are vying for management's attention. The tools of comparison are Net Present Value (NPV) and Internal Rate of Return (IRR). The NPV is the value in current dollars of the positive and negative cash flows over time related to the project. In order to perform this calculation, an interest rate must be selected and usually it is equivalent to the opportunity cost of your company. This cost is the rate of return that the company receives on average in most of its relatively safe projects. The IRR is the rate at which the NPV is zero. The project has a high likelihood of being accepted by your management if its NPV or IRR is higher than other projects. For financing success stories, see Chapter 10.

Chapter 8

Financing Energy Efficiency In International Markets

Financing energy efficiency has become a topic of great interest. The lack of available capital is an often-cited, but not the only, constraint on financing energy-efficiency projects. Overcoming barriers in this area depends on developing sound, bankable projects. This chapter draws on recent work conducted under USAID's Energy Efficiency Project and Hagler Bailly's extensive experience in the field of energy efficiency and conservation.

Energy efficiency investments that are embedded in new plant and equipment investments may be financed through traditional sources and make use of fiscal incentives. International markets, however, require expertise in trade finance which, for some countries, includes counter trade. Regardless, local financial institutions are an important segment of the market and deserve support for building their capability in energy efficiency financing. Removing barriers to financing also requires that project sponsors, investors and energy users know how to identify, develop and package bankable projects. Limited and non-recourse project financing may also be an excellent tool for obtaining financing for large projects in markets where significant barriers remain. The public sector has been an important source of financing for energy efficiency in the past and must continue to remain so. However, reduced government budgets require that they focus their efforts on successful and cost-effective programs.

Innovative financing, which we define as applying known financing structures to new countries, designing and implementing market interventions, encouraging new entrants to finance energy efficiency or har-

---

Presented at the International Energy & Environmental Congress 1995 by John R. Armstrong and Suzanne G. Smith.

nessing concern for global environmental issues for energy efficiency financing, must also be explored. Performance contracting and energy service companies (ESCOs) have proven to be a workable model in the U.S. market and provide one of the better ways to address several barriers to investment in energy efficiency internationally. However, the entry of new market participants, such as utilities, environmental groups and venture capitalists, may be required to fill gaps in the skills and resources of existing market participants and ESCOs. In the future, increased concern for global environmental issues may provide greater incentives for energy efficiency investments through the mechanisms of joint implementation and emissions-allowance trading. Because there is no single solution to the financing constraint, approaches that combine traditional and innovative financing with sound business development efforts are required.

INTRODUCTION

The topic of financing energy efficiency projects in international markets is one of increasing interest. Nearly every study on energy efficiency refers to the lack of financing as a major constraint. Some reports add the lament that if only a special pot of funds were available, the financing constraint would be eliminated. Our experience in energy efficiency around the globe has taught us that this is not the case. Even in instances where dedicated funds have been made available (and funds have been established in many countries), these funds are not the panacea for the lack of investment in energy efficiency. Some funds have gone unutilized or have required extensive promotion, subsidized interest rates and below-market terms before they were fully disbursed. Other funds have existed for years and have been successful.[1] The mixed results of these funds emphasize that the solution to financing energy efficiency is more complicated than simply directing that capital be made available.

We believe that part of the solution lies in improving business fundamentals—in building a more thorough understanding of the commercial markets and business opportunities for energy efficiency products and services, in developing sound business plans with experienced project sponsors, in conducting an increased amount of careful project identification and in putting together packages that meet the credit criteria of financial institutions.

Bankers say that if a project is properly structured, if it makes fundamental business and economic sense and the project's sponsors are trustworthy and experienced, the project can be financed. If this is true, why should it not apply to energy efficiency projects? Bankers also say that they cannot afford to spend time analyzing credits unless projects are ready and likely to go forward—soon. While we wish bankers would be more broad-minded, we understand why they operate the way they do.

We acknowledge that accessing traditional sources of private capital requires that energy efficiency investments be molded to meet the requirements of existing financial institutions. We are unable to estimate the portion of the market that can be financed this way, but experience leads us to believe that it is significant. This is the untapped potential of traditional methods of financing. Traditional financing includes commercial lines of credit and term loans, trade finance, consumer credit, mortgage financing and project financing. These methods may be applicable for new investments as well as energy-efficient retrofits.

Much is also heard about innovative methods of financing, and many different definitions of innovation exist. To us, innovation in energy efficiency financing means: (1) applying known financing structures, such as performance contracting, in new countries, (2) designing and implementing market interventions in instances where market forces are not working, (3) encouraging new entrants (e.g., venture capitalists, utilities, investment fund managers, multilateral financial institutions) to finance energy efficiency, or (4) devising ways in which increased concern for global environmental issues will cause energy efficiency investments to be financed through joint implementation or emissions trading credits. Not surprisingly, we cannot estimate the amount of innovative financing that will occur over the next several years or the portion of total financing that it might represent. However, our experience and judgement indicate that innovative financing is well worth pursuing.

This chapter sets forth our ideas on how to finance energy efficiency through both traditional and innovative methods. It draws on recent work conducted under the U.S. Agency for International Development's Energy Efficiency Project in which Hagler Bailly is the prime contractor. The Energy Efficiency Project is active in many developing and post-communist countries conducting policy and regulatory reforms that support energy efficiency, implementing demand-side management and integrated resource-planning techniques, assessing market developments and promoting investment and through information dissemination. The chapter also draws upon the 15 years of experience Hagler Bailly has gained from previous energy conservation work for USAID, other international agencies, foreign governments and private companies. Some of the ideas suggested here represent hypotheses, based upon our experiences, that have yet to be fully researched or tested. A more in-depth report, published by the USAID and written by Hagler Bailly, incorporates more detailed research than is included in this chapter.

TRADITIONAL SOURCES OF FINANCING

Financing for New Investments

Many energy efficiency investments are imbedded in the goods and services that constitute investments in new plant and equipment. In these instances, the energy efficiency component can account for a relatively small portion of the total investment. This occurs in new commercial and residential building construction where the latest energy management systems, windows, lighting and HVAC systems are installed, in newly constructed industrial facilities that procure energy-efficient process controls, or the simple consumer purchase of a new, more efficient, electrical appliance. In countries with rapid economic growth, such as Indonesia, Thailand and China, new investments can be in excess of half of the energy efficiency market. In these instances, the costs of energy efficiency are included in the total financing package such as the mortgage,[2] commercial bank loan, construction loan, municipal revenue bond or project loan from a multilateral financial institution such as the World Bank or a financial intermediary that borrows for on-lending. In many cases the internal financial resources of the energy user or consumer will suffice, requiring no external financing.

Generally speaking, when included as part of a new investment, financing for energy efficiency can be accomplished through the range of traditional sources of financing already mentioned. However, our studies indicate that only a small portion (less than 10-20%) of new, productive investments adopt energy efficiency technologies in project designs and procurement. For example, on a global basis, energy-efficient lighting is used in less than 1% of potential applications in new product sales.[3] In these cases, the availability of capital is not an impediment. Rather, the decision to invest in energy efficiency at all becomes the overriding market driver. The solution, therefore, lies in providing better investment incentives (e.g., fiscal incentives, market-based incentives such as unsubsidized energy prices, or buyer incentives). More aggressive marketing programs would also be helpful. Programs that increase the market penetration of energy-efficient goods and services will have a more significant impact than increasing the amount of financing which may already be available.

Some countries, such as Japan and Brazil, offer attractive fiscal incentives such as tax credits for investment in energy-efficient products or accelerated depreciation benefits which provide immediate tax benefits. Some countries (e.g., Pakistan, Morocco, Bulgaria, Romania, Egypt) provide direct grants for a portion of the energy efficiency equipment, as well as for institutional development. These programs have stimulated market development.

Trade Finance

Wherever the discussion of financing includes international markets, the subject of trade finance must be introduced. Many countries do not manufacture the necessary equipment and must import all or some of their energy-efficient equipment (e.g., heat exchangers, high-efficiency boilers, heat distribution controls, steam optimization technologies, air conditioner and chiller equipment, low-energy lamps and reflectors, advanced motors and drives). Trade in services (e.g., design engineering, management or technical) may also be desired. Whereas imports of equipment and services can be financed through a trade finance component included in an initial financial plan, or through a competitive bidding process tied to a multilateral development bank loan, excluded are the large numbers of retrofit projects or purchases of single pieces of equipment. These retrofit and individual procurements require access to trade financing mechanisms such as letters of credit confirmed through the commercial banking sector, medium- and long-term buyer's credits, trade insurance policies and relationships with financial intermediaries that work alongside commercial banks and export credit agencies. For large, capital-intensive items, it may make business sense for manufacturers to work directly with export finance divisions of commercial banks or directly with export credit agencies. For multiple sales of lower-priced items, however, companies must work through intermediaries. In addition to commercial banks with trade divisions, equipment distributors, agents and energy service companies can take on this important role as intermediary. While intermediaries can be located in the importer's or exporter's country, local intermediaries have the advantage of being much closer to their markets.

Equipment distributors, either through their manufacturers or through commercial banks, need to explore ways in which they can provide their customers with credit (e.g., short- or medium-term loans, credit enhancement guarantees or equipment leases). This type of credit could be structured similar to the commercial credit business of the Ford Motor Credit Company (for cars, trucks, tractors and unrelated lines of equipment) or the Singer Company, which provides loans throughout the world for the purchase of its sewing machines.

Vendor financing is not new, not in a trade finance context, and maybe not in the context of energy efficiency either. In Mexico, for example, in connection with a pilot industrial motors project implemented by the Fideicomiso de Apoyo al Programa de Ahorro de Energia del Sector Electrico (FIDE), a portion of the financing for the purchase and installation of the motors will come from General Electric, the manufacturer of the motors.

In addition to financing direct from manufacturers, vendor financing may require that groups of equipment distributors join together to obtain economies of scale in the development and application of trade finance expertise or in providing a large-enough amount of business to interest a commercial bank already active in trade finance. It is also possible that access to trade financing could be improved by special lines of credit for imported equipment at commercial banks. In order to justify the development of these special lines of credit, end users would have to demonstrate that sufficient demand for the product exists.

Typical trade credits range from short-term (i.e., 30 days to one year) to medium-term (i.e., about three years). Some long-term trade credits of up to ten years are also available, depending upon the life of the equipment and the country risk. Credits of shorter maturities will not be suitable for all types of equipment, especially for certain large-ticket items (e.g., waste-heat recovery systems) that may have payback periods that exceed three years. However, many low-cost/no-cost energy improvements that repeatedly show up in energy audits have rapid payback calculated in numbers of months, not years. These projects (e.g., installation of steam traps, acquisition of energy-auditing and measuring devices, combustion control systems) lend themselves well to the traditional short-term trade finance system.

In many countries, a shortage of foreign exchange, credit and also trading traditions, mean that trade finance may be conducted through barter and counter trade. The countries of Eastern and Central Europe and the former Soviet Union fall into this category. The potential market for energy efficiency in these countries is large. As countries transition to market economies, increases in energy prices provide a strong incentive to conserve energy. Also, as former state-owned industries struggle to survive, competitiveness becomes an important issue and one where energy efficiency can be an integral part of the solution to becoming competitive. Energy intensity in many former Soviet republics is about three times higher than in the United States and Western Europe, thus making the market potential quite high.

However, difficult economic and business conditions are a barrier to conducting business and obtaining financing. Where interest rates and inflation are in excess of 100%, traditional trade credits are available only for very short periods of time and must be associated with hard currency repayments. Counter trade and barter transactions represent one viable way to obtain better trade terms and the hard currency component of the transaction. We think it is good advice for those pursuing energy efficiency business in transitional economies to invest in the special expertise of barter specialists and counter trade experts.

Local Financial Institutions

Local financial institutions are arguably the most important players in financing energy efficiency. Local financial institutions have the potential take on a wide variety of important roles: (1) as financial intermediaries with international institutions, (2) as direct lenders and guarantors, (3) as lessors, (4) as mortgage and construction lenders, (5) as fund administrators and agents, and, not to be overlooked, (6) as experts in the business conditions, companies and economies of particular countries.

It is difficult to generalize about the current level of involvement of local financial institutions in financing for energy efficiency, since this varies dramatically from country to country and is the subject of additional research. However, one generalization that can be made is that there are an insufficient number of local financial institutions engaged in energy efficiency lending and this is a major barrier to implementing projects. Exceptions to this are in Hungary, where the Energy Savings Office of the Magyar Hotel Bank has been making energy efficiency loans since 1987 with funding from the German "coal aid" fund and the World Bank, or in India, where the Industrial Credit and Investment Corporation of India, Ltd. (ACACIA) has been working with the Asian Development Bank and the U.S. Agency for International Development for over five years to strengthen its management's ability to support energy efficiency projects and act as an agent bank in the development and commercialization of energy-efficient technologies.

Local financial institutions can also play an important role in financing the construction and start-up of local manufacturing capability and to upgrade the energy efficiency standards of existing products, such as residential and commercial appliances and lighting. Both of these options may require significant financing from local financial institutions. This is already occurring in Thailand, Mexico, Brazil and elsewhere.

We think that future support for the development of energy efficiency financing must focus heavily on the capability of local financial institutions. Through our work with USAID we are developing new initiatives to assist local financial institutions increase their role in energy efficiency financing. This can be done through training programs for bankers on energy efficiency loan evaluation, by developing manuals for widespread distribution, by providing financial institutions with the latest information on technologies and projects and through working with local financial institutions that act as fund administrators and intermediaries for international financial institutions. We believe that financial institutions must be carefully selected, where the size of the energy efficiency market and the level of interest of the financial institution are important selection criteria.

Project Identification and Business Development

Given our experiences over the last several years, however, we also believe that working with local financial institutions alone will not address many of the barriers that exist in energy efficiency financing. As mentioned earlier, many barriers relate to business and economic fundamentals as well as the absence of financing. Energy prices may be too low to justify an investment in energy efficiency. Imperfect market information may cause business opportunities to go unnoticed. Energy end users may be uncreditworthy or be concerned with technological risks. Import tariffs may be prohibitive and render a project uneconomic. Local partners may not have the necessary mix of skills.

Removing barriers to financing also requires that project sponsors, investors, energy users and developers know how to identify, develop and package bankable projects. One significant problem, however, is the high cost of business development and project identification. Who will invest in business development and project identification—equipment vendors, energy service companies, venture capital firms, public sector agencies, industry associations or a combination of the above? Which of these entities has the resources (managerial and technical, as well as financial) to finance business development and project identification? Until entities are willing to provide the funding for these activities, the pipeline of project opportunities will languish. We offer three suggestions for dealing with these issues.

The first suggestion is to involve the public sector in business development and project identification activities. In an era of declining public sector resources this may sound counter-intuitive, but we feel it is consistent with the public sectors desire for sustainable development, environmental improvements and resolution of global environmental problems. Business development and project identification activities are a necessary adjunct to support for policy and market reforms. It may be expected that interest in efficiency on the part of the public sector will not abate, and, wherever possible, the public sector will want to provide "market-push" interventions. We think that business development and project identification are good uses of public sector resources, provided they are properly designed, evaluated and proven effective.

The second suggestion involves encouraging energy efficiency project developers, equipment vendors and other private-sector parties to take a closer look at the current market situation and see for themselves that recent worldwide trends are making the market for energy efficiency products and services more attractive. In the 1970s, rising oil prices greatly improved the project economics for energy efficiency. In the 1980s, in the United States, utility involvement in demand-side management created new markets for energy efficiency products and services. In this decade and into the next, global and regional economic and political trends are changing the way in which energy efficiency will be implemented; again, project economics for energy efficiency investments are improving.

The movement towards increased international competitiveness and liberalized global trade, power sector restructuring, rationalization of energy prices (still at an estimated $100 billion/year subsidy) and improved technologies are good reasons for the private sector to take a closer look at energy efficiency projects. Another good reason is market size and growth. Hagler Bailly recently estimated the world market for private sector purchases of energy efficiency products and services at over $80 billion per year.[4] Estimates of market growth range from 4%-10% per year with the highest growth expected in developing countries and emerging market economies. Now is an opportune time to commit new resources to energy efficiency project development.

A third suggestion is to package or pool project identification and business development activities so that larger pieces of business may be developed. One example of this approach, currently being developed in Slovakia, is in the development of an energy efficiency retrofit program for all of the hospitals in a municipality or region under one contractual and financing umbrella. From a developer's viewpoint this approach may be preferable to developing a single hospital retrofit and then attempting to replicate it.

Project Financing (Limited and Non-Recourse)

In connection with energy efficiency, project financing can be considered to be both a traditional and nontraditional source of financing. (Repayments under project financing are made solely from the cash generated by a specific project.) It can be considered a traditional source of financing because it has been used for decades to finance capital-intensive projects. It may be considered nontraditional because its applications to energy efficiency, especially outside of the United States, have been limited. Project financing is now a popular financing mechanism for structuring privatized infrastructure investments, such as private power, water and wastewater treatment facilities, and privately owned and operated toll roads. We are not aware of similar popularity of project financing for energy efficiency, except for cogeneration projects. We think there is potential in project financing—for many more cogeneration projects, and for large-scale projects entailing the implementation of common energy efficiency measures across entire industries (i.e., "generic ECMs"). However, we also realize that the transaction costs for project financing are high and that this is not a model easily replicated for projects of a smaller size. Financial institutions engaged in project financing are generally involved in projects with total costs not less than $5-$10 million and ranging upwards to several billion dollars. Project development costs for even small projects can reach $1 million. Feasibility studies, which generally are a prerequisite to project financing, may cost at least $300,000.

For projects above a minimum threshold size, project financing can be an excellent tool for obtaining financing in markets where significant barriers remain. Project financing can be used effectively to mitigate poor credit and certain country risks. Performance contracting and shared savings are variants of project financing in that they rely upon cash flows from a project for repayment and are effective tools for dealing with credit and country risks. Performance contracting is discussed separately under the section on innovative financing, below.

Public Sector Support

For policy reasons, public sector agencies have long been involved in the financing of energy efficiency. In addition to grants, low-interest loans, guarantees, tax credits, technical assistance and legislative mandates, the public sector pulls the market by being a large market buyer and market participant. Hagler Bailly recently estimated that energy efficiency procurement of the national governments of OECD member nations is greater than $2 billion annually, or 3% to 4% of the OECD energy efficiency market.[5] The public sector in OECD nations offers additional support in the form of over $600 million of government-sponsored energy conservation research and development annually.[6]

In the United States, energy efficiency has been supported by federal energy management programs, fiscal incentives, PURPA and state-supported grants and loan programs. Elsewhere specifics vary but most European countries have supported energy efficiency programs in a similar manner. Japan has been an active supporter of energy efficiency since its experience with the oil supply shock of 1973.

Developing-country governments have also been involved, either on their own or in cooperation with international foreign assistance agencies such as USAID. In Morocco, Mexico, Egypt, Jamaica, Thailand, the Philippines, Russia and the Baltics (just to name a few), governments have set up energy conservation programs, implemented demonstration projects, conducted industrial energy audits and provided grants or loans for imports of equipment. When designed and implemented well, most of this support has been very successful.

In this era of reduced public sector budgets and encouragement of increased private sector activities, what role can governments play in energy efficiency financing? The simplistic answer is that they should continue to design and implement the types of programs that have proved successful, but do so with fewer resources. It means that new programs will need to be considered in terms of their sustainability, their leveraging capabilities and the impact that they have on encouraging increased private sector involvement. This could mean more tax incentives or investments in project development, it could mean developing new partnerships with project sponsors and local financial institutions, it could mean providing seed capital for special-purpose revolving funds. This chapter does not attempt to list or describe the wide range of public sector options available to support energy efficiency.

INNOVATIVE FINANCING

The different skills necessary to conduct business internationally, combined with the peculiarities of the energy efficiency business, require that innovative approaches to doing business be developed. Innovative financing for energy efficiency entails the development of new financing mechanisms tailored to the specific needs of energy efficiency. It recognizes that with energy efficiency there is a large number of small projects and a small number of large projects, requiring the packaging or pooling of investments. Innovative financing for energy efficiency also recognizes that the market for energy efficiency financing is driven by policy and economic conditions—most obviously by energy prices, the pace of commercialization and reform in the energy sector, the level of government concern for environmental issues and commitments to solving global environmental problems. The players in innovative financing for energy efficiency thus include all types of institutions, some of which are not typically associated with financing: energy service companies, utilities, governments, venture funds, energy users, local financial institutions and industry associations. The field continues to be open to new ideas, market entrants and models of innovation. Three types of innovation are discussed below.

Performance Contracting and ESCOs

Recent efforts to apply known financing structures for energy efficiency to new countries have focused on support for energy service companies (ESCOs) and the transfer of performance contracting and shared savings. ESCOs take on the multiple roles of energy services contractor, risk mitigator and arranger of financing. In the United States, the business of performance contracting and shared savings has proven successful and been replicated numerous times. Utility involvement in demand-side management programs has helped sustain the growth of the market for ESCOs. However, even in the United States, performance contracting is not a billion-dollar market; most of the market participants are small businesses (although several market participants are subsidiaries of electric utilities). In Europe, where performance contracting is generally referred to as "third-party financing," examples of successful ESCO businesses also exist.

Efforts to apply the ESCO model of performance contracting in developing countries are still relatively new and perhaps it is too soon to predict its long-term applicability and replicability. Over the short term, results have been mixed. Several companies have committed to investments only to pull out of them at a later date. Other ESCOs have conducted initial business development and concluded that the development costs were too high, the financing unavailable or the risks unmanageable. Some companies have been successful and have executed performance contracts—EPS in the Czech Republic is one example, Intesco in India,is another.

One hypothesis for why the success stories are few is the difference in skill mix and resources between ESCOs in the United States and the requirements of doing business internationally in developing countries and emerging economies. Few of the ESCOs in the United States have in-depth experience in international business, including knowledge of how to obtain international financing and find the best local partners. While international business skills can be built or acquired, as small businesses many ESCOs do not have the financial resources to sustain high market-entry costs. Another hypothesis is that the market is currently not large enough to justify investing in business development and project identification.

We are certainly not ready to write off ESCOs as a viable financing model for energy efficiency. To the contrary, we believe that performance contracting, together with ESCO development, represents one of the better ways to address several of the most frequently mentioned barriers to investment. Performance contracting transfers the technology and management risks away from the end user to the ESCO. For energy users reluctant to invest in energy efficiency, a performance contract could be a powerful incentive to implement a project. As energy management contractors, ESCOs possess the technical and managerial skills needed to perform on specific contracts. More importantly, ESCOs possess the experience needed by local partners in building viable businesses able to respond to multiple opportunities and to assistance in promoting energy efficiency projects.

For these reasons we feel that support for ESCOs is a good idea. We suggest that one way to support ESCOs is general support for sustained growth in energy efficiency markets. While this may not lower the market entry costs, improving business fundamentals will enhance the ability of ESCOs to develop new partnerships and alliances with a variety of participants, providing them with many of the skills they currently lack.

New Market Entrants: Utilities, Environmental Groups, Venture Capitalists

The entry of new market participants is an important area of innovation because it serves to fill gaps in the skills and resources of existing market participants.

The example discussed above with ESCOs deserves greater elaboration. ESCOs, as small businesses, lack the financial resources to invest heavily in business development and project identification. One suggestion is for an ESCO to assemble a business plan demonstrating the potential for profits and present it to the venture capital community for financing. While the venture capital community is not known for its expertise in international financing, this is changing quickly. Several funds have been started that have stated interests in environmental investing, including energy efficiency. Funds established to invest in renewable energy will also consider energy efficiency investments. Venture capital funds won't provide 100% of the financing required, but based upon the successful track record of an ESCO in the United States or elsewhere, and its demonstrated commitment to the business plan, venture capitalists may be willing to provide a significant portion of the required financing in return for a share of the profits.

Other entities could make good financial partners. Private power developers are learning quickly how to develop international projects; some of them have considerable financial resources. Local private sector companies are also good partners for development in that they understand the markets much better than foreign companies and they frequently have access to capital from local financial institutions.

One other category of market participant needs to be mentioned here. Certain market participants can create market demand and "pull" the market. These participants are important to identify. In the United States, through demand-side management programs, utilities have created market demand. There is no reason why they cannot do the same outside the United States. Involvement by energy efficiency suppliers with utility demand-side management programs is an important approach to creating more business overseas.

Governments also create market demand, through legislative mandates, sector reforms, institutional reforms and investment incentives. Support and participation in the wide range of government efforts will certainly help create markets for energy efficiency suppliers.

Environmental interest groups, concerned about local and global environmental problems, can also help pull the market. Some of these groups have significant financial resources but are not knowledgeable about how to translate their policy objectives into viable business opportunities. We believe that efforts to channel their resources to specific projects or towards creating a framework that supports market growth will have a positive effect on energy efficiency financing. We encourage energy efficiency suppliers to play an active role in helping environmental groups understand how best to allocate their resources.

Global Environmental Issues as a Source of Financing

In the future, we believe that environmental issues will drive energy efficiency markets in new ways. Today, concern about global warming causes a great deal of discussion about the level of damage and methodologies for treating it, but we believe that, in the future, global warming concerns will mandate that investments be made to alleviate the problem. This will cause energy efficiency markets to continue their sustained growth, again requiring new methods of financing.

One of these methods is known as joint implementation. Joint implementation refers to programs of international cooperation on climate change in which industrialized countries seek to "balance" their emission of greenhouse gases in conjunction with developing countries by using emissions reduction efforts and the enhancement of greenhouse gas sinks. Specific mechanisms for using joint implementation as a source of financing for energy efficiency have not yet been developed, although discussions have begun.

A major market driver for the implementation of greenhouse gas mitigation strategies, and energy efficiency measures to the extent that they contribute to this goal, is the potential for an emissions-trading program. The goal of such an approach is to harness market forces to achieve environmental targets at the lowest cost to society. While traditional environmental regulation prescribes a fixed emission limit for each regulated facility, allowance-trading programs set emission reduction goals for the regulated community as a whole. The regulatory authority then allocates allowances to the various facilities. Those facilities that have invested in efficient technologies and can reduce their emissions more cost-effectively have an incentive to over-control their emissions and sell their unused allowances to entities that face greater emission-reduction costs. As such, the private sector decides how to most efficiently reduce overall emissions to the mandated level.

In the U.S., an allowance-trading program has been used to reduce the emissions that cause acid rain. The prospect of wider application of allowance-trading programs around the world for a variety of emissions still remains subject to future developments. However, allowance-trading permits regulated entities to capture the value of emission abatement and may provide a powerful incentive for private enterprises to invest in energy efficiency and other emissions-abating technology.

CONCLUSIONS

There is no single solution for addressing the financing constraint for energy efficiency investments. Approaches that combine traditional sources of financing and business development with innovative approaches are required. Traditional sources of financing combined with investment incentives are likely to be a good source of financing for increased market penetration for new projects. This requires more expertise in trade financing and possibly counter trade. Vendor financing programs which provide commercial credits may be suitable for individual product programs, such as industrial motors, steam traps and measurement equipment. More active participation from the local financial community is critical because they are in the best position to serve multiple roles. Where local financial institutions are undeveloped, project financing and continued government interventions can fill the gap in the short term.

One of the most important causes of the unavailability of financing for energy efficiency is not lack of capital, but imperfections in the market. We believe that by supporting market developments, many of the financing constraints will be solved without intervention. The energy efficiency market is already heading in this direction, spurred on by the reform of energy pricing, privatization and restructuring, and, in some countries, serious energy shortages.

Innovation requires new entrants, an eye towards tracking future trends and support for performance contracting and ESCOs. It requires new and unusual pairings—ESCOs teaming with local financial institutions who understand industrial lending, or governments getting into the business of project identification and business development.

Of course, we're not sure we're on the right track. There are many issues for further research, some of which are ongoing. How will we know if we are on the right track? Only if widespread gains in energy efficiency and profits for energy efficiency suppliers result from our suggestions will we be convinced that they are valid.

Endnotes

1. For example, the USAID-supported Technology Transfer for Energy Management (TTEM) Demonstration Loan Fund (DLF) in the Philippines has proven successful. After changes were made in DLF terms, including raising the loan ceiling from $100,000 to $200,000, all funds under the DLF were committed in less than two years. Because of the revolving nature of the loan fund, the DLF is replenished as loans are repaid. However, not all programs are this successful. As part of the World Bank's first Energy Sector Loan, $5 million was set aside for energy conservation projects identified by the USAID-sponsored ENERCON program in Pakistan. Although ENERCON identified numerous attractive projects and completed feasibility studies on ten, these funds were never utilized. Potential borrowers cited a high interest rate, cumbersome application procedures and uncertainty regarding the value of energy conservation as an investment.
2. In the United States, residential mortgage financing has also been used in innovative ways to encourage energy efficiency financing. Programs exist where the total mortgage can be increased to cover the costs of an energy conservation component and the borrower's qualifying ratios are lowered to accommodate the increase in total monthly payments.
3. Henri-Claude Bailly. *The Global Market for Energy Efficiency.* Presented to the World Energy Efficiency Association, Istanbul Finance Seminar, April 11-12, 1995. Arlington, VA: RCG/Hagler Bailly, Inc., 1995.
4. *Ibid.*
5. *Ibid.* This $2 billion figure excludes a great proportion of total national government energy budgets which may encompass additional energy-efficiency components, and may exclude expenditures below the national level for some countries. Also excluded are demand reduction programs of electric and gas utilities, on which spending totaled $2 billion in the U.S. alone in 1992 (IEA, 1994).
6. International Energy Agency. *Energy Policies of IEA Countries: 1993 Review.* Paris: OECD, 1994.

Chapter 9

Utilities: Emerging Opportunities In Performance Contracting

G.W. Wood*

Many utilities are struggling to decide on the types of services they want to offer their large commercial and industrial customers. A&C Enercom is often consulted to help utilities develop packages of services for these customers. Many of the utility services being developed for the large commercial and industrial sectors include those offered by a performance contractor. These services, including engineering assessments, design, construction management and financing, can be offered by utilities in a total turnkey approach with repayment on a paid-from-savings, or performance contracting, basis.

Historically, utilities have used performance contracting companies (also referred to as energy service companies or ESCOs) to fulfill demand-side management obligations to regulators. Although many of these contracts have been successful, the utilities often have mixed (if not outright negative) feelings toward the performance contractors. These often-adversarial relationships stem in part from the lack of utility control over an ESCO's relationship with its largest commercial and industrial customers. In many cases, a utility is concerned that the ESCO will follow up the lighting or motor retrofit program with a cogeneration or other load reduction program. There is, however, an opportunity for utilities to take the knowledge and experience of the performance contracting industry they helped create and apply it toward the challenges

---

*George Wood is a Director of TRITECH, a division of A&C Enercom, serving utilities and their large commercial and industrial customers.

they face in the emerging competitive marketplace.

What type of performance contracting could utilities offer their customers? What customers are the ideal target for such services? What kind of culture change is needed within utilities to support the development of these services? How could such services complement existing utility products/services such as electricity sales, metering and equipment maintenance? Although no one can know just where the utilities and their regulators will take the industry, nearly everyone agrees that it will become more competitive.

UTILITIES: THE CURRENT SITUATION

It's happening now—utilities' largest C&I customers are buying their electricity elsewhere. Some of these customers are installing their own power generation equipment (cogeneration), some are partnering with their communities and "municipalizing" and still others are already wheeling power from other sources. Utilities find themselves in a defensive and reactive position. They are hastily forming new rate options or contesting the latest cogeneration offer or, in the worst case, struggling to hold on to power sales for just a portion of a facility.

A&C Enercom, through its TRITECH business group, possesses a great deal of experience in helping utilities that face these often defensive positions. Our experience gives us a unique insight into why utilities have been surprised, how this situation might be avoided in the future, and how to turn this experience into the development of a new business. A brief examination of the causes of market changes and utilities' past lack of preparedness would be helpful in understanding the platform from which utilities can launch new business opportunities.

UTILITIES:
A DRAMATICALLY CHANGING MARKETPLACE

Increasing competition is forcing the utility industry to undergo enormous changes. The factors driving this increased competition include regulation, alternate power generation and customer demands. These changes are forcing utilities to wrestle with their current core competencies and identify those needed for future survival.

Regulation

Recent government deregulation started with the Federal Energy Policy Act of 1992, which allowed wholesale transmission of power. This action was followed by the impending retail wheeling experiments in a number of states. A form of retail wheeling is already occurring in a number of communities. In these cases, a large industrial customer joins with the local town officials to create a new municipal electric or gas utility serving only the city or town. Once declared a municipal utility by vote of the town and regulators, the new utility can buy "wholesale" power that can result in as much as a 30% reduction in energy costs.

Alternate Power Generation

Cogeneration equipment sited at a customer's facility often produces a lower cost for electricity and the required thermal energy. The increasing number of operating and planned cogeneration facilities at customer sites has caused many utilities to develop special rate options—sometimes as much as 20% less than current industrial rates.

Customer Demands

Virtually every major industry journal is writing about the coming age of retail wheeling and the new-found power of their readers, who can now pressure their utilities for rate concessions. Our utility clients get customer calls every day asking for "competitive" rate options. Many utilities now have these rate options for customers who can document a serious intention to install cogeneration or enter a municipalization agreement with their local community.

Each of these factors contributes to customers' unhappiness with their existing utility service. The principal reasons utilities are often surprised by their customers' unhappiness include the lack of solid relationships with a customer's decision makers and an inadequate understanding of the customer's needs. Until recently, the message from utility management to field sales representatives has been to service each account, not "mine" the account for business opportunities. This atmosphere has led to the current "reactive" sales culture.

The reactive service representative responds only to the call for metering, billing, power quality or substation problems. Rarely does this representative probe for other concerns and opportunities. In addition, they often have great relationships at the wrong level in the customer management hierarchy. The utility representative typically knows the facility maintenance supervisor, the shift foreman or billing manager very well. The representative has been on a first-name basis with these folks for years—but rarely has this rep attempted to develop relationships with the key decision makers at senior levels. This type of customer relationship is no longer acceptable in a competitive marketplace. It is clearly not an acceptable relationship for utilities wanting to develop new products and services.

Changing the Culture and What's Next?

There is now a broad-reaching attempt in nearly every utility across the country to radically change the culture within the utility sales and customer service departments. Utilities want to change the business atmosphere from a reactive, non-competitive one to an aggressive, sales-driven culture.

Nearly all utilities have restructured sales management, restaffed their sales departments and developed new sales training programs. There is no question that attention to customer needs, especially in the large C&I sectors, is improving. Although the culture change is far from complete, work has begun.

Once utilities begin to make these changes, they start to focus on the next step: "How do we capitalize on our relationships with customers and diversify our business into other products and services?" *A&C Enercom/TRITECH believes that performance contracting (or its component parts) can be one of many value-added services that utilities offer to their customers... and those in other utility territories.* Why? Because all of these new activities lead to new business opportunities, including increased energy sales.

WHY SHOULD UTILITIES OFFER "VALUE-ADDED" SERVICES?

Utilities that begin the transformation **NOW** *from* regulated monopolies *to* competitive service providers will be better prepared for the upcoming competitive power/service marketplace. Some other benefits of adding services include:

- Increased sales of energy
- New significant center of revenue and profit
- Support retention efforts

- Aid in culture change for utility
- Begin staff training before the arrival of full access with current customers

The remainder of this chapter explains how performance contracting or its component services can enhance utilities' ability to create and offer successful value-added services. In addition, they will build customer relationships, increase revenue and possibly create new, unregulated services. First, we should define what we mean by performance contracting.

Performance Contracting: A Definition

Performance contracting involves the turnkey installation and financing of equipment with extended payment for these services based on "performance." The equipment, often installed to reduce net energy costs or improve productivity, will have some previously agreed-upon mechanism to document results. The performance results, whether estimated or metered regularly, will form the basis for calculating periodic payments to the contractor by the customer.

For example, a utility wants to retrofit a customer's large lighting project to energy-efficient lighting. This would result in a 3,000,000 kWh reduction in electricity consumption, giving the customer the same light level. If the utility engaged a performance contractor, this company would typically bear all costs for this installation and take the risk that the energy savings levels will remain for as many as 10 years. The utility will pay the performance contractor only on a measured basis—the measurement frequency and methods upon by the utility, the customer and the performance contractor. Should the savings level drop, then payments to the performance contractor drop (hence the terms "performance" and "paid from savings"). This approach has been applied to more complicated technologies, such as major process improvements, variable frequency drives on motors and energy management systems.

Performance contracting includes many discrete parts that can be custom-packaged for each customer. These services, in a rough chronological order, include:

- Preliminary and detailed facility assessments
- Cash flow analysis
- Meter installation

- Performance specification
- Detailed engineering design
- Bidding
- Equipment installation
- Commissioning
- Financing
- Maintenance
- Ongoing metering and billing

Utilities have an opportunity to apply these services, as a package or individually, to support their *reactive* customer retention activities as well as the *proactive* sales growth opportunities. Understanding the current relationship between performance contractors and utilities is helpful when thinking about how a utility department or unregulated subsidiary could offer these services.

PERFORMANCE CONTRACTORS AND UTILITIES: THE CURRENT RELATIONSHIP

Hundreds of megawatts of energy efficiency have been installed as a result of the "block bid" DSM programs of the 1980s and 1990s. In these relationships, the performance contractor or ESCO contracts with the utility to install a block of energy efficiency projects and agrees to be paid based on documented kWh or kW savings. These contracts, typically lasting 5 to 10 years, are structured around meeting annual minimum construction milestones and annual savings levels. Once an ESCO has installed a block of energy savings in multiple-customer facilities, the utility then has the rough equivalent of a power generating station—except megawatts of savings are being "generated" at a lower cost than a new power plant could offer.

It has been this kind of contract that has nurtured the growth of the performance contracting industry, which is now responsible for hundreds of millions of dollars in construction. Some would argue that independent of performance contracting offered by energy management controls companies, the utility block bids are the sole reason for the existence of ESCOs.

Although often successful in meeting the DSM targets set by regulators, utilities have found the relationship with ESCOs to be less than satisfying in advancing the development of strong relationships with their large C&I customers. Some critics would say that the utility representative has been excluded from the transaction between the customer and the ESCO (see figure below). This, in many cases, has created an adversarial relationship between the utility and the ESCO. There is now an opportunity, however, to build on the experience gained from "Block Bid" ESCO services and craft a series of programs to be offered by the utility.

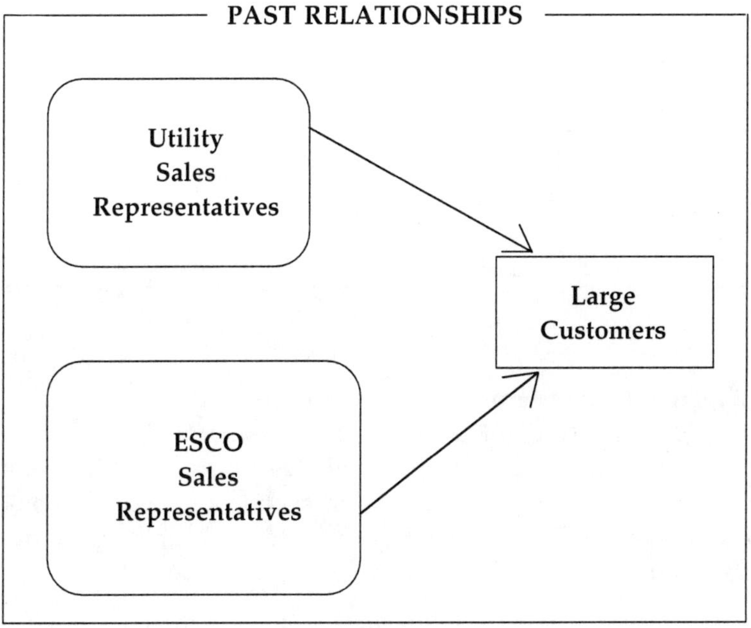

We often recommend that our utility clients repackage the beneficial parts of the ESCO "bundle of services" and offer these components to customers directly and eliminate the ESCO as a "middleman" (see following figure). Utilities *can* offer assessment, design, financing and construction services *themselves to* promote load growth, retention, energy efficiency and ultimately, profitability. The experience gained in building just such services will also prepare a utility's sales and service staff members to compete for customers in other utility service territories.

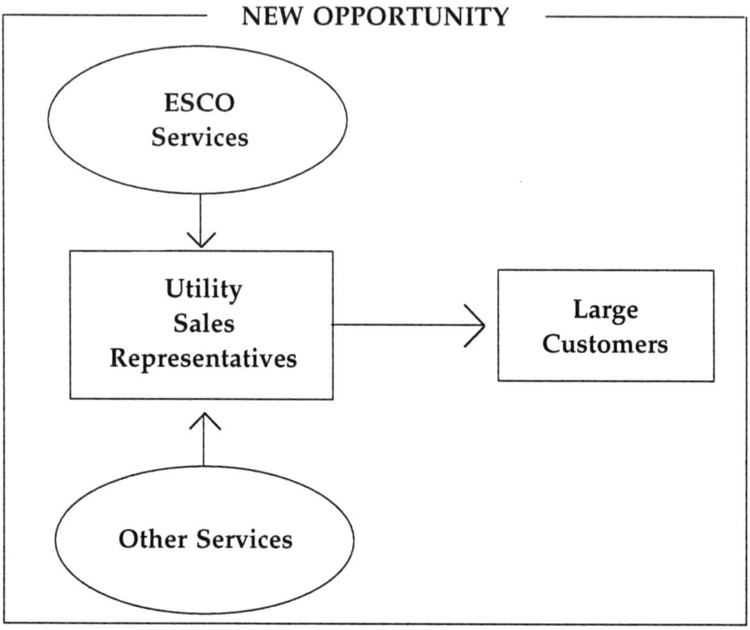

The Emerging Opportunity: Utilities Will Meet Their Customers' Needs Via Performance Contracts

Market research conducted by A&C Enercom's TRITECH, coupled with a heavy dose of common sense and experience, reveal a number of large customer needs that could be fulfilled by a utility performance contracting-type service. These customer needs include:

- Identifying, designing and installing electric and gas technologies
- Load management services, such as standby generation and direct load control
- Manufacturing and environmental assessments
- End-use pricing
- Power services, such as power conditioning and power factor correction
- Multiple metering services

- Energy brokering (in the future)
- Financing
- District heating and cooling systems

These services could be provided by a team of technologists, project managers, financiers and developers. This team can take a customer need, prequalify the opportunity and develop a full-service package. While this package of services is attractive, many are already being offered in the marketplace.

What is the Utilities' Competitive Edge?

Utilities enjoy a unique position in the overall energy marketplace, one that provides a strong competitive advantage for offering a full range of services to large C&I customers. These competitive advantages include:

- Customer name recognition and trust
- Utilities are (and some portion will continue to be) regulated
- Most utilities already own their power generators; other players do not
- Access to capital
- Existing support services infrastructure—they are ready to go
- Lead generation (customers already asking for services)
- "One-stop shopping"—utilities are able to offer diverse but related services from one company

Marketing Performance Contracting Services through Two Channels

Utilities could use two approaches to deliver a performance contracting-type service: the first is through the regulated side of the business (usually residing in the C&I sales and/or service departments). In this approach, a customer need, such as chiller replacements, would be identified by the utility sales representative and brought to a team of engineers and developers from both in-house and third party. This team, generally a separate group under the sales and/or marketing department, would assess the opportunity for equipment replacement and enhanced electric or gas sales and present a preliminary proposal to the customer.

The second approach, increasingly popular with utilities, is to create a separate company under the unregulated side of the company that would offer the same bundle of services but not as part of the regulated sales team. Here, the sales representative would follow a similar in-house path in developing a prospective opportunity, but the technical and development support would come from a separate corporate entity. There are several advantages to this approach including:

- Separate, competitive culture from the regulated utility
- Separate accounting system
- Freedom to develop projects off the system that could be future utility customers
- Company is not ratepayer-based and as such not subject to certain regulations
- Freedom to develop many types of end-use pricing "bundling"
- Easily integrated into the utility "family" of services

HOW DOES A UTILITY START A PERFORMANCE CONTRACTING COMPANY?

With a lot of help! There is a great opportunity, a strong customer need and a very optimistic outlook for future growth in servicing the large C&I sector with performance contracting services. Three strategies can be employed to develop this capability: 1) grow the business "organically," or from the ground up, 2) manage from within the utility but use outside services during start-up and 3) acquire companies that currently have performance contracting capabilities and a track record.

Organically grown performance contracting will require from 12 to 30 months before beginning to turn a profit. Our biggest concern with this approach is that while the utility is developing this capability, other

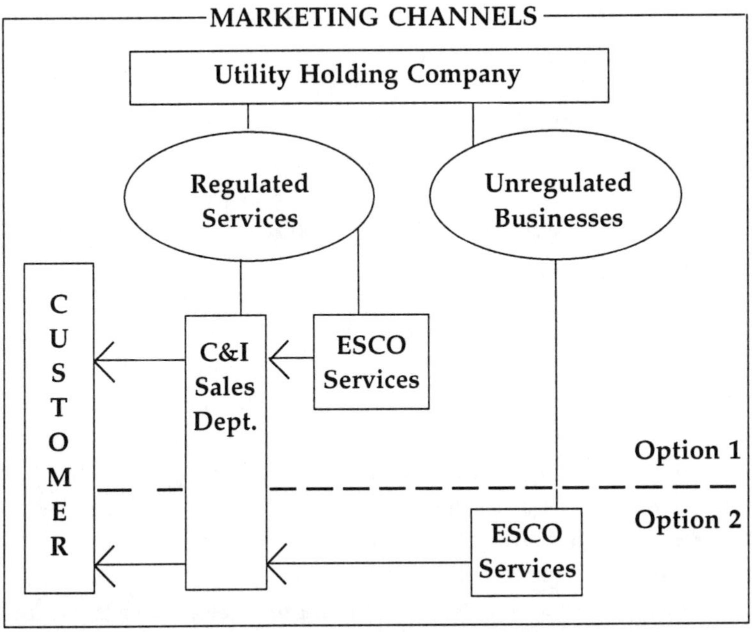

utilities may be stealing their customers with their own team of developers, technologies and contractors.

The second strategy, that of developing a performance contracting capability through creative partnerships with select service firms, could begin to turn a profit in as little as 12 months. In this scenario, the utility maintains control of the new service or company but can, through a service provider, offer full performance contracting capability on day one. These third parties can also conduct a methodical, account-by-account review to assist the sales representative in "mining" the opportunities in each of the larger customers. A program that is custom-designed for each utility's number and type of large C&I customers is often the most successful approach. Here, the utility staff works closely with the third parties and both learn through formal classroom training and on-the-job training. Keeping the service or new business flexible enough to respond to changing demands of the marketplace is also imperative.

The third strategy, development through acquisition, offers the same "day one" delivery capability and the potential for immediate profitability, depending on the terms and cost of acquisition. Care must be taken to carefully integrate new services into existing ones and to maintain proper control over the newly acquired company.

However the new service is developed, it must be flexible enough to respond to diverse needs. One industrial customer may need manufacturing assessments for expanded production capacity, another customer may need power quality testing, another is interested in fuel switching for paint drying and another may be studying cogeneration. The challenge is to offer this wide range of services while maintaining profitability.

The Customer Energy Service Company—CESCo

The opportunities are exciting—utility representatives are now walking into industrial customers' facilities with the capability of **identifying, installing** and **financing** process, environmental, resource efficiency or HVAC improvements. Utilities are profiting by offering these turnkey services **and** by frequently obtaining additional electricity or gas sales. These same utilities are fine-tuning their energy services and contracting with large customers in other service territories. The time to develop energy services capability for large C&I customers is now... **while utilities still have their customers.**

Chapter 10

Financing Energy Projects: Success Stories

An independent school district in Texas needed to renovate its air conditioning system and improve air quality. The project was estimated at $1,000,000 with a simple payback of five years (estimated savings of $200,000 per year). Although the public wanted the schools in the district to undergo renovations, a bond-issue referendum was defeated. This left the district's board of directors in a difficult situation: the equipment was over 15 years old but there were no funds in the budget or available through a bond issue.

Solution: The school district applied for a municipal lease. This program offered an interest rate similar to a bond issue (6%) and allowed for 100% of the project to be financed. With a term of 10 years, the lease had payments of $133,225, which the district could easily afford out of project savings.

A U.S. Army base was faced with having to minimize costs in order to avoid being closed due to federal budget cutbacks. Since the base had large numbers of residences and an outdated HVAC system, a retrofit project was identified. The base's budget did not include the $15,000,000 necessary to complete the work.

Solution: ESPC was chosen as the financing methodology. An ESCO was selected. It did not have the borrowing capacity to handle the job but had developed financing relationships with various investment companies. Those companies analyzed the project and the ESCO's track record with similar projects. An approval was issued for a loan which represented 80% of the total project costs, or $12,000,000, at an interest rate of 9.75% for 15 years. The ESCO contributed the difference ($3,000,000) and both the ESCO and the investment companies were compensated by the savings generated from the upgraded HVAC system.

An airplane-parts-manufacturing facility was facing an increasing volume of business with an outdated lighting system throughout various hangars. Although the CFO of the manufacturing company indicated that the capital was available to perform the project, at the last minute he was forced to use the funds for an investment in a foreign facility. The project seemed doomed to be delayed until the following budget year despite having a simple payback of only 2.5 years.

Solution: The ESCO that made the original audit and savings estimates suggested a capital lease. The leasing company reviewed the project and determined the manufacturing company had placed numerous bonds in the public markets. Standard & Poor's rated that debt at "A$^-$." Based on this rating and the latest publicly available information about the manufacturing company's financials (its stock traded on the New York Stock Exchange), the leasing company approved a four-year lease at 8.5%. Although the CFO was satisfied with the rate and approved the lease obligation, he was surprised to discover one year after the project had been installed that not only were the savings projections met, but the improved lighting caused a substantial rise in quality control and employee productivity.

A hospital built in the early 1970s in the Northeast was suffering from persistently overheated and underheated sections of floors and intermittent variations of water pressure. Hospital management recognized the national trend of hospital acquisitions by large health-care companies and wanted to prepare for a possible bid from an interested buyer. Unfortunately, the hospital's balance sheet was weighed down with debt from the construction of a new wing. A bank loan or lease seemed out of the question, but the hospital's basic operations were starting to suffer. A possible patient lawsuit was a real threat to the hospital.

Solution: After issuing a Request For Proposals, hospital management selected an ESCO which offered to purchase the HVAC and water supply plant from the hospital and contract to sell the system's services back to the hospital. The ESCO set up a single-purpose entity to own and operate the plant. An investment company reviewed the project and issued an approval for non-recourse project financing. The loan was for 10 years at 12%. The plant generated enough revenues from the long-term contract to sell the services and water to the hospital, pay the debt service and leave a profit margin for the ESCO.

Appendix A
Leasing Options

Guideline Lease/True Lease/Tax-Oriented Lease

Compliance with all of the IRS guidelines including those listed below is required:

(a) The total lease term (including extensions and renewals at a predetermined, fixed rate) must not exceed 80% of the estimated useful life of the equipment at the start of the lease, i.e., at the end of the lease the equipment must have an estimated remaining useful life equal to at least 20% of its originally estimated useful life. Also, this remaining useful life must not be less than one year, thereby limiting the maximum term of the lease.

(b) The equipment's estimated residual value at the expiration of the lease term must equal at least 20% of its value at the start of the lease. This requirement limits the maximum lease term and the type of equipment to be leased.

(c) No bargain purchase option.

(d) The lessee cannot make any investment in the equipment.

(e) The equipment must not be "limited-use" property. Equipment is limited-use property if no one other than the lessee or a related party has a use for it at the end of the lease agreement.

(f) Tax-oriented leasing is 100% financing. Guideline leases (tax-oriented leases) may be classified as either a capital lease or an operating lease for reporting purposes under Financial Accounting Standards Board Rule #13.

Capital Lease

A capital lease is one that fulfills any ONE of the following criteria:

(a) The lease transfers ownership of the property to the lessee by the end of the lease term.

(b) The lease contains a bargain purchase option which is less than fair market value.

(c) The lease term is equal to 75% or more of the estimated economic life of the leased property.

(d) The present value of the minimum lease payments equals or exceeds 90% of the fair market value of the leased property.

The capital lease shows up on the lessee's balance sheet as an asset and a liability.

Operating Lease

An operating lease does not meet ANY of the above criteria for a capital lease. An operating lease is not booked on the lessee's balance sheet but is recorded as a periodic expense on the income statement.

Capital or Operating Lease?

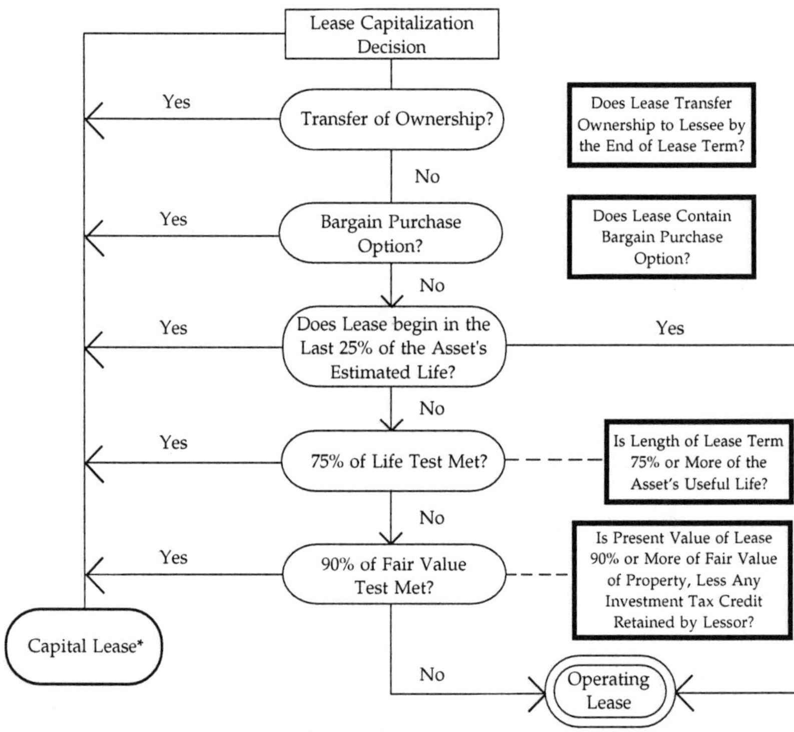

*Lessor must determine that additional criteria are met to account for the lease as a capital lease:
1. Collection of minimum lease payments is reasonably predictable.
2. No important uncertainties exist for unreimbursable costs to be incurred by the lessor.

Sources:

Accounting For Leases, Financial Accounting Standards Board
Leasing and Tax Reform: A Guide Through The Maze, General Electric Credit Corporation
Handbook of Leasing, General Electric Credit Corporation
Miller Comprehensive GAAP Guide, 1988 by Martin A. Miller, Harcourt Brace Jovanovich, Section 30.18

Appendix B
Model Energy Service Agreement

Listed below are key sections of an energy service agreement between the customer and the ESCO. Lenders will pay particular attention to various sections based on the type of financing program that is being applied to the project. Note: Consult your legal counsel before authorizing any legal agreement.

I. TERM SHEET.

The purpose of the Agreement is the evaluation, engineering, design, procurement, installation, financing and monitoring by ESCO of Energy Conservation Measures ("ECMs") at **Customer's** facility(ies) identified in Appendix A attached hereto ("Premises"). ESCO and **Customer** (the "Parties") agree to the following terms pursuant to which this Agreement shall be performed:

1. EXECUTION DATE: ______________________

2. TERM OF AGREEMENT: __________ (years after ECM Commencement Date)

3. LOCATION: ______________________

4. OWNERSHIP OF PREMISES: ______________________

PROJECT FIGURES

	Preliminary Engineering Estimates	Final Installed Figures
5. ESCO PERCENTAGE OF ENERGY SAVINGS:	______	______
6. PROJECT TOTAL CAPITAL COST ($ 000):	______	______
7. ECM COMMENCEMENT DATE:	______	______

8. VALUE OF FIRST YEAR ENERGY SAVINGS ($): ________ ________

9. **CUSTOMER** FIRST YEAR PAYMENT TO ESCO (line 5 × line 8) ($): ________ ________

10. **CUSTOMER** FIRST YEAR MONTHLY PAYMENT (line 9÷12) ($): ________ ________

11. **CUSTOMER** FIXED MONTHLY PAYMENT (OPTIONAL): ________ ________

Please Initial: ESCO______ CUSTOMER________

Final Installed Figures appearing in the right-hand column above shall be completed in accordance with Section V and the Parties shall then re-execute the Agreement below and enter their initials above.

As indicated in TERM #11 above, upon completion of any determinations required by Section V, **Customer** shall have the option to fix its monthly payment to **ESCO**, as required under Section VI, by multiplying One Hundred and _____ Percent (1__%) of ESCO's Percentage of Energy Savings by the Final Installed Figure for the Value of First Year Energy Savings and paying such product to ESCO each year in twelve (12) monthly payments ("**Customer** Fixed Monthly Payment"). **Customer** shall indicate its exercise of such option by initialing the appropriate line below. Except as set forth in this Agreement, such monthly payments shall be due and payable each month of this Agreement from and after the month in which the Commencement Date occurs and shall be made by **Customer** without regard to the amount of Energy Savings in any such month or year. Any excess of Energy Savings over the Final Estimated Value of First Year Energy Savings shall be retained by **Customer**. **Customer's** Fixed Monthly Payment shall not be revised except as may be specifically required in accordance with the terms of this Agreement.

ESCO: **Customer:**

____

____

(business organization) (business organization)

By:____________________________________

By:____________________________________

(entity) (entity)

By:____________________________________

By:____________________________________

(name) (name)

Its:____________________________________

Its:____________________________________

(title) (title)

ACCEPTANCE OF FINAL INSTALLED FIGURES

ESCO: **Customer:**

____

____

(business organization) (business organization)

By:____________________________________

By:____________________________________

(entity) (entity)

By:______________________________

By:______________________________

(name) (name)

Its:______________________________

Its:______________________________

(title) (title)

CUSTOMER ACCEPTANCE OF FIXED MONTHLY PAYMENT

Accepted Not Accepted

______ __________ (please initial)

II. DEFINITIONS.

When used in this Agreement, the following terms shall have the meaning specified:

2.1 **Agreement:** This Agreement between **Customer** and **ESCO.**

2.2 **ESCO's Percentage of Energy Savings:** The percentage of Energy Savings ESCO shall receive as compensation for its services under this Agreement, subject to **Customer's** option to fix monthly payments set forth in Section I, paid to ESCO in accordance with Section VI and Section VII.

2.3 **Current Market Value of Energy Savings:** The total market value rate expressed in r/kWh of electrical energy use and/or $/kW of electrical demand imposed by the Utility company in the current monthly period then occurring, or in any future monthly period then being considered, including applicable taxes, surcharges and franchise fees, if applicable.

2.4 **ECM Commencement Date:** The ECM Commencement Date shall be the date on which the installation of the ECMs is substantially complete.

Prior thereto, **ESCO** shall give **Customer** a Notice of Substantial Completion and shall therein identify the ECM Commencement Date, which shall occur no sooner than fifteen (15) days after such notice.

2.5 **Energy Audit Report ("EAR"):** The analysis performed by ESCO of the electric energy use by **Customer** at the Premises, and the potential for electric energy savings. Such analysis includes, without limitation, the ECMs recommended by ESCO and agreed to by the Parties for installation at the Premises and the Measurement Plan for measuring the savings estimated to result from such ECMs, all as attached hereto and made a part hereof as Appendix B.

2.6 **Energy Conservation Measure ("ECM"):** The various items of equipment, devices, materials and/or software as installed by **ESCO** at the Premises, or as repaired or replaced by **Customer** hereunder, for the purpose of improving the efficiency of electric consumption, or otherwise to reduce the electric utility costs of the Premises.

2.7 **Energy Savings:** Electric energy reduction (expressed in kilowatt-hours of electric energy and/or kilowatts of electric demand and measured in accordance with the Measurement Plan) achieved through the more efficient utilization of electricity resulting from the installation of the ECMs agreed to by the Parties under this Agreement.

2.8 **Measurement Plan:** The plan for measuring Energy Savings under this Agreement which shall be in accordance with the requirements of the Utility Agreement and shall be a part of the Energy Audit Report attached as Appendix B hereto.

2.9 **Monthly Period:** A span of time covering approximately 30 days per month, corresponding to **Customer's** billing period from its electric utility.

2.10 **Party: Customer** or ESCO. **Parties** means **Customer** and **ESCO.**

2.11 **Premises:** The buildings, facilities and equipment used by **Customer,** as identified in Section I and as more fully described in the attached Appendix A, where **ESCO** shall implement the Project under this Agreement.

2.12 **Project:** The complete range of services provided by **ESCO** pursuant to this Agreement, including evaluation, engineering, procurement, installation, financing and monitoring of ECMs at the Premises.

2.13 **Uncontrollable Circumstances:** Any event or condition having a material adverse effect on the rights, duties or obligations of **ESCO**, or materially adversely affecting the Project, if such event or condition is beyond the reasonable control, and not the result of willful or negligent action or omission or a lack of reasonable diligence, of **ESCO**; provided, however, that the contesting by **ESCO** in good faith of any event or condition constituting a Change in Law shall not constitute or be construed as a willful or negligent action, or a lack of reasonable diligence. Such events or conditions may include, but shall not be limited to, circumstances of the following kind:

a. An act of God, epidemic, landslide, lightning, hurricane, earthquake, fire, explosion, storm, flood or similar occurrence, an equipment failure or outage, an interruption in supply, an act or omission by persons or entities other than a Party, an act of war, effects of nuclear radiation, blockade, insurrection, riot, civil disturbance or similar occurrences, or damage, interruption or interference to the Project caused by hazardous waste stored on or existing at the Project site;

b. strikes, lockouts, work slowdowns or stoppages, or similar labor difficulties, affecting or impacting the performance of **ESCO** or its contractors and suppliers;

c. a change in law or regulation or an act by a governmental agency or judicial authority.

2.14 **Utility Agreement:** The agreement entered into by **ESCO** with __________, a ________ public utility company ("Utility"), pursuant to which ESCO is required to install certain ECMs at facilities such as **Customer's** Premises and in accordance with the terms of which ESCO has entered into this Agreement.

III. ECM COMMENCEMENT DATE AND TERM OF AGREEMENT.
The term of this Agreement shall commence as of the date on which this Agreement is executed and shall continue, unless sooner terminated in accordance with the terms hereof, for the period of years after the ECM Commencement Date set forth in Section I.
Upon receipt of the Notice of Substantial Completion identifying the

ECM Commencement Date, **Customer** shall provide **ESCO**, within fifteen (15) days, any comments and requests for work or corrections. **ESCO** shall make all commercially reasonable efforts to respond to such comments and requests within thirty (30) days of the ECM Commencement Date, which shall occur on the identified date.

Upon the expiration or termination of this Agreement, the provisions of this Agreement that may reasonably be interpreted or construed as surviving the expiration or termination of this Agreement shall survive the expiration or termination for such period as may be necessary to effect the intent of this Agreement.

At the end of the term of this Agreement, **Customer** shall purchase the ECMs and **ESCO** shall transfer title to **Customer**, free and clear of all liens and encumbrances, all as set forth in Section 6 of the General Terms and Conditions.

IV. SCOPE OF ESCO'S SERVICES.

Subject to and in accordance with the terms and conditions of this Agreement, **ESCO** shall provide the evaluation, engineering, design, procurement, installation, financing and monitoring of the ECMs set forth in the Energy Audit Report. The ECMs in the EAR may be modified pursuant to Section V. **ESCO** shall use reasonable commercial efforts to achieve the ECM Commencement Date estimated in Section I, line 7, column entitled, "Preliminary Engineering Estimates."

ESCO agrees to extend its funds and install such ECMs in return for **Customer**'s agreement to perform hereunder and, in particular, to pay **ESCO**'s Percentage of Energy Savings pursuant to Section VI. The Parties anticipate the measurement of Energy Savings, but notwithstanding this expectation or any provision of this Agreement which may suggest to the contrary, in light of the factors affecting savings which are beyond **ESCO**'s reasonable control, **ESCO** assumes no obligation that any particular level of savings shall materialize due to its services hereunder.

ESCO warrants that the ECMs which have a lifetime greater than one (1) year shall be, and shall remain, free of defects for one (1) year after the ECM Commencement Date. In addition, **ESCO** agrees to assign all manufacturers' warranties for such ECMs to **Customer** for the period of manufacturer warranty, subject to all exclusions and limitations as may be set forth therein.

V. DETERMINATION OF FINAL TERMS.

After the execution of this Agreement, if the Parties agree to revise the ECMs listed in the EAR, and to make all associated revisions to this Agreement, including, without limitation, to the numbers or dates, as the case may be, entered in the column entitled, "Preliminary Engineering Estimates," appearing in lines 5 through 11 of Section I, the Parties shall, upon such agreement, make any associated revisions and enter in the column entitled, "Final Installed Figures," the agreed-to numbers and/or dates. The Parties shall then re-execute Section I upon the entry of such numbers and/or dates and **ESCO** shall revise the Termination Values and the Final Installed Project Total Capital Cost then appearing in line 6 of Section I above.

All such revisions shall be voluntary and the Parties shall not be required, absent mutual consent, to revise the ECMs listed in the EAR after the execution hereof; provided, however, **ESCO** shall not be required to install ECMs affected by Uncontrollable Circumstances. Such ECMs shall be deleted from the Project unless the Parties agree to all necessary changes to the Project numbers and/or dates required to adjust to such circumstances. Changes to Project numbers and/or dates in connection with Uncontrollable Circumstances shall be entered by **ESCO** in the column entitled, "Final Installed Figures," in lines 5 through 11 of Section I. The Parties shall then re-execute Section I and **ESCO** shall revise the Termination Values and the Final Installed Project Total Capital Cost then appearing in line 6 of Section I.

VI. COMPENSATION.

From and after the ECM Commencement Date, except as provided in Sections I and VII with respect to **Customer's** option to fix monthly payments, **Customer** shall pay **ESCO** an amount equal to **ESCO**'s Percentage of Energy Savings, as set forth in Section I above, multiplied by the applicable Current Market Value of Energy Savings. **ESCO** shall prepare and send **Customer**, and **Customer** agrees to pay, a monthly invoice calculated pursuant to Section VII below.

VII. BILLING.

ESCO will submit monthly invoices to **Customer** in amounts determined in accordance with Section VI. In the event **Customer** exercises its option under Section I to fix its monthly payments to **ESCO**, **Customer** shall pay to **ESCO** the **Customer** Fixed Monthly Payment defined in Section I. If

Customer does not exercise such option, monthly payments shall be estimated as set forth in this Section VII and paid on such estimated basis, subject to reconciliation as provided hereinafter. Monthly invoices shall be paid within thirty (30) calendar days following receipt. Reconciliation payments, or refunds, as the case may be, shall be due within thirty (30) calendar days of receipt of a reconciliation invoice.

Subject to reconciliation, invoice amounts shall be estimated for each year following the ECM Commencement Date. In the first such year, monthly payments shall be one-twelfth of the product of **ESCO**'s Percentage of Energy Savings and the Current Market Value of Energy Savings expected in such first year as set forth in line 8 of the column entitled, "Final Installed Figures" in Section I above. Such fixed amounts shall be reconciled and adjusted as necessary every six (6) months based on the difference between the Current Market Value of Energy Savings and the estimated value then applicable. **ESCO** shall prepare and submit to **Customer** a reconciliation invoice within thirty (30) calendar days of the expiration of each such six-month period. The estimate of the value of Energy Savings then in effect at the end of any year shall continue until replaced with reconciled amounts in the succeeding year. In the second year and following years, the estimate of the Current Market Value of Energy Savings expected in such year shall equal the expected value (in nominal dollars for the billing year in question) of the Energy Savings actually delivered in the prior year.

VIII. NOTICES.

All notices to be given by either Party to the other shall be in writing and must be delivered or mailed by registered or certified mail, return receipt requested, or sent by a courier service which renders a receipt upon delivery addressed as set forth above in Section I or such other addresses as either Party may hereinafter designate by a Notice to the other. Notices are deemed delivered or given and become effective upon mailing if mailed as aforesaid and upon actual receipt if otherwise delivered to the addresses set forth in Section I.

IX. APPLICABLE LAW.

This Agreement and the construction and enforceability thereof shall be interpreted under the laws of the state of ______________ .

X. FINAL AGREEMENT.

This Agreement, together with its appendices and attachments, shall constitute the full and final Agreement between the Parties, shall supersede all prior agreements, communications and understandings regarding the subject matter hereof and shall not be amended, modified or revised except in writing. **This Agreement shall bind the Parties as of the date on which it was executed.** Any inconsistency in this Agreement shall be resolved by giving priority in the following order: (a) Amendments to the Agreement, in reverse chronological order; (b) the Agreement, Sections I through XI; (c) General Terms and Conditions of the Agreement; (d) Appendices to the Agreement; and (e) Attachments to, or other documents incorporated into, the Agreement.

XI. INCORPORATION OF GENERAL TERMS AND CONDITIONS.

THIS AGREEMENT IS SUBJECT TO THE GENERAL TERMS AND CONDITIONS ATTACHED HERETO AND MADE A PART HEREOF. IN ADDITION, THE PERFORMANCE OF THIS AGREEMENT BY ESCO IS SUBJECT TO THE PROVISIONS OF THE UTILITY AGREEMENT.

IN WITNESS WHEREOF, and intending to be legally bound, the Parties hereto subscribe their names to this instrument herein above in Section I as of the date of execution first written above.

SUMMARY OF GENERAL TERMS AND CONDITIONS

Section 1.	Operating the Premises and Maintaining the Use of the ECMs.
Section 2.	Equipment Location and Access.
Section 3.	Construction.
Section 4.	Ownership of ECMs.
Section 5.	Condition of ECMs.
Section 6.	Transfer of ECMs.
Section 6.1.	Material Shortfall in Energy Savings.
Section 6.2.	Customer Purchase Option.
Section 6.3.	Purchase Upon Customer Default.
Section 6.4.	Expiration of the Term.

Section 6.5.	Transfer Without Encumbrance.
Section 7.	Requirements of Utility Agreement.
Section 8.	Insurance.
Section 8.1.	Insurance on the Premises and the ECMs.
Section 8.2.	Risk of Loss of the ECMs.
Section 8.3.	ESCO's Insurance Requirements.
Section 9.	Hazardous Materials and Activities.
Section 10.	Events of Default.
Section 10.1.	Events of Default by Customer.
Section 10.2.	Events of Default by ESCO.
Section 11.	Remedies Upon Default.
Section 11.1.	Remedies Upon Default by Customer.
Section 11.2.	Remedies Upon Default by ESCO.
Section 11.3.	Limitation of Remedies.
Section 12.	Representations and Warranties of Both Parties.
Section 13.	Compliance With Law and Standard Practices.
Section 14.	Assignment.
Section 15.	Taxes.
Section 16.	Severability.
Section 17.	Effect of Waiver.
Section 18.	Usage of Customer's Records.
Section 19.	Air Emission Rights, Credits or Allowances.

Appendix C

North American Energy Measurement and Verification Protocol

ACKNOWLEDGMENTS

Participating Organizations

Canada
Canadian Association of Energy Service Companies (CAESCO)

Mexico
Comision Nacional Para El Ahorro De Energia (CONAE)
Fideicomiso De Apoyo Al Programa De Ahorro De Energia Del Sector Electrico (FIDE)

United States
American Society of Heating, Refrigerating and Air-Conditioning Engineers (ASHRAE)
National Association of Energy Service Companies (NAESCO)
National Association of Regulatory Utility Commissioners (NARUC)
National Association of State Energy Officials (NASEO)
U.S. Department of Energy (DOE)
U.S. Environmental Protection Agency (EPA)

Technical Subcommittee

Cary Bullock
KENETECH Corporation

John Brennan
Natural Resources Canada

David Claridge
Texas A&M University

Margaret Fels
Princeton University

Anne Gumerlock-Lee
Pacific Gas and Electric Company

Brad Gustafson
Lawrence Berkeley National Laboratory

Jeff Haberl
Texas A&M University

Jim Halpern, Technical Subcommittee Chair
Measuring & Monitoring Services

Kerri Herrity, Technical Editor
KENETECH Corporation

Steve Kromer
Lawrence Berkeley National Laboratory

David McGeown
Public Service Conservation Resources Corporation

George Reeves
George Reeves Associates

Larry Ringel
Synergistic Control Systems, Inc.

Steven R. Schiller
Schiller Associates

Policy Committee

Cary Bullock, Policy Committee Cochair
KENETECH Corporation

David Claridge
Texas A&M University

Howard Coogler
South Carolina Energy Office

Lynn Fischer
Commission for Environmental Cooperation

Brad Gustafson
Lawrence Berkeley National Laboratory

Jeff Haberl
Texas A&M University

Jim Halpern
Measuring & Monitoring Services

Salvador Herrera-Gonzales
Fideicomiso De Apoyo Al Programa De Ahorro De Energia Del Sector Electrico

John Horn
Iowa Department of Natural Resources

Greg Kats, Policy Committee Cochair
U.S. Department of Energy

Robert Krizman
Consultant

Steve Kromer
Lawrence Berkeley National Laboratory

Henry Kurth
Illinois Department of Commerce and Community Affairs

Tom Livers
Montana Department of Environmental Quality

Mark Martinez
ENvest

David McGeown
Public Service Conservation Resources Corp.

Rick Morgan
U.S. Environmental Protection Agency

Tom Philbin
HEC, Inc.

Jim Redden
HEC, Inc.

George Reeves
George Reeves Associates

Dave Rinebolt
National Association of State Energy Officials

Larry Ringel
Synergistic Control Systems, Inc.

Arthur Rosenfeld, Policy Committee Cochair
U.S. Department of Energy

Steven R. Schiller
Schiller Associates

Terry Singer
National Association of Energy Service Companies

Barry Solomon
formerly U.S. Environmental Protection Agency

Roya Stanley
Iowa Department of Natural Resources

Catherine Strickland
International Institute for Energy Conservation

John Sugar
California Energy Commission

Tom Tamblyn
Tescor Energy Services, Inc.

Mateo Trevino
Fideicomiso De Apoyo Al Program De Ahorro De Energia Del Sector Electrico

Financial Advisory Committee

John Augustine
Lehman Brothers

Robert Bergstrom
Allison Williams Capital, Inc.

Bryan Colwell
Goldman, Sachs & Co.

Joe Euphrat
EA Capital Solutions

John Gibb
Sallie Mae

Jeffrey Heinz
Sallie Mae

Matthew Heller
Energy Capital Partners

Dave Kuncio
Citicorp

John MacLean
Energy Efficiency Capital Corporation

Scott Mills
Lehman Brothers

Steve Petty
LINC Anthem

Curtis Probst
Salomon Brothers, Inc.

Donald Reed
Excelsior Capital Corporation

Lisa Rogers
Goldman, Sachs & Co.

Glenn Tobias
Banque Paribas

For Additional Information Regarding a Specific Topic

Below is a list of the authors responsible for specific sections of this protocol. Section authors encourage readers to contact them directly for additional information on a given topic.

Section 1.0
B. James Halpern, *Technical Subcommittee Chair*
Measuring & Monitoring Services, Inc.
TEL: 800.942.2703
FAX: 908.576.8067
E-mail: bjhalpern@aol.com

Section 2.0
Steve Kromer
Lawrence Berkeley National Laboratory
TEL: 510.486.6619
FAX: 510.486.5394
E-mail: jskromer@lbl.gov

Cary Bullock
Kenetech Energy Management, Inc.
TEL: 617.273.5894
FAX: 617.273.1732
E-mail:cbull21415@aol.com

Section 3.0
Steven R. Schiller
Schiller Associates
TEL: 510.444.6500
FAX: 510.444.6502
E-mail: steves@schiller.com

Section 4.0
Jeff Haberl, Ph.D.
David Claridge, Ph.D.
Texas A&M University
TEL: 409.845.6065
FAX: 409.862.2457
E-mail: jhaberl@esl-net1.tamu.edu

Section 5.0
Lawrence Ringel
Synergistic Control Systems, Inc.
TEL: 504.885.8180
FAX: 504.885.1180
E-mail: synergis@ix.netcom.com

Please direct comments and suggestions related to this protocol to: Mr. B. James Halpern, Measuring & Monitoring Services, Inc., 620 Shrewsbury Avenue, Tinton Falls, New Jersey 07701.

To Obtain the NEMVP or the FEMP Guidelines, or Information on ASHRAE Guideline 14:

1) To obtain the NEMVP:
   A) As a book: call the "Efficiency and Renewable Energy Clearing House (EREC)" **1 (800) DOE-EREC** or fax your name, address & telephone number to EREC at **(703) 893-0400**, ask for the "North American Energy Measurement and Verification Protocol" and include the code "NEMVP"
   B) Electronically via E-mail: access EREC at **doe.erec@nciinc.com**
   C) Electronically via the World Wide Web: access the U.S. Department of Energy's "Energy Efficiency and Renewable Energy Network (EREN)" Homepage **http://www.eren.doe.gov**
   The Protocol is listed under the topic "Building Systems and Community Programs."
2) To obtain the Federal Energy Management Program's Guidelines and the NEMVP together:
   A) As a book: call EREC at **1(800) DOE-EREC**
   B) Electronically via E-mail: **MVinfo@lbl.gov**
   C) Electronically via the World Wide Web: access EREN at: **http://www.eren.doe.gov**
   The Guidelines are listed under the topic "Building Systems and Community Programs."
3) For information about ASHRAE Guideline 14 via the World Wide Web: Access ASHRAE's Homepage at: **http://www.ashrae.org**
   Search for "Guideline 14. Measurement of Energy and Demand Savings."
4) To access reference documents via Internet:
   Access Energy Systems Laboratory at Texas A&M server at: **ftp:/info.tamu.edu/doe**

TABLE OF CONTENTS

Section

2.4 Procuring Performance Contracts
2.5 Selecting the Best M&V Option

3.0 Overview of Measurement and Verification
3.1 General Approach to M&V
3.2 Verifying ECM Performance Potential
3.3 Verifying ECM Performance (Actual Savings)
3.4 M&V Implementation Procedures
3.5 M&V Issues
3.6 Defining the Appropriate Level of M&V
3.7 Measurement and Verification Options

4.0 Three M&V Options with Examples
- 4.1 Option A: End-Use Retrofits Measured Capacity, Stipulated Consumption Approach
  - 4.1.1 Confirming Installed Equipment Performance
  - 4.1.2 Examples
  - 4.1.3 Expected Accuracy
  - 4.1.4 Expected Cost
- 4.2 Option B: End-use Retrofits - Measured Capacity, Measured Consumption Approach
  - 4.2.1 Confirming Installed Equipment Performance
  - 4.2.2 Examples
  - 4.2.3 Expected Accuracy
  - 4.2.4 Expected Cost
- 4.3 Option C: Whole-Facility or Main Meter Measurement Approach
  - 4.3.1 Utility Billing Methods
  - 4.3.2 Whole-Facility or Main Meter Hourly Before/After Analysis
  - 4.3.3 Expected Accuracy
  - 4.3.4 Expected Cost

5.0 Other M&V Issues
5.1 Report Format and Invoicing
5.2 M&V Professionals
5.3 Instrumentation and Measurement Techniques
5.4 Verification of Project Maintenance
5.5 Special Notes on New Construction
5.6 Minimum Energy Standards (Federal, State and Local)
5.7 Dealing with Data Collection Errors and Lost Data
5.8 Commissioning Projects
5.9 Interactive Effects Between Measures
5.10 Dispute Resolution
5.11 Special Notes on Residential Projects
5.12 Calibration of Instrumentation
5.13 Calculating Uncertainty

6.0 Definitions of Terms

7.0 References

SECTION 1.0: PURPOSE AND SCOPE OF DOCUMENT

Introduction

In 1995, North Americans installed $5 billion in efficiency equipment in their buildings in order to save money and conserve energy and water. But this covers only a small fraction of the existing cost-effective opportunities for energy-savings investments. If all cost-effective efficiency investments were made in public and commercial buildings, the United States would save $20 billion per year on energy bills, create over 100,000 jobs, and significantly cut pollution.

When firms invest in energy efficiency, they naturally want to know how much they have saved and how long their savings will last. If the installation had been made to generate energy, measurements would be trivial—install a meter. But to measure savings is a challenge and requires both metering and a methodology, known as a measurement and verification protocol.

To determine energy savings, the parties (the building owner, the installer and perhaps the financier) must first agree on the "base case" (what the building used before retrofit), and then must measure energy use after retrofit. They may want to adjust the savings for variations in the weather or changes in occupancy or work schedules. And they should keep up the measurements to ensure that their savings persist.

In early 1994, our financial advisors complained that existing protocols (and those under development) create a patchwork of inconsistent, sometimes unreliable efficiency installation and measurement practices that prevent development of new forms of lower-cost financing. We believe that the North-American Energy Measurement and Verification Protocol (NEMVP) will help overcome these impediments to the growth of the efficiency industry.

The NEMVP Protocol is the result of a remarkable collaborative effort between federal and state agencies and experts in the energy and efficiency industries in America, Canada and Mexico. This year-long effort, initiated by the U.S. Department of Energy, has been largely driven by industry and reflects a broad industry consensus. The work was drafted by the NEMVP Technical Subcommittee and reviewed and guided by the NEMVP Policy Committee with financial guidance from the NEMVP Financial Advisory Committee. In addition, comments were received from over 250 corresponding members.

It is our hope that this Protocol will be adopted throughout North

America and internationally. Our expectation is that, by providing greater and more reliable savings and a common approach to efficiency installation and measurements, the financial markets will respond with financial products allowing the securing of energy and water efficiency projects. Ultimately, we hope this will lead to the development of a secondary market for efficiency investments, with increased availability of low-cost and off-balance-sheet financing, allowing the efficiency industry to grow much more rapidly, resulting in widespread benefits in the form of increased employment, lower energy and water bills and reduced damage to the environment.

1.1 Purpose

The long-term success of energy management projects has been hampered by the ability of project partners to agree on an accurate, successful M&V plan. This M&V protocol discusses procedures that, when implemented, allow buyers, sellers and financiers of energy projects to quantify energy conservation measure (ECM) performance and energy savings. By using one of the different M&V options discussed in this document, readers can allocate various risks associated with achieving energy cost savings to either the buyer or seller of the project. Additionally, this protocol:

- is designed to be consistent with EPA Acid Rain Program verification protocols for conservation and renewable energy reserve allowances,
- gives buyers, sellers and financiers a basis to discuss key M&V project-related issues,
- helps ensure the accurate verification of projects, with respect to anticipated versus achieved savings, and
- uses procedures which i) are consistently applicable to similar projects throughout all geographic regions, and ii) are nationally accepted, impartial and reliable.

Two basic aspects of ECM performance verification are addressed in this document:

1. Verification of: i) the accuracy of baseline conditions as specified in the contract between buyer and seller, and ii) the complete installa-

tion and proper operation of new equipment/systems specified in the contract.

2. Verification of the quantity of energy savings and/or energy cost savings that occur during the term of the contract.

This M&V protocol is not intended to prescribe contractual terms between buyers and sellers. Once other contractual issues are decided, this document may be used to select the verification plan that best matches: i) project costs and savings magnitude, ii) technology-specific requirements, and iii) risk allocation between buyer and seller, i.e., which party is responsible for installed equipment performance and which party is responsible for achieving long-term energy savings.

The protocol provides an overview of current techniques available for verifying aspects of third-party-financed energy projects. It may also be used by building operators to assess and improve facility performance.

Coordinated with the efforts of the ASHRAE GPC 14P Committee, this document will be maintained and revised under the sponsorship of the Department of Energy on a biannual basis, based on recommendations developed by a consensus group of facility owners/operators, financiers, contractors or energy service companies (ESCOs) and other stakeholders. As the document is updated, it will incorporate new measurement techniques as they become available. Changes, modifications and additions will be reviewed on a regular basis and, if approved, will be incorporated directly into this document, so that buyers, sellers and financiers of energy projects are kept abreast of the latest options.

1.2 Scope

The scope of this M&V protocol includes:

- addressing the energy and cost-savings verification needs of participants in third-party-financed energy projects—financiers, sellers, buyers and technical consultants,

- defining the role of verification in third-party-financed energy project contracts and implementation,

- discussing procedures, with varying levels of accuracy and cost, for verifying: i) baseline and project installation conditions, and ii) long-term energy savings performance,

- creating a *living* document that includes a set of methodologies and procedures that enable the document to evolve over time,
- designing M&V procedures for a variety of facilities including residential, commercial, institutional and industrial buildings,
- providing techniques for calculating "whole-facility" savings, individual technology savings and stipulated savings,
- looking at a variety of ECMs, including gas and electric, fuel switching, load shifting and other measures which involve the installation of equipment and result in energy cost savings,
- reviewing methods of measuring energy savings for retrofit as well as new construction,
- providing procedures for the investigation and resolution of ECM performance issues, and
- designed to verify the installation of renewable energy technologies in a way that best optimizes their system benefits.

1.3 Target Audience

The target audience for this M&V protocol includes:

- Facility Energy Managers
- Consultants
- Researchers
- Government Agencies
- Utilities
- ESCOs
- Financiers
- Any Other Party Needing to Determine the Value of Energy Improvements

1.4 Relationship to ASHRAE GPC 14P

This document is designed to be complementary to the work of the ASHRAE GPC 14P Committee, currently writing *Measurement of Energy and Demand Savings.* In contrast to the ASHRAE document, which focuses on the relationship of the measurement to the equipment being verified, this M&V protocol discusses a variety of M&V topics as they relate to actual contracts for energy services. By design, portions of these documents overlap. It is advised that the reader use both documents, as well as others referenced herein, to formulate a successful M&V plan.

1.5 Relationship to EPA's Conservation Verification Protocol

This M&V protocol has been written to be compatible with Environmental Protection Agency (EPA) protocols. EPA's Conservation Verification Protocols (CVPs) are designed to verify energy (electricity) savings from utility Demand Side Management (DSM) programs for the purpose of awarding sulfur dioxide allowances under EPA's Acid Rain Program. EPA's protocols are designed to verify energy savings, and not specific environmental savings, although the latter could be easily accomplished with additional calculations.

To this end, the CVPs are designed to provide EPA with sufficient confidence that the savings from utility DSM programs are real, without placing an undue burden on the utility. The CVPs are intended mainly for utility-run DSM programs, as opposed to performance contracting. The CVPs emphasize actual measurement of savings over engineering estimates. However, discounted stipulated savings are permitted for certain measures and procedures.

Energy savings verified from performance contracting are potentially eligible for bonus allowances under EPA's Acid Rain Program, provided the measures are paid for in part by an electric utility. This M&V protocol could also be used in that context to verify performance contracting energy savings, in conjunction with the CVPs or other verification procedures used by state utility commissions.

Copies of EPA's CVPs are available from the EPA Acid Rain Division (6204J), 401 M Street, SW, Washington, D.C. 20460.

1.6 Relationship to Federal Energy Management Program Guidelines

The Department of Energy's Federal Energy Management Program (FEMP) was established, in part, to reduce energy costs to the government to operate federal facilities more efficiently. FEMP assists federal energy managers by identifying and procuring energy-saving projects. Part of this assistance includes preparing a document called the "Standard Procedures and Guidelines for Verification of Energy Savings Obtained Under Federal Energy Savings Performance Contracting Programs."

The FEMP M&V Guidelines are an application of the NEMVP Protocol for Federal sector projects only and are intended to be fully compatible and consistent with the NEMVP Protocol. The FEMP Guideline is intended to be used by Federal procurement teams consisting of contracting and technical specialists. The content of the FEMP Guidelines includes those items necessary to draft an RFP and evaluate responses. The focus of the FEMP Guidelines is on choosing the M&V option and method most appropriate for specific projects. Contractors responding to RFPs may refer to the FEMP Guidelines for more information on specific procedures referenced in Federal RFPs. Although the FEMP Guidelines strive for compatibility, discrepancies may exist between the FEMP Guidelines and the NEMVP Protocol. In these cases, please send E-mail to MVinfo@lbl.gov.

SECTION 2.0: PERFORMANCE CONTRACTING: WHY MEASUREMENT AND VERIFICATION

2.1 Definition and Role of Performance Contracts

The term "Energy Savings Performance Contract" (ESPC) covers a broad range of contracts where the cost of ECM implementation is recovered through savings created by the ECMs. An ESPC can be used to accomplish any or all of the following objectives: upgrade capital equipment, provide for maintenance of existing facilities, save energy and/or save money. These contracts range in complexity from simple projects, such as lighting upgrades, to more detailed projects involving all aspects of energy consumption.

Because the basis for ESPCs is the performance of the ECMs, contracts must include a clear method of: i) assessing project performance, and ii) determining savings distribution. The role of M&V is to provide that method—a means of quantifying installed ECM performance, and calculating the savings that accrue as a result of increased efficiency.

This M&V protocol formalizes the language and techniques of M&V used in many existing contracts and provides several options for assessing energy savings. This protocol is not meant to prescribe an M&V option for every type of retrofit, but to state options available and help clarify the relationship of various M&V options to the associated risks assumed under an ESPC.

Typically, a number of performance risk issues are addressed in the ESPC. Many are listed below. In the interest of brevity, some contractual issues, such as maintenance and operation of the ECMs, are not covered by this document.

2.2 Participants in Performance Contracts

An ESPC usually requires the participation of several parties. While energy projects that are directly financed typically involve only an owner and a contractor (or ESCO), ESPC projects may also include financiers and third-party M&V professionals.

2.2.1 Owner

The role of the owner is to determine project objectives and resources and to understand the options that are available in the ESPC to meet those objectives. For example, some owners are interested in replacing old, inefficient equipment (capital renewal), while other owners may be interested in saving energy and still others in saving money.

2.2.2 Contractor/Energy Services Company

The role of the contractor is to provide assistance in identifying and capitalizing on energy-saving opportunities and/or to implement the ECMs that are specified in the contract. Contractors with the resources to package engineering, financing and construction of these projects are referred to as energy service companies (ESCOs).

2.2.3 Financiers —Banks, Utilities, Etc.

ESPCs require that one party be responsible for paying the initial cost of the ECMs to be installed. There are several financing options. The optimal source of financing is dictated by how risks are shared in the contract.

2.2.4 Independent M&V Professionals

The application of concepts and procedures presented in this protocol requires the skills of professionals familiar with measurement techniques, data manipulation and technology performance. In some circumstances, it may be preferable that a third party be obtained by the owner to judge whether agreements are being met.

In order to adequately understand the implications of various measurement strategies, the M&V professional should have a thorough understanding of the ECMs being installed.

2.2.5 ESCO/Owner/Financier Relationship

The relationship between the parties is directly related to the different contract types and risks. When the owner is at risk for project payments, the relationship between the owner and the financier is as shown below:

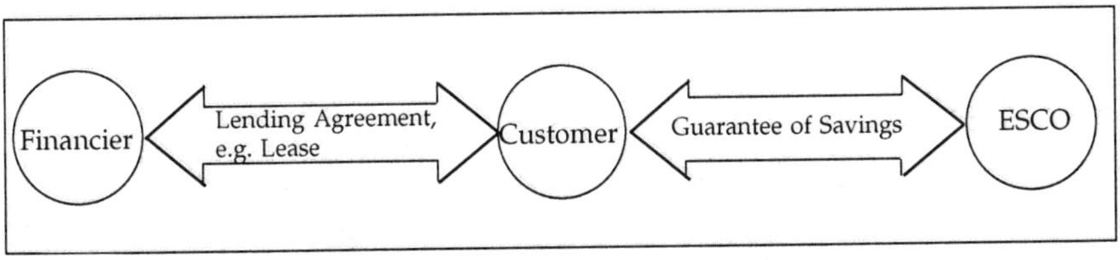

Facility owners often erroneously believe that in a "guaranteed savings arrangement," the ESCO guarantees the financial obligation assumed by the owner. In fact, the ESCO guarantees *a level of savings which is generally adequate to cover the financial obligation*. Hence, the owner still bears the financial risk associated with performance if, for example, the ESCO is unable to deliver on the guarantee, e.g., in the case of bankruptcy.

Shared savings, pay from savings or chauffage contracts use an arrangement shown in the following figure:

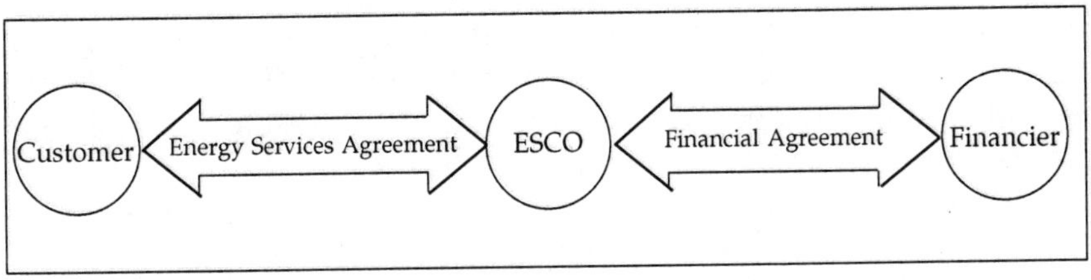

Clearly this group of contracts requires that additional services be provided by the ESCO. And the services, consequently, are more costly to provide. Not addressing the operations and maintenance (O&M) services normally attendant to these ESPCs could cause the ESCO to assume the risk of project performance and/or arrange for the equity aspect of the project. In these projects, financing must be obtained solely on either the project's ability to perform or the strength of the ESCO's guarantee. In a guaranteed savings arrangement, money is raised solely on the strength of the owner's credit.

2.3 M&V Options

The purpose of defining several M&V options is to allow the reader flexibility in the cost and method of assessing savings. Therefore, the M&V options, described briefly below and in more detail in Section 4.0, vary in accuracy and in cost of implementation. The reader should first review and understand the three general approaches (Options A, B and C) and then define a specific measurement plan based on the option that makes the most sense for the intended project.

It is important to note that all methods of defining savings are estimates. ***Performance can be measured, savings cannot be measured.*** The options described in this document are created to meet the needs of a wide range of contracts that use savings to determine financial payments. It is vital that the reader understand the limitations as well as the strengths of each method presented.

2.3.1 Option A: Performance Verification, End-Use Retrofits — Measured Capacity, Stipulated Consumption Approach

The verification techniques for Option A determine savings by measuring the performance of a system before and after a retrofit, and multiplying the difference by an agreed-upon, or "stipulated," factor, such as hours of operation. Option A is best applied to individual loads or systems within a building, such as a lighting system or chiller. This method is appropriate for projects where both parties will agree to a payment stream that is not subject to fluctuation due to changes in the operation of the equipment. Payments could be subject to change based on periodic measurements of system performance.

Option A relies on the direct measurement of affected end uses. For projects where the baseline is well understood, and operating hours are not expected to change, only the "change in equipment performance" is needed in order to calculate savings.

2.3.2 Option B: Savings Verification. End-Use Retrofits— Measured Capacity, Measured Consumption Approach.

Verification techniques for Option B are designed for projects where long-term continuous measurement of performance is desired.

Under Option B, individual loads are continuously monitored to determine performance, and this measured performance is compared with a baseline to determine savings. Option B M&V techniques provide long-term persistence data on the operation and performance of the ECMs. These data can be used to improve or optimize the operation of the equipment on a real-time basis, thus improving the benefit of the retrofit. Option B also relies on the direct measurement of affected end uses.

2.3.3 *Option C: Whole Building or Main-Meter Measurement Approach*

Verification techniques for Option C determine savings by studying overall energy use in a facility and identifying the effects of energy projects from changes in overall energy use patterns. Option C methods are required when measuring interactions between energy systems is desired, and when determining the impact of projects that cannot be measured directly, such as insulation or other envelope measures, is necessary.

2.4 Procuring Performance Contracts

This section is intended to help readers of this protocol understand the forms of performance contracts that exist and to relate the most appropriate measurement needs to each form or type of contract. ESPC types are typically designed to deal with certain contracting problems or to capitalize on specific owner opportunities. Most owners are understandably skeptical of performance financing of energy projects. This skepticism usually stems from a lack of familiarity with the contract types, as well as a natural distrust of financially complex deals.

There are several different types of contracts used for project financing within the performance contracting industry. Some of these require M&V, while others do not. In cases where M&V is not contractually required, it may still provide valuable information to the owner.

To a significant degree, different types of financing contracts can be organized by determining whether the project financing is on the owner's balance sheet[1] or not, and whether the owner's payment for

[1]On balance sheet means that the entire financial obligation is on the owner's balance sheet. The current portion of the obligation would be listed as a current liability, while the long-term portion would be listed as a long-term liability. If the financial obligation is "off balance sheet," only the current portion (i.e., the portion paid during the designated accounting period, usually one year) appears on the balance sheet of the owner. This distinction is important to owners, and is almost always important to their lenders.

services or amortization of the financing is contingent on project performance. This organization is shown in the table below:

	On Owner's Balance Sheet	Not on Owner's Balance Sheet
Net Owner Payment Contingent on Performance	Guaranteed Savings (Long-Term)	Shared Savings Pay from Savings Chauffage
Net Owner Payment Not Contingent on Performance	Guaranteed Savings Loans (Short-Term) Capital Leases	Certain Municipal Leases Operating Leases

When Owner Payment is Not Contingent on Project Performance

When the owner takes the long-term project risk, which occurs in most guaranteed savings contracts, the most cost-effective way to finance the project is through the use of leases or loans. When the owner needs further assurances, which may be required by law (e.g., in some states, a third-party guarantee of savings is a prerequisite for a local government agency to enter a performance contract relationship), a savings guarantee creates a guaranteed savings contract.

In essence, the project finance portion of the transaction is supported by a debt-like vehicle (loan, capital lease, municipal lease, operating lease) which is fully backed by the owner. The owner is willing to provide the capital, even though it may be arranged by the performance contractor, because the owner receives a guarantee from the performance contractor that the savings level will meet or exceed a specified minimum, which is generally greater than the amount required to service the assumed debt. With this approach, the financier does not rely on the savings guarantee, and the guarantee's presence or absence has little effect on the cost of funds.

These types of arrangements may result in an "on balance sheet" transaction in the case of capital leases or loans, or an "off balance sheet" transaction in the case of operating leases and municipal leases.

Guaranteed savings contracts are useful in some particular market niches. In the local public sector, i.e., cities, towns and counties, interest paid is generally tax-exempt for most investors.[2] Since the interest spread between tax-exempt and taxable interest is generally 250 to 300 basis points, using tax-exempt financing can be of considerable advantage, often amounting to a present-value benefit of 10-20% of the amount financed. On a $1,000,000 project, the benefit may be as much as $150,000. Since these projects are often multimillion-dollar projects, all parties have a significant incentive to use tax-exempt financing wherever possible.

When the Owner Payment is Contingent on Project Performance

This type of financing arrangement is typically how most owners view performance contracts. It includes such contracts as shared savings, pay from savings and chauffage. These contracts are more easily explained than guaranteed savings contracts, but are generally much more complex in structure. Shared-savings contracts are arrangements where one party provides the capital (and generally the technical expertise), while the other party provides the facility. The parties measure savings in accordance with a defined protocol, like this one, and share the project savings on a defined basis. The basis can be as simple as fifty/fifty, or perhaps variable or stepped percentages.

Pay-from-savings contracts are similar to a loan with a variable term based on the level of savings, i.e., more savings reduce the financial obligation more quickly. In this instance, the contractor/ESCO may loan the owner the money to build the project, while charging an elevated interest rate (to recognize the increased risk). The owner then pays the contractor/ESCO from the savings generated by the ECMs installed. Moreover, the contractor/ESCO guarantees that the savings will be adequate to pay off the obligation within a specified time period, typically five to ten years.

Chauffage contracts provide for a radically different structure. In essence, end uses, themselves, are sold. For example, a contractor might offer lighting from a specified fixture type on a dollars-per-hundred-hours-of-usage basis. Or, chilled water might be sold on a MMBTU basis (gallons delivered at a specified temperature). These contracts are different from others discussed above in two respects: i) they typically involve lengthy contract periods (20-30 years), and ii) the contractor provides all associated O&M support during the contract.

---

[2]If the investor is a bank, the local entity must also be "bank qualified." This requirement simply sets a limit on long-term, tax-exempt financing issued by the local entity in the year of the financing.

Why Contract Differences?

Each type of contract solves certain problems or capitalizes on specific owner advantages. Guaranteed-savings contracts are most useful in the local public sector, since they provide the assurance needed for this market niche to be comfortable securing financing. This approach also offers great ease in using tax-exempt financing vehicles. The reader should be advised that these types of transactions are often required to meet exacting internal investment criteria of the target company, and often they do not.

Shared-savings contracts are useful where the owner cannot or does not want to use borrowing capacity. For example, many subsidiaries of large companies do not generally secure debt independently. For them, shared savings is a useful approach since the transaction structure ensures that the owner will never pay more than the savings, and the obligation will likely be off the balance sheet. Hence, this structure guarantees that, absent a contract breach, the obligation can be retired from current funds. This is not true in the case of a guaranteed-savings contract (e.g., suppose the guarantor refuses to pay or goes bankrupt).

Shared-savings contracts are also useful because they are most often regarded as new equity and may not be required to meet internal investment criteria. For example, if the internal hurdle rate of the target company is 30%, and the expected APR of a shared-savings contract is 18%, the shared-savings contract becomes attractive, since it is much less expensive than internal funds. By contrast, a guaranteed-savings approach would likely not be acceptable for this transaction. As well, shared-savings contracts generally provide the owner O&M responsibilities if the owner desires.

Pay-from-savings contracts would appear to have the same risk characteristic for the owner as shared savings, but these contracts lend themselves to a much more "open book" approach. This is one reason they have historically been popular in situations where cost-based construction is desired (e.g., public sector, not tax-exempt and institutional).

Chauffage contracts are useful where the owner wishes to "outsource" facility services and investment.

The contract types discussed in this section are listed in the table on the following page in order of cost. To the extent cost reflects value, i.e., each form is used in its most appropriate application, this table also represents the relative value to the owner, from the least valuable to the most valuable.

Contract Type	Order of Cost 1 = least 4 = most
Guaranteed Savings	1
Pay from Savings	2
Shared Savings	3
Chauffage	4

2.5 Selecting the Best M&V Option

M&V is used for a variety of reasons:

1. To define how much an owner pays a contractor/ESCO.
2. To help operate a facility more efficiently.
3. To assess, by the owner and where no third-party guarantees are present, whether a particular investment is performing.

If the owner chooses an M&V plan to define how much the owner will pay a contractor/ESCO, then:

- Depending on the size of the project, an independent third party is sometimes useful in providing an M&V framework for a project. Among the reasons for using a third party should be independence and objectivity, although in many projects the contracting parties can perform these tasks themselves.
- The type of measurement described in this protocol in its current form is appropriate primarily for guaranteed-savings contracts, pay-from-savings contracts and shared-savings contracts. Chauffage contracts, by contrast, usually define payment in terms of a positive utilization rather than a negative utilization (e.g., pounds of steam used instead of savings). The measurement pro-

fessional can use concepts developed in this protocol as a basis for developing payment formulae under chauffage contracts.

- Cost-effective measurement approaches and contract types generally match up as follows:

Short-Term Guaranteed Savings	Option A
Long-Term Guaranteed Savings	Option B
Shared Savings	Option B, C
Pay From Savings	Option B, C

- An M&V plan should be selected so that M&V costs do not consume the savings. In general, M&V costs should be approximately less than 20% of the anticipated net savings benefit to the owner.

SECTION 3.0: OVERVIEW OF MEASUREMENT AND VERIFICATION

3.1 General Approach to M&V

The basic approach to determining energy savings involves comparing energy use associated with a facility, or certain systems within a facility, both before and after ECM installation. The "before" case is called the baseline. The "after" case is called post-installation. In general:

Energy Savings = Baseline Energy Use – Post-Installation Energy Use

Exceptions to this simple equation are:

- new construction projects where baseline energy use has to be determined by methods other than pre-installation inspections or measurements and
- projects where the baseline is determined from other similar facilities, not from the facility where the retrofit actually occurred.

Baseline and post-installation energy use can either be constant during the term of a contract, or one (or both) can vary with time. The following are three examples:

- **Baseline and post-installation energy use constant during term of agreement**:
  Example: Lighting project where lamps and ballasts in office building are changed, and the operating hours of the lights do not change during term of agreement.

- **Baseline and post-installation energy use vary during term of agreement**:
  Example: HVAC project where new chillers are installed, and the occupancy of the building changes during the term of the agreement.

- **Baseline energy use remains constant, and post-installation energy use varies during term of agreement**:
  Example: Lighting controls project where occupancy sensors are installed, and the operating hours of the lights with occupancy sensors change during term of agreement.

Verifying baseline and post-installation conditions involves inspections, spot measurement tests and/or commissioning activities. Commissioning activities include:

- documentation design assumptions for the ECM design,

- documentation of the design intent for use by contractors, owners and operators,

- functional performance testing and documentation necessary for evaluating the ECM for acceptance, and

- adjusting the ECM to meet actual needs within the capability of the system.

For projects based on "pay for performance," each ECM or site will have a separate verification process to determine its savings. For each site or project, the baseline and post-installation energy use will be defined using a combination of metering, billing analysis and/or engineering calculations (including computer simulation). In addition, values for certain factors which affect energy use and savings, and which are beyond the control of the contractor/ESCO, may be stipulated by the owner.

After each project is completed, the contractor/ESCO submits a report that defines projected energy savings for the first year. Typically, first-year payments to the ESCO will be based on projected savings values submitted in the report. This post-installation report must be accepted and approved by the owner.

For the remaining years of the contract, the contractor/ESCO provides annual (or at some other regular interval) "true-up" reports. These reports include inspection documentation of the installed equipment/systems and, perhaps, updated savings values using data obtained and analyzed during each year of the contract. Previous payments would be reconciled as necessary based on results of the periodic report, and future-year payments would be calculated based on information in the periodic report. This true-up and payment reconciliation would not apply if the contract specifies fixed payments.

3.2 Verifying ECM Performance Potential

3.2.1 Baseline Verification

Baseline conditions may be defined by either the owner or the contractor/ESCO. If the baseline is defined by the owner, then the contractor/ESCO will have the opportunity to verify the baseline. If the baseline is defined by the contractor/ESCO, then the owner will have the opportunity to verify.

Baseline physical conditions such as equipment counts, nameplate data, energy consumption rate and control strategies will typically be determined through surveys, inspections and/or spot or short-term metering activities.

3.2.2 Post-Installation Verification

One aspect of post-installation M&V is verification by the ESCO and the owner that the proper equipment/systems were installed, are operating correctly and have the potential to generate the predicted savings. Verification methods may include surveys, inspections, and/or spot or short-term metering. System/equipment commissioning is expected to be completed by the contractor/ESCO. Current editions of ASHRAE's GPC-13 can be the basis for commissioning activities.[3]

---

[3]Guidelines for Commissioning of HVAC Systems, ASHRAE Guideline 1-1989.

3.2.3 Regular Interval Post-Installation Verification

The contractor/ESCO and owner, at defined intervals during the contract term, verify that the installed equipment/systems have been properly maintained, continue to operate correctly and continue to have the potential to generate predicted savings.

3.3 Verifying ECM Performance (Actual Savings)

Either once after the ECM is installed, continuously or at regular intervals, the contractor/ESCO and owner determine energy savings in accordance with an agreed-upon M&V approach as defined in a site-specific M&V plan.

3.3.1 M&V Techniques

Baseline energy use, post-installation energy use and energy (and cost) savings can be determined using one or more of the following M&V techniques:

- Engineering Calculations
- Metering And Monitoring
- Utility Meter Billing Analysis
- Computer Simulations, e.g., DOE-2 Analysis
- Agreed-Upon Stipulations by the Owner and the Contractor/ESCO

3.3.2 Energy-Use Stipulations

There are numerous factors that can affect energy savings during the term of a contract, such as weather, operating hours for lighting projects, and ton-hours and heat exchanger fouling for chiller replacement projects. In general, but not always, one contract objective may be to adjust the baseline energy use up or down for factors beyond the control of the contractor/ESCO (e.g., building occupancy, weather) and adjust the post-installation energy use for contractor/ESCO-controlled factors (e.g., maintenance of equipment efficiency).

Therefore, in order to calculate energy savings the owner may, under certain circumstances, stipulate the value of factors which may vary during the term of the contract. For example, for a lighting project, the owner (or contractor/ESCO) measures the baseline and post-installation lighting-fixture power draw and then stipulates the operating hours of the facility. Another example, for a chiller replacement project, the owner verifies the baseline and post-installation chiller performance factors (e.g., kW/ton, % of rated load, etc.) and then stipulates the ton-hours of cooling at the facility for calculation of annual energy savings

However, for other projects, continuous or regular interval measurements throughout the term of the contract will be compared against baseline energy measurements to determine energy savings. For example, for a "constant-speed motor to variable-speed-drive motor" conversion project, post-installation motor energy use may be continuously metered and compared against baseline measurements of motor energy use.

If any values are stipulated, a "reality check" is suggested, such as comparing total predicted savings against utility bills.

3.4 M&V Implementation Procedures

M&V activities can be divided into the following tasks:

- Define a general M&V approach for inclusion in the agreement between buyer and seller of energy services (the owner and the contractor/ESCO).

- Define a site-specific plan for the particular project being installed once the project has been fully defined, usually after the agreement is signed.

- Define pre-installation baseline, including: i) equipment/systems, ii) baseline energy use and iii) factors which influence baseline energy use—this could simply include site surveys; spot, short-term or long-term metering; and/or analysis of billing data.

- Define post-installation, including: i) equipment/systems, ii) post-installation energy use, and iii) factors which influence post-installation energy use—this could simply include site surveys; spot, short-term or long-term metering; and/or analysis of billing data.

- Calculate energy savings for the first year or all of the remaining years of a contract.

- Calculate first-year payments.

- Conduct annual M&V activities to verify operation of the installed equipment/systems and/or calculation of current-year energy savings (if required in the contract).

- Calculate annual payments.

3.5 M&V Issues

3.5.1 Independent Reviewer

Often the contractor/ESCO has more expertise and experience than the owner in dealing with performance contracts and ECM savings. Therefore, it is almost always more cost-effective and beneficial for the owner to utilize ESCOs or, where appropriate, independent professionals to assist with defining M&V site-specific plans and analyzing the results. This helps provide a "level playing field" for negotiation and determination of savings and payments to the ESCO.

M&V professionals are typically engineering consultants with experience and knowledge in verifying ECM savings, ECM technologies and performance contracting.

3.5.2 Metering and Monitoring Issues Common to All Projects

Metering is just one part of a successful M&V program. Other key components include:

- Properly defining the project and critical factors which affect energy consumption in order to prepare an appropriate M&V plan. These factors may include minimum energy standards.

- Completely defining the baseline conditions.

- Defining the analysis equations and required confidence in the savings calculations in order to determine: i) the data which must be collected, ii) period of time for data collection and iii) the required accuracy of the data collection and analysis technique(s).

- Calculating the value of the project in order to define a cost-effective level (accuracy) of M&V; address the relative value of the M&V information.

- Use qualified staff and/or contractors to collect and analyze data.

- Define the data reporting and archiving requirements.
- Have an appreciation for Murphy's Law.

A note regarding Murphy's Law: Because performance contracts require a multitude of tasks be completed simultaneously, Murphy's Law applies. In other words, "if something can go wrong, it will." Performance contracts require skilled project management. Attention to detail is important for successful execution of a performance contract.

3.5.3 Metering and Monitoring Protocols

A site-specific M&V plan should demonstrate that any metering and monitoring will be done in a consistent and logical manner. Metering and monitoring reports must address exactly what was measured, how, with what meter, when and by whom. Calibration of sensors and meters to known standards is required to ensure that data collected are valid. Project information and metered data must be maintained in usable formats. Both "raw" and "compiled" data should be submitted to the owner with the post-installation and regular interval reports.

The duration of metering and monitoring must be sufficient to ensure an accurate representation of the average amount of energy used by the affected equipment both before and after project installation. The measurements should be taken at typical system outputs within a specified time period, such as one month. These measurements can then be extrapolated to determine annual and time-of-use period energy consumption.

The required length of the metering period depends on the type of project.

- If, for instance, the project is a system that operates according to a well-defined schedule under a constant load, such as a constant-speed exhaust fan motor, the period required to determine annual savings could be quite short. In this case, short-term energy savings can be extrapolated easily to the entire year.
- However, if the project's energy use varies across both day and season, as with air-conditioning equipment, a much longer metering or monitoring period may be required to characterize the system. In this case, long-term data are used to determine annual and time-of-use period energy savings.

- For some types of projects metering time periods may be uncertain. For example, there is still controversy over how long lighting operating hours must be measured in office buildings to determine a representative indication of annual operating hours. For these situations, the project parties must determine the appropriate measurement period for the ECM under consideration.

If energy consumption varies by more than 10% from one month to the next, sufficient measurements must be taken to document these variances. Any major energy consumption variances due to seasonal activity increases or periodic fluctuations must also be monitored. If these variances cannot be monitored, for whatever reason, they must be included in the annual energy consumption figure through a mathematical adjustment agreeable to both parties.

Extrapolation can be used by measuring and normalizing energy consumption as a function of some independent parameter, such as temperature, humidity, product type or production quantity. Once the relationship between the energy consumption of the equipment and the parameter(s) is established, then extrapolation can be done by extending the relationship over a one-year period. Therefore, a site-specific M&V plan should identify critical variables, explain how they will be measured or documented and discuss how they will be used in the extrapolation. Additionally, assumptions and mathematical formulas used in the M&V plan must be clearly stated.

Any auxiliary energy-consuming equipment must be metered or accounted for if its energy consumption changes as a result of the project installation.

3.5.4 Energy Costs

For some projects, contract payments will be based on energy savings, e.g., kWh, kW, therms, etc. For other projects payments will be based on energy *cost* savings. When required, energy cost savings will be calculated using energy savings and the appropriate cost of energy. In most cases, the cost of energy will be based on the servicing utility's energy rate schedules. The cost of energy that will be used in calculating energy cost savings must be defined in sufficient detail in the contract to allow calculation of energy cost savings using each of the factors which affect cost savings. These factors include items such as kWh saved, kW saved, power factor, kW ratchets, energy rate tiers, etc.

3.5.5 Minimum Energy Standards

When a certain level of efficiency is required either by law or by the owner's standard practice, savings may be based on the difference between the energy usage of the new equipment and minimum standard equipment. In these situations the baseline energy and demand consumption may be determined to be equal to or less than any applicable minimum energy standards.

3.5.6 Interactive Effects

It is commonly understood that various ECMs interact with each other. Reduced lighting loads, for example, can reduce air-conditioning energy consumption but increase heating consumption. However, the detailed relationship between most dissimilar but interactive ECMs is not known, and the methods for measuring interactive effects are not cost-effective for most applications. For these reasons, payments for ECM projects with interactive effects will typically:

- be made on savings directly related to the ECM being evaluated,
- include some stipulated interactive factors, or
- be calculated based on Option-C-type analyses.

3.6 Defining the Appropriate Level of M&V

The level of certainty required for verifying an ECM's performance potential and actual performance will vary from project to project. The confidence which is appropriate for establishing savings is a function of the value of the project and the cost-effectiveness of increasing or decreasing confidence in savings. Factors which will affect the level of effort, i.e., cost, are:

- Value of ECM in terms of projected savings
- Complexity of ECM
- Number of ECMs at a single facility and the degree to which savings are interrelated
- Number of interrelated ECMs
- Uncertainty of savings

- Risk allocation between the contractor/ESCO and the owner for achieving savings
- Other uses for M&V data and systems

With respect to value of ECM, suppose a project has an expected savings of $100,000 per year, and that it was believed that this estimate was good ±20% or $20,000 per year. Thus, it may be reasonable to spend $10,000 per year on M&V to bring the actual determination of savings to within an accuracy of ±10%. However, it would not be appropriate to spend $30,000 per year as the value of the results would not be worth the price paid.

Factors which will typically affect M&V accuracy and costs are (some of these are interrelated):

- Level of detail and effort associated with verifying baseline and post-installation surveys
- Sample sizes (number of data points) used for metering representative equipment
- Duration and accuracy of metering activities
- Number and complexity of dependent and independent variables which are metered or accounted for in analyses
- Contract term
- Confidence and precision levels specified for energy-savings analyses

Discussions and definitions of site-specific M&V plans should include consideration of accuracy requirements for M&V activities and the importance of relating M&V costs and accuracy to the value of the ECM savings. For certain types of projects, a statistical definition of accuracy could be included in a contract. For other types of projects, it may be only possible to define a subjective accuracy range or percent of payment budget for M&V.

- *Value of ECM in terms of projected savings.* Scale of a project, energy rates, term of contract, comprehensives of ECMs, benefit sharing arrangement and magnitude of savings can all affect the value of the project. The M&V effort should be scaled to the value of the project so that the value of information provided by M&V activity is appropriate to the project value. "Rule of thumb" estimates put M&V costs at 1-10% of typical project construction cost.

- *Complexity of ECM.* More complex ECM projects may require more complex and expensive M&V methods to determine energy savings. However, this is not always the case. In general, the complexity of savings isolation is the critical factor. A complicated HVAC measure may not be difficult to assess if there is a utility meter dedicated to the HVAC system.

  When defining the appropriate M&V requirements for a given project it is helpful to place projects in one of the following categories (listed in order of increasing M&V complexity):
  - Constant load, constant operating hours
  - Constant load, variable operating hours
    - Variable hours with a fixed pattern
    - Variable hours without a fixed pattern, i.e., weather-dependent
  - Variable load, variable operating hours
    - Variable hours or load with a fixed pattern
    - Variable hours or load without a fixed pattern, i.e., weather-dependent

- *Number of ECMs at a Single Facility and the Degree to Which Their Savings are Interrelated.* If there are multiple ECMs being installed at a single site, the savings from each measure may be, to some degree, related to the savings of other measure(s) or other non-ECM activities at the facility, e.g., interactive effects between lighting and HVAC measures, or HVAC control measures and a chiller replacement. In these situations it will probably not be possible to isolate and measure one system in order to determine savings. Thus, for multiple, interrelated measures Option C is almost always required.

  Uncertainty of Savings. The importance of M&V is often tied to the uncertainty associated with estimated energy or cost savings. ECMs with which the facility staff is familiar may require less M&V than other, less common ECMs. In addition, if a given ECM project is similar to other projects which have documented savings, M&V results may be applied from the other project. If the ESCO specifies the baseline, it may be more appropriate to use M&V Options B or C to verify savings.

- *Risk Allocation Between the Contractor/ESCO and the Owner.* If a contractor/ESCO's payments are not tied to actual savings, M&V is not typically required. Likewise if a contractor/ESCO is not held responsible for certain aspects of project performance, these "aspects" do not need to be measured or verified. The contract should specify how payments will be determined and exactly what needs to be verified. For example, variations in facility operating hours during the contract term may be a risk the owner takes. Consequently, operating hours need not be continuously measured for purposes of payment. In this example, the Option A approach may be appropriate.

- *Other Uses for M&V Data and Systems.* Often the array of instrumentation installed and the measurements collected during M&V can be used for other purposes. These include: commissioning, system optimization and fine tuning, diagnostics, alarms and control. Such uses can be more cost-effective if combined with the objectives of the M&V activities. In addition, there is the possible interest in quantifying savings beyond the requirements of the performance contract. Information may be desired for cost allocation between facility tenants, for future projects or for research purposes.

3.7 Measurement and Verification Options

Three M&V options (A, B and C) are defined in this protocol for use with performance-based projects. Any one option is not necessarily better or more/less expensive than another. Each M&V option is applicable to different types of performance contracts. The three options are described below. The owner and contractor/ESCO should select an M&V option and method for each project and then prepare a site-specific M&V plan that incorporates project-specific details. The M&V options have been defined to help organize selection. The following table provides a quick overview of the options:

M&V Option	Metering	Cost	Accuracy
Option A: Verifying ECM has potential to perform & generate savings	None or short-term periodic	Dependent on no. of measurement points. Approx. 1-5% of construction cost.	Performance accuracy dependent on metering. Energy savings accuracy dependent on estimate of stipulated hours.
Option B: Verifying ECM has potential to perform; verifying actual performance by end use	Continuous in post-installation at system level	Dependent on no. of systems measured. Typically 3-10% of construction cost.	Performance accuracy dependent on metering. Energy savings accuracy dependent on baseline assumptions and metering.
Option C: Verifying ECM has potential to perform; verifying actual performance (whole bldg. analysis)	Continuous in post-installation at whole-facility level	Dependent on no. of relative parameters. Typically 1-10% of construction cost.	Energy savings accuracy dependent on baseline assumptions and selection of relevant variables.

M&V costs depend on many factors, such as the:

- M&V option method selected
- complexity of the ECM
- number of exterior factors affecting its performance
- number of similar ECMs in a single project or program
- accuracy requirements
- duration of contract
- reporting requirements
- experience of the people conducting verification

As a general rule, M&V costs should fall within the ranges listed in the table above. Percentages listed are representative of a percentage of construction costs for the project.

3.7.1 Option A

Option A is a verification approach designed for projects where the potential to perform needs to be verified, but the actual performance (savings) can be stipulated based on the results of the "potential to perform and generate savings" verification and engineering calculations. Option A involves procedures for verifying that:

- Baseline conditions have been properly defined.

- The equipment and/or systems that were contracted to be installed have been installed.

- The installed equipment/systems meet the specifications of the contract in terms of quantity, quality and rating.

- The installed equipment is operating and performing in accordance with the specifications in the contract and meeting all functional tests.

- The installed equipment/systems continue, during the term of the contract, to meet the specifications of the contract in terms of quantity, quality and rating, operation and functional performance.

This level of verification is all that is contractually required for certain types of performance contracts. For example, baseline and post-installation conditions (e.g., equipment quantities and ratings such as lamp wattages, chiller kW/ton, motor kW or boiler efficiency) represent a significant portion of the uncertainty associated with many projects.

The potential to perform may be verified through inspections and/or spot or short-term metering conducted immediately before and/or immediately after project installation. Annual (or some other regular interval) inspections may also be conducted to verify an ECM's continued potential to perform and generate savings.

With Option A, actual achieved energy or cost savings are not verified; they are predicted using engineering or statistical methods that do not involve long-term measurements. All end-use technologies can be verified using Option A. Within Option A various methods and levels of accuracy in verifying performance are available. The level of accuracy involves moving from an inventory method of ensuring nameplate data and quantity of installed equipment to short-term measurements for verifying equipment ratings, capacity and/or efficiency.

Performance can be quantified using any number of methods, each depending on the accuracy requirements of the contract. Performance of equipment can be obtained either directly, i.e., through actual measurement, or indirectly, i.e., through the use of manufacturer data. There may be sizable differences between published information and actual operating data. Where discrepancies exist, or at least are believed to exist, field operating data should be obtained. This could be spot measurement for a constant load application. Short-term M&V can be used if the application is not proven to be a constant load.

Baseline and post-installation equipment should be verified with the same level of detail. Either formally or informally, all equipment baselines should be verified for accuracy and for concurrence with stated operating conditions. Actual field audits will almost always be required.

3.7.2 Option B

Option B is for projects where: i) the potential to perform and generate savings needs to be verified and ii) actual performance during the term of the contract needs to be measured (verified). Option B involves procedures for verifying the same items as Option A plus actual achieved energy savings during the term of the contract. Performance verification techniques involve engineering calculations with metering and monitoring. Option B:

- confirms that the proper equipment/systems were installed and that they have the potential to generate the predicted savings

- determines an energy (and cost) savings value using measured data taken throughout the term of the contract

All end-use technologies can be verified with Option B. However, the degree of difficulty and costs associated with verification increases proportionately as metering complexity increases.

Energy-savings-value accuracy is defined by the owner or negotiated with the contractor/ESCO. The task of measuring or determining energy savings using Option B can be more difficult and costly than with Option A. However, the results will typically be more precise.

Methods employed in this option will involve the use of long-term measurement of one or more variables. The use of long-term measurement accounts for operating variations and will more closely approximate actual energy savings than the use of stipulations as defined for Option A. However, under certain circumstances there is no inherent increase in accuracy.

Measurement of all end-use operating systems may not be required through the use of statistically valid sampling. Examples of this include measurement of operating hours for a selected group of lighting fixtures or power draw of certain motors which have been predetermined to operate in a similar manner.

3.7.3 Option C

Option C may be employed for projects where: i) the potential to perform needs to be verified and ii) actual performance during the term of the contract needs to be verified. Option C involves procedures for verifying the same items as Option A plus actual achieved energy savings during the term of the contract.

Performance verification techniques involve utility whole-facility meter analysis and/or computer simulation calibrated with utility billing data. Option C:

- confirms that the proper equipment/systems were installed and that they have the potential to generate the predicted savings

- determines an energy-savings value using measured utility meter data taken throughout the term of the performance contract

All end-use technologies can be verified with Option C. This option may be used in cases where there is a high degree of interaction between installed energy conservation systems and/or the measurement of individual component savings is difficult. Accounting for changes (other than those caused by the ECMs) is the major challenge associated with Option C, particularly for long-term contracts.

SECTION 4.0: THREE M&V OPTIONS WITH EXAMPLES

4.0.1 Documenting Baseline/Installed Equipment

Energy-consuming equipment to be replaced or modified as part of an energy-conservation project requires a thorough documentation of the installed equipment operating during the baseline and post-installation periods. Many reports, papers and text books address the specific information that needs to be gathered and provide procedures and methods for accomplishing this task. The following are sources of additional information: ASHRAE (1990), Dukelow (1991), Dyer and Maples (1981), Dubin and Long (1978), Dubin et al. (1976), Dutt and Harrje (1988a; 1988b), DOE (1980), EPA (1993), Fracastoro and Lyberg (1983), Haberl et al. (1990; 1992), Haberl and Komor (1989), Harrje (1982), Harrje (1986), IES (1987), Jilar (1990), Lyberg (1987), MacDonald et al. (1989), SMACNA (1985), Stein and Reynolds (1992), Ternes (1987), Turner (1993) and Witte et al. (1988).

Procedures

At a minimum, the following procedures are recommended to characterize and document the installed equipment during the baseline and post-installation periods:

- Record the location and count of equipment to be retrofitted so that it can easily be located on a set of plans. Indicate the facility, room and location of the equipment within the room.

- Take photographs and/or videotapes of the equipment to accurately document its condition. Each piece of equipment, or equipment lots, should have the manufacturer's model number, serial number and nameplate information recorded. This information is usually necessary when contacting the manufacturer to obtain equipment performance specifications.

- Obtain an accurate count of the systems to be replaced if the retrofit involves multiple units of one type of equipment. This should also be accompanied by a location diagram that indicates where new/different equipment is to be located.

- If a lighting retrofit is being considered, measure baseline and post-installation lighting conditions using standard Illuminating Engineering Society (IES) measurements. Determine lighting fixture operating schedules including: general, task, hallway and exterior lighting.

- If a heating/cooling equipment retrofit is being considered, determine system setpoints and operating schedules including: thermostat setpoints; system temperature settings, i.e., cold deck temperature, boiler temperature/pressure; on/off schedules for air-handler units, pumps, air conditioners, chillers, boilers, etc. An assessment of thermal comfort and/or indoor air quality (IAQ) may also prove useful in cases where the new system does not perform as well as the old, inefficient system.

4.1 Option A: End-Use Retrofits—Measured Capacity, Stipulated Consumption Approach

Option A, the first approach to M&V presented in this protocol, is intended for energy conservation retrofits where end-use capacity, demand or power level can be measured or estimated with manufacturer's measurements, and energy consumption, or hours of operation, are known in advance, stipulated or agreed upon by both parties. Option A usually involves a one-time measurement of the instantaneous energy use before the retrofit (baseline), and a one-time measurement of the instantaneous energy use after the retrofit (post-installation). In certain circumstances, representative measurements can be made in place of *in-situ* measurements where multiples of identical units are being installed. Periodic equipment inspections may also be warranted. Estimated energy consumption is calculated by multiplying the measured end-use capacity, i.e., the kW, Btu/hr or kJ/hr) by the stipulated hours of operation for each characteristic mode of operation, i.e., weekday/weekend hourly profiles.

4.1.1 Confirming Installed Equipment Performance

Option A performance verification is estimated by multiplying the representative energy capacity by the hours of operation. The capacity, demand or power level, i.e., kW, Btu/hr or kJ/hr, needs to be measured using one-time, *in-situ* end-use measurements. This may be estimated with representative sample measurements, representative manufacturer's measurements or representative baseline power levels. The hours of operation are either known in advance, stipulated or agreed upon by both parties. Each of these methods is described below.

One-Time, In-Situ *End-Use Measurements*

One-time, *in-situ,* end-use measurements are measurements taken at the site using calibrated instrumentation. Information regarding calibration and instrumentation can be found in Section 5.0 of this document. One-time, *in-situ* measurements are appropriate for energy-consuming equipment that does not vary significantly in load, i.e., by more than ±5%. Types of one-time, *in-situ* energy measurements that can be taken include: electrical energy use measurements (i.e., watt measurements using an RMS Wattmeter), thermal product energy use measurements (i.e., Btu-thermal—Btu-t, or Joule-thermal—Joule-t) and thermal fuel input energy use (i.e., Btu-f or Joule-f).

For electrical loads, this type of measurement usually requires isolating the device to be measured and measuring the electrical power (RMS Wattage) that the device draws on all phases. This can be accomplished at the electrical distribution panel or at the plug with a specially modified extension plug that allows access to individual wires in the branch circuit. Section 5.3.1 of this document discusses this issue in depth.

Thermal product energy-use measurements are measurements taken after the energy fuel (electricity, natural gas) has been converted into thermal energy (steam, hot or chilled water). Thermal product energy-use measurements usually require a volumetric flow rate per unit time (m), a specific heat value (cp) and a temperature difference (delta-T). Steam measurements require a steam flow rate (m), temperature (T) and pressure (P) of the steam, and temperature of the boiler feedwater.

Section 5.3.6 of this document discusses this issue in depth.

Thermal fuel energy-use measurements are measurements of the weight, mass or quantity of fuel being consumed by the energy conversion device including: electricity, coal, wood, biomass, natural gas, oil and/or various forms of liquid petroleum.

In general, estimates of energy savings using the Option A approach may be adversely affected by the following factors:

- Measured capacity, stipulated consumption savings estimates may vary if there are changes to the equipment during the course of the retrofit that affect equipment operating efficiency.

- Measured capacity, stipulated consumption savings estimates may vary if operational settings that affect facility system performance are changed after measurements are taken.

- Measured capacity, stipulated consumption savings estimates of chillers will vary if the chiller evaporator or condenser temperature operates at different temperatures than those which occurred during the *in-situ* tests. Measured capacity, stipulated consumption estimates of chiller retrofits should only be used in cases where the owner and contractor/ESCO accept the uncertainty due to the large variation that occurs in chiller performance, given varying operating conditions.

- Measured capacity, stipulated consumption savings estimates of boilers may vary if the boiler operates at different temperatures, loads or combustion settings than those which occurred during the *in-situ* tests. Tests will, therefore, need to be conducted over a range of operation to characterize performance. Measured capacity, stipulated consumption estimates for boiler retrofits should only be used in cases where the owner and contractor/ESCO accept the uncertainty due to the large variation that occurs in boiler performance, given varying conditions.

- Measured capacity, stipulated consumption savings estimates of pumps and blowers may vary if the pump or blower operates at different pressure settings or flow rates than occurred during the *in-situ* tests. Tests will, therefore, need to be made over a range of operation to characterize the performance.

- Measured capacity, stipulated consumption savings estimates of lighting retrofits may vary if there is a significant number of lamp outages, or if the actual operating schedule varies significantly from the stipulated operating schedule.

- Measured capacity, stipulated consumption savings estimates of air conditioners may vary if the air conditioner operates at different condenser temperatures or evaporator temperatures than those which occurred during the *in-situ* tests. Air conditioner efficiency may also be adversely affected by compressor degradation, low/high refrigerant charge, and/or insufficient evaporator or condenser air flow. Measured capacity, stipulated consumption estimates for air conditioner retrofits should only be used in cases where the owner and contractor/ESCO accept the uncertainty due to the large variation that occurs in air conditioner performance, given varying conditions.

All of these operating conditions should be noted carefully during both the baseline and post-installation periods.

Representative Sample Measurements

Representative sample measurements are measurements that are taken with calibrated instrumentation on a representative sample of equipment being installed. Representative sample measurements are appropriate for energy-consuming equipment that does not vary significantly in load, i.e., by more than ±5%, and must be taken on similar equipment model types. Types of representative sample measurements that may be taken include: electrical energy use measurements (i.e., watt measurements using an RMS Wattmeter), thermal product energy use measurements (i.e., Btu-t or Joule-t) and thermal fuel input energy use (i.e., Btu-f or Joule-f).

Estimates using representative sample measurements and stipulated consumption may be adversely affected by the same factors as Option A: one-time, *in-situ* measurements as listed previously.

Representative Manufacturer's Measurements

Representative manufacturer's measurements are measurements

taken and published by the manufacturer. In order for such measurements to be valid, they should be taken with calibrated instrumentation on a representative sample of equipment being installed. Representative manufacturer's measurements are appropriate for energy-consuming equipment that does not vary significantly in load, i.e., by more than ±5%, and must be taken on similar equipment model types. Types of representative sample measurements that may be taken include: electrical energy-use measurements (i.e., watt measurements using an RMS Wattmeter), thermal product energy use measurements (i.e., Btu-t or Joule-t) and thermal fuel input energy use (i.e., Btu-f or Joule-f).

Estimates using manufacturer's sample measurements and stipulated consumption may be adversely affected by the same factors as Option A: one-time, *in-situ* measurements as listed previously.

Representative Baseline Power-Level Profiles

Representative baseline power-level profiles are either hourly or 15-minute measurements taken at the site usually at the whole-facility level or subpanel level using portable monitoring equipment. These measurements represent an aggregate end-use load, e.g., all motors or lighting loads in a facility. Representative baseline power-level profiles capture the *in-situ* 24-hour profiles of a group of equipment operating during weekday or weekend modes. Such measurements are appropriate for non-weather-dependent energy-consuming equipment loads that vary within a 24-hour period, but do not vary from day to day by more than ±10%.

Examples include: weekday/weekend, whole-facility lighting loads and motor control center loads that include only constant-load motors. In general, representative baseline power-level profiles can be used to measure weather-independent loads. Representative baseline power-level profiles for weather-dependent loads should include measurements taken over a long-enough period to adequately characterize the schedule, i.e., weekday/weekend and weather-dependent characteristics of the end-use load. Examples of weather-dependent, day-type profiling can be found in Katipamula and Haberl (1991), Akbari et al. (1988), Hadley and Tomich (1988) and Bou Saada and Haberl (1995a).

These types of measurements include: continuous 15-minute electrical energy-use measurements (i.e., watt measurements using an RMS Wattmeter), thermal product energy-use measurements (i.e., Btu-t/hr, or kJ-t/h), and thermal fuel input energy use (i.e., Btu-f/h or kJ/h).

In general, estimates of energy savings using Option A: representative baseline power-level profiles can be adversely affected by the same factors as Option A: one-time, *in-situ* measurements as listed previously.

4.1.2 Examples

Lighting Efficiency and/or Controls Project. Savings resulting from lighting efficiency and/or lighting controls retrofits can be estimated using the Option A approach, provided both the owner and contractor/ESCO are willing to accept the uncertainty that accompanies stipulated consumption. Existing baseline conditions should be documented according to the procedures outlined in Section 4.0.1.

Lighting projects require the capacity, demand or power level (i.e., kW, Btu/hr or kJ/hr) be measured using one-time, *in-situ,* end-use measurements, representative sample measurements, representative manufacturer's measurements or representative baseline power levels. The hours of operation are either known in advance, stipulated or agreed upon by both parties.

Electricity savings, due to reduced lighting energy, are calculated by multiplying the difference between baseline and post-installation measured capacity by the stipulated consumption or hours of operation. Electric demand reductions can also be analyzed, provided representative baseline and post-installation measurements have been taken.

Electricity savings, due to reduced cooling load, are not included in Option A estimates. As well, negative savings, which account for heating increases due to reductions in internal heating load, are not included.

Calculating Electricity Savings. Electricity savings resulting from lighting retrofits can be estimated in the following fashion: First, measure the baseline capacity of the facility's lighting load using one-time, *in-situ,* end-use measurements, representative sample measurements, representative manufacturer's measurements or representative baseline power levels. Second, estimate energy savings by multiplying the difference between baseline and post-installation measurements by the stipulated hours-of-use or hourly profiles. Both the owner and contractor/ESCO should understand that this analysis provides an energy-savings estimate which may or may not represent actual energy savings from the lighting project.

Calculating Electric Demand Reductions. Electric demand reductions resulting from a lighting retrofit can be estimated in the following fashion: First, develop a baseline demand measurement using the methods previously described. Second, calculate retrofit electric-demand savings by comparing baseline demand to measured post-installation demand, where the demand is measured using one-time, *in-situ*, end-use measurements, representative sample measurements, representative manufacturer's measurements or representative baseline power levels. Both the owner and contractor/ESCO should understand that this analysis provides a demand-savings estimate which may or may not represent actual demand savings from the lighting project.

Calculating Interactive Cooling Savings. An estimate of interactive cooling savings is not included in the Option A approach. Estimates of interactive cooling savings may be stipulated as part of the contract; however, both the owner and contractor/ESCO should understand that these estimates may or may not reflect actual interactive cooling savings from a particular retrofit.

Calculating Interactive Heating Increases. An estimate of interactive heating increases is not included in the Option A approach. Estimates of interactive heating increases may be stipulated as part of the contract; however, both the owner and contractor/ESCO should understand that these estimates may or may not reflect actual interactive heating increases from a particular retrofit.

Limitations of Calculating Retrofit Savings From Lighting and/or Lighting Controls Projects Using Option A. Savings resulting from lighting efficiency and/or lighting controls projects that are calculated using Option A can be adversely affected by the following factors:

- Savings estimates may vary if there are equipment changes during the retrofit that affect equipment operating efficiency.

- Savings estimates may vary if operating settings that affect facility system performance are changed after measurements are taken.

- Savings estimates may vary if there is a significant number of lamp outages, or if the actual operating schedule varies significantly from the stipulated operating schedule.

- Savings estimates calculated using the Option A approach do not measure cooling interaction or increases in heating load due to reductions in internal heating caused by improved lighting system efficiency.

All of the above operating conditions should be noted carefully during both the baseline and post-retrofit periods.

Constant Load Motor Replacement Project. Savings resulting from constant load motor replacement projects can be estimated using the Option A approach, provided both the owner and contractor/ESCO are willing to accept the uncertainty that accompanies stipulated consumption or hours of operation of the motor. Existing baseline conditions should be documented according to the procedures outlined in Section 4.0.1.

Load motor replacement projects require the capacity, demand or power level, i.e., kW, Btu/hr or kJ/hr, be measured using one-time, *in-situ,* end-use measurements estimated with representative sample measurements, representative manufacturer's measurements or representative baseline power levels. The hours of operation are either known in advance, stipulated or agreed upon by both parties.

Electricity savings, due to reduced motor load, are calculated by multiplying the difference between baseline and post-installation measured capacity by the stipulated consumption or hours of operation. Electric demand reductions can also be analyzed, if representative baseline and post-installation demand measurements have been taken.

If the motors being replaced are used to deliver chilled or hot water, downsizing the motor may reduce thermal flows to the facility, which may cause a reduction in cooling or heating. Savings resulting from reduced cooling or heating load are not included in Option A estimates because only baseline and post-installation motor electrical loads are being multiplied by the run-time.

Calculating Electricity Savings. Electricity savings resulting from a constant load motor replacement can be estimated in the following fashion: First, measure the baseline capacity of the motor(s) to be replaced using one-time, *in-situ,* end-use measurements, representative sample measurements, representative manufacturer's measurements or representative baseline power levels. Second, estimate the energy savings by multiplying the difference between baseline and post-installation capacity measurements by the stipulated hours of use or hourly profiles. Both the owner and contractor/ESCO should understand that this analysis provides an energy-savings estimate which may or may not represent actual energy savings from the constant load motor replacement project.

Calculating Electric Demand Reductions. Electric demand reductions resulting from a constant load motor replacement can be estimated by using the same method described for lighting projects in this Section 4.0.

Calculating Cooling or Heating Savings. Estimate of cooling or heating savings due to the downsizing of a motor used to deliver thermal energy is not included in the Option A approach. Cooling or heating savings may be stipulated as part of the contract; however, both the owner and contractor/ESCO should understand that these estimates may or may not reflect actual interactive cooling or heating savings from a particular retrofit.

Limitations of Calculating Retrofit Savings from Constant Load Motor Retrofits Using Option A. Savings from a constant load motor replacement that are calculated using Option A can be adversely affected by the following:

- Savings estimates may vary if there are equipment changes during the retrofit that affect equipment operating efficiency.
- Savings estimates may vary if operating settings that affect facility system performance are changed after measurements are taken.
- Savings estimates may vary if there is a change in the load placed on the motor, e.g., if there is a significant increase in the pressure drop across the motor due to a valve closure in the piping system.
- Savings estimates using Option A do not measure cooling/heating savings due to downsizing in the pump that may be delivering thermal energy to a facility.

All of the above operating conditions should be noted carefully during both the baseline and post-retrofit periods.

4.1.3 Expected Accuracy

Option A is meant to serve as a contractual substitute for measuring post-installation savings. Option A substitutes baseline and post-installation measured capacity multiplied by a stipulated hours-of-use number, for actual measured energy retrofit savings. Accuracy of expected savings is dependent on the accuracy of the one-time, baseline and post-installation *in-situ* measurements and the stipulated hours of use or baseline and post-installation consumption estimates.

If significant attention is paid to measurement accuracy, and if the estimates of run-time or load profiles are collaborated with *in-situ* measurements, the accuracy of such tests can be ±20% of the actual performance. However, any inaccuracies in estimated annual run-time profiles can severely affect the savings estimates. In the worst case, errors of 100 to 200 percent have been observed.

4.1.4 *Expected Cost*

Option A costs will generally fall between 1-5% of construction costs. This includes any periodic reports made over the payback period of the retrofit. For example, if a $100,000 retrofit was installed, roughly between $1,000 and $5,000 should be allocated to estimate savings and produce the appropriate reports.

4.2 Option B: End-Use Retrofits— Measured Capacity, Measured Consumption Approach

Option B is intended for energy conservation retrofits where the end-use capacity, demand or power level can be measured before the retrofit (baseline), and the continuous energy consumption of the equipment or subsystem can be measured after the retrofit (post-installation) for a selected period of time. Option B can involve a continuous measurement of energy use both before and after the retrofit for the specific equipment or energy end use affected by the retrofit for a limited period of time necessary to determine retrofit savings. Periodic inspections of the equipment may also be warranted.

Energy consumption is calculated by developing statistically representative models of the energy end-use capacity (i.e., the kW or Btu/hr) and consumption (i.e., the kWh or Btu). As with Section 4.1, this section includes information regarding installed equipment performance confirmation and includes examples, expected accuracy and cost information.

4.2.1 *Confirming Installed Equipment Performance*

The primary difference between Options A and B is that Option A uses *one-time* baseline and post-installation "snapshot" measurements, whereas Option B involves portable monitoring equipment installed in a facility for a *period of time* to measure the *in-situ*, baseline and post-installation performance of the specific equipment being replaced. Time allotted for installing portable metering devices during the baseline and post-installation periods depends on the type of equipment being measured.

For example, the *in-situ* measurement of constant-load motor replacements may take only a few hours or a few days before the retrofit and some period of time after the retrofit. Measurement of the 24-hour profile of whole-facility lighting loads may take several weeks to one month to determine average weekday and weekend use (before and after the retrofit). Option B does not include measurement of whole-facility heating or cooling loads which would be necessary to calculate heating-cooling interaction of a lighting retrofit.

Specific tests may need to be performed on the equipment to force it through all possible operating modes while input-output efficiency measurements are being taken. Examples of this type of testing include chiller efficiency tests (Gordon and Ng 1994; Anderson and Breene 1995; Phelan et al. 1995), boiler efficiency tests (CEUE 1995; Dyer and Maples 1981; Dukelow 1991) and tests regarding pumps and fans (Phelan et al. 1996).

4.2.2 Examples

Lighting Efficiency and/or Controls Project. Savings resulting from efficiency and/or lighting controls projects can be measured using Option B, provided both the owner and contractor/ESCO are willing to accept the uncertainty that accompanies estimates made to extrapolate sample measurements, so that one year of lighting consumption is represented. Existing baseline conditions should be documented according to the procedures outlined in Section 4.0.1.

The capacity, demand or power level (i.e., kW, Btu/hr or kJ/hr) and consumption are measured during the baseline period using portable hourly or 15-minute monitoring equipment for a period deemed sufficient to characterize lighting system performance during all operational periods, i.e., weekday, weekend, etc. After the retrofit, the measurements are repeated to develop post-installation, 24-hour profiles of lighting system energy consumption. Continuous post-installation measurements can also be taken.

Electricity savings due to reduced lighting energy consumption are calculated by analyzing the difference between measured, 24-hour consumption profiles for the baseline period and the post-installation period. Electric demand reductions can also be analyzed, provided representative baseline and post-installation demand measurements have been taken.

Electricity savings due to reduced cooling load are not included in

Option B estimates. As well, negative savings, which account for increased heating due to reduced internal heating load, are not included.

Calculating Electricity Savings. Electricity savings due to reduced lighting energy consumption are calculated by analyzing the difference between measured 24-hour consumption profiles for the baseline and post-installation periods. Care should be taken to adequately capture the correct number of day-type profiles to accurately represent the facility's baseline electricity use during weekday, weekend and holiday periods. In some cases, additional profiles may be needed to capture lighting energy use during secondary schedules. For example, in educational facilities there is often a significant difference between school-year and summer-vacation-period loads. In some instances baseline, weekday/ weekend profile measurements may be necessary during both school-year and summer-vacation periods. Electric demand reductions can also be analyzed, provided representative baseline and post-installation demand measurements have been taken. Post-installation measurements can either be taken continuously throughout the payback period, or for a representative sample period. Savings can be projected with statistical projections.

Calculating Electric Demand Reductions. Electric demand reductions resulting from a lighting retrofit can be calculated in this fashion: First, develop an hourly baseline demand measurement profile using the methods previously described. Second, calculate retrofit electric demand savings by comparing baseline demand to measured, post-installation demand. Both the owner and contractor/ESCO should understand that this analysis provides an estimate of the demand savings which may or may not represent actual demand savings from the lighting project.

Calculating Interactive Cooling Savings. Interactive cooling savings estimates are not included for Option B because whole-facility cooling measurements that correspond to the baseline and post-installation periods are not normally taken. Estimates of interactive cooling savings may be stipulated as part of the contract; however, both the owner and contractor/ESCO should understand that these estimates may or may not reflect actual interactive cooling savings from a particular retrofit.

Calculating Interactive Heating Increases. Interactive heating savings estimates are not included for Option B. Estimates of interactive heating increases may be stipulated as part of the contract; however, both the owner and contractor/ESCO should understand that these estimates may or may not reflect actual interactive heating increases from a particular retrofit.

Limitations of Calculating Retrofit Savings from Lighting and/or Lighting Controls Projects Using Option B. Savings resulting from lighting efficiency and/or lighting controls projects that are calculated using Option B can be adversely affected by the following factors:

- Savings calculated using Option B are intended to be estimates of electricity savings which utilize representative one-time samples of baseline electricity use and either continuous or representative samples of post-installation electricity use. Therefore, measurement accuracy is completely dependent on how well representative profiles match actual baseline and/or post-installation lighting profiles in the facility.

- Savings estimates may vary if there are equipment changes during the retrofit that affect equipment operating efficiency.

- Savings estimates may vary if operating settings that affect facility system performance are changed after measurements are taken.

- Savings estimates may vary if there is a significant number of lamp outages, or if the actual operating schedule varies significantly from the stipulated operating schedule.

- Savings estimates calculated using the Option B approach do not measure cooling interaction or increases in heating load due to reductions in internal heating caused by improved lighting system efficiency.

All of the above operating conditions should be noted carefully during both the baseline and post-retrofit periods.

Constant Load Motor Replacement Project. Savings resulting from constant load motor replacement projects can be using the Option B approach, provided both the owner and contractor/ESCO are willing to accept the uncertainty that accompanies short-term measurements. Existing baseline conditions should be documented according to the procedures outlined in Section 4.0.1.

Baseline capacity, demand or power level (i.e., kW, Btu/hr or kJ/hr) needs to be measured using short-term, *in-situ*, end-use measurements estimated with representative sample measurements (Biesemeyer and Jowett 1994; Phelan et al. 1996). These measurements are then repeated post-installation to determine any change in the energy use of the motor. Depending upon the type of system or load, these measurements can either be representative measurements (for constant-speed, constant-load systems) or continuous measurements (for constant-speed, varying-load systems).

Electricity savings due to reduced motor load are calculated by analyzing the difference between baseline and post-installation measured capacity multiplied by the estimated hours of operation. Electric demand reductions can also be analyzed, provided representative baseline and post-installation demand measurements have been taken.

If the motors being replaced are used to deliver chilled or hot water, downsizing the motor may reduce thermal flows to the facility, which may cause cooling or heating reductions. Savings due to any cooling or heating load reductions are not included in Option B estimates and will need to be either stipulated or measured using other methods.

Calculating Electricity Savings. Electricity savings resulting from constant load motor replacement can be estimated in the following fashion: First, measure the baseline capacity of the motor(s) to be replaced using short-term, *in-situ* measurements during the baseline period. Next, either repeat the measurements one time or continuously during the post-installation period. Calculate energy savings by analyzing the difference between baseline and post-installation measured electricity use. When sample measurements are used to calculate savings, statistical models of electricity use will need to be created and energy use projected using the appropriate load profiles. Both the owner and contractor/ESCO should understand that this analysis provides an energy-savings estimate which may or may not represent actual energy savings from the constant load motor replacement project.

Calculating Electric Demand Reductions. Electric demand reductions resulting from constant load motor replacement can be calculated in the following fashion: First, develop a baseline demand measurement for the electric load of the motor(s) to be replaced. Second, calculate the retrofit electric demand savings by comparing the baseline demand to the measured post-installation demand. Both the owner and contractor/ESCO should understand that this analysis provides a demand-savings estimate which may or may not represent actual demand savings from the constant load motor replacement project.

Calculating Cooling or Heating Savings. An estimate of cooling or heating savings due to downsizing a motor delivering thermal energy is not included in Option B. Estimates of cooling or heating savings may be stipulated as part of the contract; however, both the owner and contractor/ESCO should understand that these estimates may or may not reflect actual interactive cooling or heating savings from a particular retrofit.

Limitations of Calculating Retrofit Savings from Constant Load Motor Retrofits Using Option B. Savings resulting from a constant load motor retrofit calculated using the Option B approach can be adversely affected by the following:

- Savings measured using the Option B approach are intended to be estimates of electricity savings which utilize representative baseline and post-installation, one-time measurements or short-term measurements of the installed electric motors. Therefore, the accuracy of the measurements is completely dependent on how well representative measurements match actual motor electricity consumption over an annual period.

- Savings estimates may vary if there are equipment changes during the retrofit that affect equipment operating efficiency.

- Savings estimates may vary if operating settings that affect facility system performance are changed after measurements are taken.

- Savings estimates may vary if there is a change in load being placed on the motor, e.g., there is a significant increase in the pressure drop across the motor due to a valve closure in the piping system.

- Savings estimates calculated using Option B do not measure cooling or heating savings due to downsizing in the pump delivering thermal energy to a facility.

All of the above operating conditions should be noted carefully during both the baseline and post-retrofit periods.

Variable-Speed Drive Motor Project. Savings resulting from variable-speed drive motor replacement projects can be estimated using the Option B approach, provided both the owner and contractor/ESCO are willing to accept the uncertainty that accompanies short-term measurements. Existing baseline conditions should be documented according to the procedures outlined in Section 4.0.1.

Baseline capacity, demand or power level (i.e., kW, Btu/hr or kJ/hr) needs to be measured using short-term, *in-situ*, end-use measurements estimated with representative sample measurements (Biesemeyer et al. 1993; Phelan et al. 1996). Short-term measurements are then repeated post-installation to adequately characterize the motor's variable electricity use. Continuous measurements are taken in cases where it is not possible to predict the varying loads on the motor. A statistical model can be used to extrapolate this variable electricity use over an entire year.

Electricity savings due to reduced motor load are calculated by analyzing the difference between the motor's measured constant baseline electricity use and either: i) actual measured electricity use in the post-installation period or ii) electricity use predicted by the statistical model of the post-installation electricity use. Electric demand reductions can also be analyzed, provided representative baseline and post-installation demand measurements have been taken.

If the motor(s) being replaced is used to deliver chilled or hot water, downsizing it may reduce thermal flows to the facility, causing reduced cooling or heating. Savings due to any cooling or heating load reduction are not included in Option B estimates because only measured baseline and post-installation motor electrical loads are used in the calculation.

Calculating Electricity Savings. Electricity savings resulting from variable-speed motor replacement can be estimated in the following fashion: First, measure the baseline capacity of the motor(s) to be replaced using short-term, *in-situ* measurements in the baseline period. These measurements should adequately characterize the 24-hour, seven-day-per-week electricity use. Second, in the post-installation period, take either continuous measurements or short-term measurements to characterize the variable electricity use (Phelan et al. 1995). In cases where short-term measurements are used, electricity use variability should be analyzed and correlated to a predictor variable (such as ambient temperature), so that an hourly statistical model can be developed to extrapolate variable electricity use for an entire year. When continuous measurements are used, only the baseline period calculation requires a statistical model be developed for predicting constant speed energy use.

Savings are determined to be significant if the difference between baseline and post-installation energy use is greater than model error as determined by calculations of the R^2, RMSE and CV(RMSE). Equations for determining model error are included in Section 5.13. Uncertainty equations for measuring *in-situ* performance can be found in Phelan et al., 1996.

Calculating Electric Demand Reductions. Electric demand reductions from a constant load motor replacement project can be estimated by comparing measured peak hourly baseline electricity use with peak hourly electricity use measured in the post-installation period or peak electricity use predicted by the post-installation statistical model. Both the owner and contractor/ESCO should understand that this analysis provides a demand-savings estimate which may or may not represent actual demand savings from the motor replacement project.

Calculating the Cooling or Heating Savings. Cooling or heating savings estimates due to downsizing a motor used to deliver thermal energy are included in Option B. Estimates of cooling or heating savings may be stipulated as part of the contract, however, both the owner and contractor/ESCO should understand that these estimates may or may not reflect actual interactive cooling or heating savings from a particular retrofit.

Limitations of Calculating Retrofit Savings from Variable-Speed Motor Retrofits Using Option B. Savings resulting from variable-speed motor retrofits calculated using the Option B approach can be adversely affected by the same issues which may affect load motor replacement projects, with the exception of the following:

- Savings estimates of variable-speed motor retrofits are dependent on the accuracy of baseline constant-speed measurements and either continuous post-installation electricity use or the post-installation statistical model. Therefore, care should be taken to develop a model(s) that accurately characterizes performance in both the baseline and post-installation periods.

HVAC and/or EMCS Project. Savings resulting from HVAC systems and/or Energy Management Control System (EMCS) projects can be analyzed using Option B, providing a calibrated engineering model is developed for each HVAC system to adequately assess performance in the baseline period and either continuous measurements are made in the post-installation period or a calibrated model is developed in the post-installation period (Knebel 1983; Katipamula and Claridge et al. 1991; Liu et al. 1995). Annual savings are calculated by comparing energy use predicted by the model(s) for the agreed-upon standard operating schedule and ambient conditions. Such models are capable of determining electricity and thermal savings, as well as electric demand reductions.

Calculating Electricity Savings. Electricity savings resulting from HVAC and/or EMCS retrofits can be calculated using calibrated baseline and post-installation engineering models of the system. To develop such models, each major HVAC system in the facility must be inspected and analyzed, and a separate baseline psychrometric model developed to predict existing system energy use. This normally includes short-term measurements of *in-situ* performance of the HVAC system (Phelan et al. 1996, Balcomb et al. 1993 and Liu et al. 1994). In the post-installation period, either continuous energy use is measured or post-installation HVAC system models are developed that reflect post-installation operational changes. These post-installation models also need to be calibrated to measure short-term data.

Savings are determined to be significant if the difference between the model-predicted baseline and post-installation energy use is greater than model error as determined by the RMSE.

Calculating Electric Demand Reductions. Electric demand reductions are also estimated by comparing the difference between projected baseline electricity use and electricity use predicted by the post-installation model. Care should be taken to ascertain the appropriate demand billing intervals that agree with those charged by the local utility.

Calculating Cooling Savings. Cooling energy savings can also be calculated if calibrated baseline and post-installation simulation models are used. Cooling savings are estimated by comparing the post-installation projections of the baseline HVAC cooling use to the HVAC cooling use predicted by the post-installation model. Appropriate calculations need to be made to determine the effect of the primary cooling system efficiency, i.e., kW/ton of the chillers for varying loads.

Calculating Heating Savings. As with cooling savings calculations, heating energy savings can be calculated if calibrated baseline and post-installation simulation models are used. Heating savings are estimated by comparing post-installation projections of the baseline HVAC heating use to the HVAC cooling use predicted by the post-installation model. Appropriate calculations need to be made to determine the effect of the primary heating system efficiency, i.e., Btu/lb.-steam or input/output boiler efficiency for varying loads.

Limitations of Calculating Retrofit Savings from HVAC and EMCS Retrofits Using Option B. Estimated savings from HVAC and/or EMCS projects calculated using Option B can be adversely affected by the following factors:

- Savings measured using the Option B approach are intended to be estimates of electricity savings which utilize representative baseline and post-installation, one-time measurements or short-term measurements of the installed HVAC electricity and thermal performance. Therefore, the accuracy of the measurements is completely dependent on how well representative measurements match actual HVAC electricity and thermal consumption over an annual period.

- Estimated savings using Option B may be affected if HVAC system operating characteristics do not complement representative schedules used to drive the models.

- Estimated savings from Option B may be affected if EMCS programming is significantly different than the representative schedule used to drive the models, i.e., setpoint temperatures, schedules, etc.

- Changes in cooling savings may be affected by procedures used to operate the cooling systems. In particular, the average chiller kW/ton ratio is affected by the rate of the cooling load on a particular chiller. Chillers that are loaded below 50% of their capacity tend to have significantly higher kW/ton ratios which can increase overall electricity consumption.

- Changes in heating savings may be affected by procedures used to operate the heating systems. Boilers or furnaces run at low loads can cycle excessively, which decreases fuel conversion efficiency.

All of the above operating conditions should be noted carefully during both the baseline and post-retrofit periods.

Chiller Project. Savings resulting from chiller retrofit projects can be estimated using Option B if calibrated baseline and post-installation chiller models are developed (Phelan et al. 1995; Gordon and Ng 1994;

Anderson and Breene 1995). Such models are sensitive to differences in chilled water supply temperatures, condenser water return temperatures (or refrigerant return temperatures for air condensers) and chiller loads.

To calibrate such models, chiller thermal output, chiller electricity use, chilled water supply temperature and condenser water return temperatures need to be measured over the expected range of operation. Measurements are repeated post-installation. Annual savings are then calculated by driving the chiller models with an agreed-upon schedule of chiller loads, chilled water supply temperatures and condenser temperatures, and comparing the differences predicted.

Limitations of Calculating Retrofit Savings from Chiller Retrofits Using Option B. Estimated chiller retrofit savings calculated using the Option B approach can be adversely affected by the following factors:

- Savings calculations are intended to be estimates of electricity savings which utilize representative baseline and post-installation, one-time measurements or short-term measurements of installed chiller performance over varying conditions. Therefore, the accuracy of the measurements is completely dependent on how well representative chiller measurements match actual chiller performance over an annual period.

- Estimated savings using Option B may be affected if chiller operating characteristics do not complement the representative schedules used to drive the models.

All of the above operating conditions should be noted carefully during both the baseline and post-retrofit periods.

Boiler Project. Boiler retrofit savings can be estimated using Option B if input-output boiler efficiency tests, or combustion efficiency tests, are taken before and after the retrofit (Dukelow 1991; Dyer 1981; Babcock and Wilcox 1992). In smaller boilers other test methods can be used, i.e., the "time to make steam" test (Center for Energy and Environment, CEE). In order to be effective, these boiler efficiency tests should be taken under varying operating conditions in order to capture boiler efficiency over its expected operating range, temperature and pressure. The results of these tests should yield a set of performance curves that can then be applied to an agreed-upon histogram of annual operating hours to establish the annual boiler performance. Retrofit savings are then calculated by comparing the differences between baseline annual boiler performance and post-installation annual boiler performance. Continuous post-installation measurements can also be taken. Savings may be calculated by comparing these measurements to the baseline measurements.

Limitations of Calculating Retrofit Savings from Boiler Retrofits Using Option B. Boiler retrofit savings calculated using the Option B approach can be adversely affected by the following factors:

- Savings calculated using Option B are intended to be estimates of electricity and/or fuel savings which utilize representative baseline and post-installation, one-time measurements or short-term measurements of installed boiler performance over varying conditions. Therefore, the accuracy of the measurements is completely dependent on how well representative boiler measurements match actual boiler performance over an annual period.
- Estimated savings from Option B may be affected if boiler operating characteristics do not complement the representative schedules used to drive the models.

All of the above operating conditions should be noted carefully during both the baseline and post-retrofit periods.

4.2.3 Expected Accuracy

As previously stated, retrofit savings calculated using the Option B approach are intended to be estimates of electricity and/or thermal energy savings which utilize representative short-term, baseline measurements and either representative short-term or continuous post-installation measurements of the installed equipment over varying conditions. Measurement accuracy is completely dependent on how well representative one-time measurements and agreed-upon hours of operation match actual equipment performance over an annual period.

If significant attention is paid to measurement accuracy, and continuous post-installation measurements are taken, the accuracy of such tests can be ±10-20% of actual performance. However, any inaccuracies in the estimated annual run-time profiles can severely affect savings estimates. In the worst case, errors of 100-200% have been observed.

4.2.4 Expected Cost

The expected cost of Option B should be 3-10% of the installed retrofit cost. For example, if a $100,000 retrofit is installed, roughly $3,000 to $10,000 should be allocated for estimating savings and producing the necessary reports. If continuous post-installation monitoring is planned, savings recording and reporting for the second and subsequent years should not exceed 1% of the cost of the retrofit each year.

The use of continuous post-installation monitoring may help identify O&M problems in a facility. Results from several studies have shown that O&M savings as high as 5-15% of annual energy costs can be identified using data from hourly data loggers (Claridge et al. 1994; Haberl et al. 1995a).

4.3 Option C: Whole-Facility or Main-Meter Measurement Approach

Option C encompasses whole-facility or main-meter verification procedures that provide retrofit performance verification for those projects where whole-facility baseline and post-installation data are available to measure savings. Option C usually involves a continuous measurement of whole-facility energy use before the retrofit (baseline), and a continuous measurement of the whole-facility energy use after the retrofit (post-installation). Periodic inspections of the equipment may also be warranted.

Energy consumption under Option C is calculated by developing statistically representative models of whole-facility energy consumption, i.e., the kWh, Btu or kJ. This section contains information concerning M&V using utility billing data methods and methods that use hourly whole-facility baseline and post-installation analysis. As with previous Option discussions, this section includes examples, expected accuracy and cost information.

4.3.1 Utility Billing Methods

Utility billing methods calculate the savings from an energy-conservation retrofit by establishing a baseline or baseline model using 12 or more months of whole-facility utility billing data. In general, this type of savings calculation procedure is intended for projects where savings are expected to be 20% or more of the monthly utility bill, and where the size of the project or metering budget is too small to justify installing an hourly data logger.

Data Requirements. Normally, 12 months or more of monthly baseline data are required to establish baseline energy consumption (Fels 1986). This includes the following information: i) the date of the meter readings, ii) daily average temperature data from a nearby airport (i.e., NWS min/max data) and iii) the amount of energy consumed during the utility billing period (i.e., the period between the current and previous month's readings).

For each billing period, the average temperature should be calculated. The appropriate statistical model is determined by regressing the billed utility data against the average billing period temperature. If several different meters are read on separate days, then a separate analysis will need to be performed on each meter having a unique billing period. The results will then need to be combined after the analysis.

Differences in billing period length can be accounted for by calculating average daily energy use in the billing period and multiplying by the number of days in the post-installation utility billing period. A small amount of error can occur due to differences between the number of weekdays and weekends in the baseline and post-installation periods, and/or differences in holiday schedules.

Developing a Baseline Energy Use Using an Inverse (Regression) Model. This procedure requires an analysis be conducted on the empirical behavior of the facility as it relates to one or more driving forces or parameters. This approach is referred to as "a system identification, parameter identification or inverse modeling approach." Using the inverse modeling approach, certain characteristics of the facility or system being studied are assumed, and the most important parameters are identified through the use of statistical analysis (Rabl 1988; Rabl and Rialhe 1992). The simplest form of an inverse model is a steady-state inverse model of a facility's energy use. The simplest steady-state inverse model can be calculated by statistically regressing monthly utility consumption data against average billing-period temperatures.

Although simple in concept, the most accurate methods use sophisticated change-point statistical procedures that simultaneously solve for several parameters including a weather-independent base-level parameter, one or more weather-dependent parameters, and the point or points at which the model switches from weather-dependent to non-weather-dependent behavior. In its simplest form, the 65°F (18.3°C) degree-day model is a change-point model that has a fixed change point at 65°F. Examples include the three- and five-parameter Princeton Scorekeeping Method, PRISM (i.e., where the three parameters include weather-independent, base-level use; change-point temperature, and a slope of the line fitted to the points above or below the change point [Fels 1986]); and a four-parameter model (4P) developed by Ruch and Claridge (1991), where the four parameters include a change point, a slope above the change point, a slope below the change point and the energy use associated with the change point.

Figure C-1 shows steady-state, single-variable models appropriate for commercial facility energy use as follows:

- (a) One-Parameter Model
- (b) Two-Parameter Model Shown for Cooling Energy Use
- (c) Three-Parameter Heating Energy Use Model (Heating)
- (d) Three-Parameter Cooling Energy Use Model (Cooling)
- (e) Four-Parameter Heating Energy Use Model (Heating)
- (f) Four-Parameter Cooling Energy Use Model (Cooling)
- (g) Five-Parameter Heating and Cooling Energy Use Model (With Distinct Heating and Cooling Modes)

Figure C-1 shows several types of steady-state, single-variable inverse models. Figure C-1a shows a simple, one-parameter, or constant model, and Equation C-1 gives the equivalent notation for calculating the constant energy use using this model. Figure C-1b shows a steady-state two-parameter model where Bo is the y-axis intercept and B1 is the slope of the regression line for positive values of x, where x represents the ambient air temperature. Figure C-1c shows a three-parameter, change-point model. This is typical of natural gas energy use in a single-family residence that utilizes gas for space heating and domestic water heating. In Equation C-3, which is given for the three-parameter model, Bo represents the baseline energy use, B1 is the slope of the regression line for values of ambient temperature less than the change point B2. In this type of notation, the exponent (+) indicates that only positive values of the parenthetical expression are considered. Figure C-1d shows a three-parameter model for cooling-energy use, and Equation C-4 gives the appropriate expression for analyzing cooling-energy use with a three-parameter model.

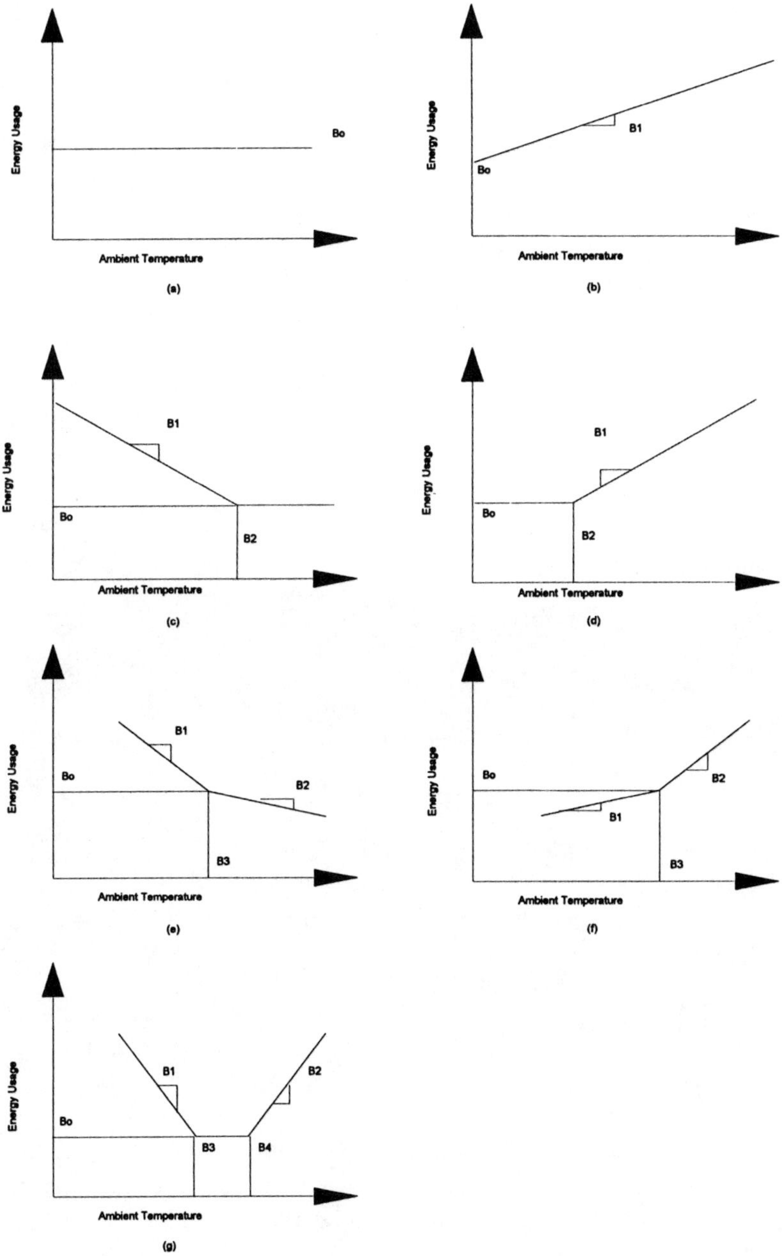
Energy Usage
Bo
Ambient Temperature
(a)
Energy Usage
B1
Bo
Ambient Temperature
(b)
Energy Usage
B1
Bo
B2
Ambient Temperature
(c)
Energy Usage
B1
Bo
B2
Ambient Temperature
(d)
Energy Usage
B1
Bo
B2
B3
Ambient Temperature
(e)
Energy Usage
Bo
B2
B1
B3
Ambient Temperature
(f)
Energy Usage
B1
B2
Bo
B3
B4
Ambient Temperature
(g)

Figures C-1e and C-1f illustrate four parameters for heating and cooling, respectively. Equations C-5 and C-6 indicate the respective expressions for calculating heating (Figure C-1e) and cooling (Figure C-1f) energy use using a four-parameter model. In a four-parameter model, Bo represents the baseline energy exactly at the change point B3. B1 and B2 are the lower and upper region regression slopes for ambient air temperature below and above the change point B3.

Equation C-7 gives the expression for calculating a five-parameter model where there are separate change points for heating and cooling energy use as might be expected in an all-electric heat pump facility for cases where the change point B3 < B4. For cases where there is simultaneous heating and cooling, i.e., B3 > B4, the base-level Bo will be artificially high, and sub-metering is recommended to differentiate between heating and cooling.

$$\text{Eperiod} = \text{Bo} \qquad \text{(Eq. C-1)}$$

$$\text{Eperiod} = \text{Bo} + \text{B1(T)} \qquad \text{(Eq. C-2)}$$

$$\text{Eperiod} = \text{Bo} + \text{B1(B2} - \text{T)+} \qquad \text{(Eq. C-3)}$$

$$\text{Eperiod} = \text{Bo} + \text{B1(T} - \text{B2)+} \qquad \text{(Eq. C-4)}$$

$$\text{Eperiod} = \text{Bo} + \text{B1(B3} - \text{T)+} - \text{B2(T} - \text{B3)+} \qquad \text{(Eq. C-5)}$$

$$\text{Eperiod} = \text{Bo} - \text{B1(B3} - \text{T)+} + \text{B2(T} - \text{B3)+} \qquad \text{(Eq. C-6)}$$

$$\text{Eperiod} = \text{Bo} + \text{B1(B3} - \text{T)+} + \text{B2(T} - \text{B4)+} \qquad \text{(Eq. C-7)}$$

There are several advantages to these steady-state linear and change-point linear inverse models, including:

- The application can be automated and applied to large numbers of facilities where monthly utility billing data and average daily temperatures are available.

- It has been shown that linear and change-point linear models have physical significance to the actual heat loss/gain mechanisms that govern the energy use in most facilities (Fels 1986, Rabl and Rialhe 1992, Claridge et al. 1994 and Rabl 1988).

Disadvantages of the steady-state inverse monthly models include:

- Insensitivity to dynamic effects, e.g., thermal mass.
- Insensitivity to variables other than temperature, e.g., humidity and solar.
- Inappropriateness for certain facility types, e.g., facilities that have strong on/off schedule-dependent loads or facilities that display multiple change points. In such cases, alternative models, such as hourly or daily models, will need to be developed.

Selecting the Best Monthly Regression Model. Ideally, model selection procedures should be simple to apply and produce consistent, repeatable results. Several selection procedures have been recommended to select the best regression model. In general, these procedures calculate several regression models and select the best model depending on the match as measured by the R^2, coefficient of variation of the normalized annual consumption, i.e., CV(NAC), or coefficient of variation of the RMSE.

Additional information concerning these selection procedures can be found in Reynolds and Fels (1986) and Kissock (1994). Public domain software related to these selection procedures can be obtained from Princeton University (Fels et al. 1995), and from Texas A&M University (Kissock et al. 1994). Spreadsheet procedures have also been developed (Landman and Haberl 1995).

In certain types of facilities (such as schools) where there is a significant difference between the facility's energy use during the school year and summer break, separate regression models may need to be developed for different usage periods (Landman and Haberl 1995).

Calculating Energy Savings Using the Baseline Model. Once the appropriate baseline model has been determined for the facility, energy savings are calculated by comparing energy use predicted by baseline parameters, projected into the post-installation period by multiplying by post-installation weather and operating conditions, to measured post-installation data. In general the following steps are used to calculate the savings:

1. Determine the appropriate baseline or baseline model.
2. Project the baseline energy use into the post-installation period by driving the baseline model with the post-installation weather and operating parameters.

3. Calculate the savings by comparing the difference between energy use predicted by the post-installation model and actual energy use. Equation C-8 is the basic equation used in this analysis.

E(save,i) = E(baseline,i) – E(post,i) (Eq. C-8)

where

(Esave,i) = energy savings from the energy-conservation retrofit during period (i).

E(baseline,i) = the baseline or baseline energy use projected into the post-installation period by multiplying the parameters of the baseline model by weather and operating parameters from the post-installation period.

E(post) = the actual post-installation energy use during period (i).

In situations where significant data are missing from the post-installation period, a post-installation model can be created to fill in the missing data. Energy savings are then calculated by comparing the energy use predicted by the baseline model to the energy use predicted by the post-installation model.

Savings are determined to be significant if the difference between the baseline and post-installation energy use is greater than model error as determined by the RMSE.

Examples of Projects Analyzed with Monthly Energy Use. The example projects that follow can be analyzed with monthly utility billing data, provided that the change in baseline and post-installation energy use is larger than the inherent uncertainty in the statistical model as calculated by the RMSE. For those retrofits where the change in monthly baseline and post-installation consumption is *less* than the uncertainty in the statistical model, alternative methods of measuring the retrofits savings should be considered, i.e., Option B, or baseline and post-installation hourly or daily analysis.

Lighting Efficiency and/or Controls Project. Lighting efficiency and/or lighting control retrofit savings can be analyzed using monthly baseline and post-installation utility billing data, provided the savings are greater than the uncertainty of the regression model. Existing baseline conditions should be documented according to the procedures outlined in Section 4.0.1.

Obtaining utility billing data for 12 months prior to the retrofit is recommended. If electricity savings and electric demand are being evaluated, a separate demand analysis should be performed that compares demand for a given month with the demand for that same month in the year prior to the retrofit.

Electricity savings due to reduced lighting energy are calculated by analyzing whole-facility electricity use. Electric demand reductions can also be analyzed using monthly utility billing data.

Electricity savings due to reduced cooling load can also be determined through the selection of the appropriate baseline and post-installation modeling strategy if the energy use of the chiller or air-conditioning equipment is considered in the whole-facility utility meter. Negative savings that account for increased heating due to reduced internal heating load can also be determined if the heating system energy use is available for analysis.

Calculating Electricity Savings. Electricity savings resulting from a lighting retrofit can be determined in the following fashion: First, develop a baseline model using the methods previously described. Second, calculate the retrofit electricity savings by comparing the electricity use predicted by baseline parameters (projected into the post-installation period by multiplying by post-installation weather and operating conditions) to measured post-installation data. In situations where significant data are missing in the post-installation period, a post-installation model can be created to fill in missing data. Energy savings are then calculated by comparing energy use predicted by the baseline model to energy use predicted by the post-installation model.

Savings are determined to be significant if the difference between baseline and post-installation energy use is greater than model error as determined by the RMSE.

Calculating Electric Demand Reductions. Electric demand reductions from a lighting retrofit can be determined in the following fashion: First, develop a baseline demand model using the methods previously described. Second, calculate the retrofit electric demand savings by comparing the monthly demand predicted by the baseline model to measured post-installation demand data. Demand savings are determined to be significant if the difference between baseline and post-installation electric demand is greater than model error as determined by the RMSE.

Calculating Interactive Cooling Savings. In most lighting retrofits there will be a significant reduction in the energy required to cool the space due to internal heat reduction. The amount of cooling savings will vary by facility depending on the relative internal load proportions versus envelope loads, the type of cooling system, cost of the energy used by the cooling system and whether or not economizer or free cooling is utilized. Cooling savings can be determined from monthly data if: i) separate metering data for the energy use of the cooling system is available, or ii) the cooling energy use is part of the main meter.

If separate metering data are available for the cooling system, cooling savings can be determined by developing a baseline cooling model using the methods previously described. Cooling energy savings are calculated by comparing the electricity use predicted by baseline parameters (projected into the post-installation period by multiplying by post-installation weather and operating conditions) to measured post-installation data. In situations where significant data are missing in the post-installation period, a post-installation model can be created to fill in missing data. Energy savings are then calculated by comparing energy use predicted by the baseline model to energy use predicted by the post-installation model. Savings are determined to be significant if the difference between baseline and post-installation energy use is greater than model error as determined by the RMSE.

If cooling energy use is part of the main meter, cooling savings may be combined with the electricity reduction (due to lighting fixture retrofits) in the whole-facility statistical model. A combined electricity and cooling reduction can be determined by developing a baseline model using the methods previously described. Electricity savings plus cooling savings are calculated by comparing the electricity use predicted by baseline parameters (projected into the post-installation period by multiplying by post-installation weather and operating conditions) to measured post-installation data. In situations where significant data are missing in the post-installation period, a post-installation model can be created to fill in missing data. Electricity plus cooling energy savings are then calculated by comparing energy use predicted by the baseline model to energy use predicted by the post-installation model. Savings are determined to be significant if the difference between baseline and post-installation energy use is greater than model error as determined by the RMSE.

Calculating Interactive Heating Savings. In most lighting retrofits additional heating will be necessary to make up for internal heating loss due to the removal or replacement of inefficient lighting fixtures. The amount of additional heating required will vary depending on the relative proportions of the internal loads versus the envelope loads, the type of heating system and the cost of heating fuel. The additional heating energy required can be determined from monthly data if: i) separate metering data for the energy use of the heating system is available, or ii) the heating energy use is part of the main meter.

If separate metering data are available for the heating system, the heating reduction can be determined by developing a baseline heating model using the methods previously described. Additional heating energy is then calculated by comparing energy use predicted by the baseline model (projected into the post-installation period by multiplying by post-installation weather and operating conditions) to measured post-installation data. In situations where significant data are missing in the post-installation period, a post-installation model can be created to fill in missing data. The additional energy is then calculated by comparing energy use predicted by the baseline model to energy use predicted by the post-installation model. Additional heating energy requirements are determined to be significant if the difference between baseline and post-installation energy use is greater than the model error as determined by the RMSE.

If the heating energy use is part of the main electric meter, the additional heating will be combined with the electricity reduction (due to lighting fixture retrofits) in the whole-facility statistical model. An evaluation of reduced lighting electricity and increased heating electricity can be determined by developing a baseline model using the methods previously described. Electricity savings plus additional heating are then calculated by comparing the electricity use predicted by the baseline model (projected into the post-installation period by multiplying by post-installation weather and operating conditions) to measured post-installation data. In situations where significant data are missing in the post-installation period, a post-installation model can be created to fill in missing data. The electricity plus additional heating energy are then calculated by comparing the energy use predicted by the baseline model to the energy use predicted by the post-installation model. Combined savings are determined to be significant if the difference between baseline and post-installation energy use is greater than model error as determined by the RMSE.

Limitations of Calculating Retrofit Savings fom Lighting and/or Lighting Controls Projects Using Monthly Utility Billing Data. Lighting efficiency and/or lighting controls project savings calculated using the Option C approach should be greater than the uncertainty as calculated by the RMSE. These savings can be adversely affected by the following factors:

- Savings in the whole-facility electricity consumption can be affected by changes in electric receptacle loads.

- Savings in the whole-facility demand can be affected by additions or subtractions of major electric-consuming subsystems.

- Changes in the whole-facility interactive cooling savings may be affected by procedures used to operate the cooling systems. In particular, the average chiller kW/ton ratio is affected by the rate of the cooling load on a particular chiller. Chillers that are loaded below 50% of their capacity tend to have significantly higher kW/ton ratios which can increase overall electricity consumption.

- Changes in whole-facility interactive heating savings may be affected by procedures used to operate the heating systems. Boilers or furnaces run at low loads can cycle excessively, which decreases fuel conversion efficiency.

All of the above operating conditions should be noted carefully during both the baseline and post-retrofit periods.

Constant Load Motor Replacement Projects. Constant load motor replacement project savings can be analyzed using Option C, provided the savings are greater than the uncertainty of the regression model. Existing baseline conditions should be documented according to the procedures outlined in Section 4.0.1. In particular, care should be taken to note pressure rises across pumps or blowers because the electric demand of a pump or blower is dependent on the pressure it exerts on the fluid stream passing through the pump or blower. For such retrofits, hourly measurements of baseline and post-installation energy use and/or *in-situ* component efficiency measurements are usually required.

For constant load motor replacement projects, obtaining utility billing data for 12 months prior to the retrofit is recommended. If electricity savings and electric demand are being evaluated, a separate demand analysis needs to be performed that compares demand for a given month with demand in the same month of the year prior to the retrofit.

Calculating Electricity Savings. Constant load motor project electricity savings can be determined in the following fashion: First, develop a baseline model using the methods previously described. Second, calculate the retrofit electricity savings by comparing the electricity use predicted by the baseline model to measured post-installation data. Savings are determined to be significant if the difference between baseline and post-installation energy use is greater than model error as determined by the RMSE.

Calculating Electric Demand Reductions. Electric demand reductions from a constant load motor retrofit can be determined in the following fashion: First, develop a baseline demand model using the methods previously described. Second, calculate the retrofit electric demand savings by comparing the monthly demand predicted by the baseline model to measured post-installation demand data. Demand savings are determined to be significant if the difference between baseline and post-installation electric demand is greater than model error as determined by the RMSE.

Limitations of Calculating Savings from a Constant Load Motor Retrofit Using Utility Billing Data. Constant load motor retrofit savings measured using the Option C approach can be adversely affected by the following:

- Savings in electricity consumption can be affected when the motor being replaced no longer operates in a constant load. For example, if the pressure drop changes across a pump, the electricity use of the pump will also change.

- Savings in electric demand can be affected by additions or subtractions of major electric-consuming subsystems.

All of the above operating conditions should be noted carefully during both the baseline and post-retrofit periods.

Variable-Speed Drive Motor Project. Variable-speed drive motor retrofit savings are not easily analyzed using the Option C approach. For such retrofits, hourly measurements of baseline and post-installation energy use and/or *in-situ* component efficiency measurements are usually required.

HVAC and/or EMCS Project. HVAC system and/or EMCS savings can be analyzed using the Option C approach, provided the savings are greater than the uncertainty of the regression model. Existing baseline conditions should be documented according to the procedures outlined in Section 4.0.1. Obtaining utility billing data for 12 months prior to the retrofit is recommended. If electricity savings and electric demand are being evaluated, a separate demand analysis needs to be performed that compares the demand for a given month to the demand in the same month of the year prior to the retrofit.

Electricity savings due to the reduction in the HVAC energy use are calculated by analyzing whole-facility electricity use. Electric demand reductions can also be analyzed using monthly utility billing data. Retrofits to HVAC and/or EMCS systems can also affect the cooling and heating energy use in a facility. Such interactions can be evaluated with utility billing data in facilities with large envelope-driven loads. Buildings with large internal loads, significant schedule changes and/or simultaneous heating and cooling may require hourly baseline and post-installation measured data.

Calculating Electricity Savings. Electricity savings from an HVAC or EMCS retrofit can be determined in the following fashion: First, develop a baseline model using the methods previously described. Second, calculate post-installation electricity savings by comparing electricity use predicted by the baseline model to measured post-installation data. Savings are determined to be significant if the difference between baseline and post-installation energy use is greater than model error as determined by the RMSE.

Calculating Electric Demand Reductions. Electric demand reductions from an HVAC or EMCS retrofit can be determined in the following fashion: First, develop a baseline demand model using the methods previously described. Second, calculate the post-installation electric demand savings by comparing the monthly demand predicted by the baseline model to measured post-installation demand data. Demand savings are determined to be significant if the difference between baseline and post-installation electric demand is greater than model error as determined by the RMSE.

Calculating the Cooling Savings. In most HVAC or EMCS projects there may be significant reductions in the energy required to cool the space due to improved HVAC system efficiency. The amount of cooling savings may vary by facility depending on the relative proportions of internal loads versus the envelope loads, the type of cooling system, cost of the energy used by the cooling system and whether or not economizer or free cooling is utilized. Cooling energy savings from an HVAC retrofit can be determined from monthly data if: i) separate metering data for the energy use of the cooling system is available or ii) the cooling energy use is part of the main meter.

In either case, cooling reductions can be determined by developing a baseline cooling model using the methods previously described. Cooling energy savings are calculated by comparing the electricity use predicted by the baseline model to measured post-installation data. Savings are determined to be significant if the difference between baseline and post-installation energy use is greater than model error as determined by the RMSE.

Calculating Heating Savings. In HVAC or EMCS retrofits there may also be significant reductions in the energy required to heat the space due to improved HVAC system efficiency. The amount of heating savings will vary by facility depending on the relative proportions of internal loads versus the envelope loads, the type of heating system, etc. Heating energy savings from an HVAC retrofit can be determined from monthly data if: i) separate metering data for the energy use of the heating system are available or ii) the heating energy use is part of the main meter.

In either case, heating reductions can be determined by developing a baseline heating model using the methods previously described. Heating energy savings are calculated by comparing the energy use predicted by the baseline model to measured post-installation data. Savings are determined to be significant if the difference between baseline and post-installation energy use is greater than model error as determined by the RMSE.

Limitations of Calculating Retrofit Savings From HVAC or EMCS Retrofits Using Monthly Utility Billing Data. Savings from HVAC or EMCS retrofits that are calculated with monthly baseline and post-installation utility billing data should be greater than the uncertainty as calculated by the RMSE. These savings can be adversely affected by the following factors:

- In facilities where simultaneous heating/cooling occurs during a significant portion of the year, savings due to HVAC system modifications may require hourly baseline and post-installation measurements or *in-situ* efficiency measurements of the HVAC system.

- Savings in whole-facility electricity consumption can be affected by changes in the electric receptacle loads.
- Savings in whole-facility demand can be affected by additions or subtractions of major electric-consuming subsystems.
- Changes in whole-facility cooling savings can be affected by procedures used to operate the cooling systems.
- Changes in whole-facility heating may be affected by procedures used to operate the heating systems. Boilers or furnaces run at low loads can cycle excessively, which decreases fuel conversion efficiency.

All of the above operating conditions should be noted carefully during both the baseline and post-retrofit periods.

Chiller Project. Chiller project savings can be analyzed using monthly baseline and post-installation utility billing data, provided the savings are greater than the uncertainty of the regression model, and that chiller operating conditions have remained the same. Existing baseline conditions should be documented according to the procedures outlined in Section 4.0.1. In particular, care should be taken to note the loading of the chiller, chilled water supply temperatures, condenser return temperatures and flow rates through the chiller. This documentation is important because the efficiency of the chiller, i.e., kW/ton or COP, is dependent on the percent load on the chiller, temperature of the chilled water supply, condenser return temperature and flow rates through the chiller (Gordon and Ng 1994).

For those retrofits where such parameters are uncertain or cannot be ascertained, it may be necessary to measure the baseline and post-installation, *in-situ* chiller efficiency as outlined in Section 4.2. Hourly baseline and post-installation measurements can be used if the loading and temperature have remained relatively constant, and if the chiller output and electricity input are being measured.

For chiller replacement projects that have constant baseline and post-installation loading conditions and operating temperatures, obtaining utility billing data for 12 months prior to the retrofit is recommended. If electricity savings and electric demand are being evaluated, a separate demand analysis needs to be performed that compares the demand for a given month with the demand in the same month of the year prior to the retrofit.

Calculating Electricity Savings. Electricity savings from a chiller retrofit that has constant baseline and post-installation loading profiles and operating temperatures can be determined in the following fashion: First, develop a baseline model using the methods previously described. Second, calculate the retrofit electricity savings by comparing the electricity use predicted by the baseline model to measured post-installation data. Savings are determined to be significant if the difference between baseline and post-installation energy use is greater than model error as determined by the RMSE.

Calculating Electric Demand Reductions. Electric demand reductions from a chiller retrofit can be determined in the following fashion: First, develop a baseline demand model using the methods previously described. Second, calculate the post-installation electric demand savings by comparing the monthly demand predicted by the baseline model to measured post-installation demand data. Demand savings are determined to be significant if the difference between baseline and post-installation electric demand is greater than model error as determined by the RMSE.

Limitations of Calculating Savings from a Chiller Retrofit Using Utility Billing Data. In most cases it may not be possible to accurately assess chiller retrofit savings by comparing monthly utility billing data. This is due to the fact that most chiller retrofits are usually accompanied by changes to the chilled-water pumping systems, chilled-water setpoints, downsizing of the chillers or staging of the chillers, etc. Therefore, most chiller retrofits require that baseline and post-installation efficiency measurements and load profiles be developed.

Even in such cases where all of these variables have been held constant, savings from a chiller retrofit project can be adversely affected by the following:

- Savings in electricity consumption can be affected when the chiller operates at different baseline and post-installation chilled-water setpoint conditions or condenser temperatures, because of additional work which is required to produce colder evaporator temperatures, or shed heat in the condenser at higher temperatures.

- Savings in electricity consumption can be affected if the baseline and post-installation loading on the chiller is substantially different. This is due to the fact that chillers tend to have a non-linear increase in kW/ton ratios as the loading drops below approximately 50 percent.

- Savings in electricity consumption can also be affected by flow rates through the evaporator and/or condenser.

All of the above operating conditions should be noted carefully during both the baseline and post-retrofit periods.

Boiler Project. Savings resulting from upgrades to or replacements of large boilers can be analyzed with monthly baseline and post-installation utility billing data, provided the savings are greater than the uncertainty of the regression model and that the baseline and post-installation conditions under which the boiler operates have remained the same. Existing baseline conditions should be documented according to the procedures outlined in Section 4.0.1. In particular, care should be taken to note boiler loading, setpoint temperatures and general condition, because the efficiency of the boiler is primarily dependent upon the loading and setpoint temperatures. Useful information regarding boiler efficiency can be found in Dyer and Maples (1981), Dukelow (1991) and Babcock and Wilcox (1992).

For those retrofits where such parameters are uncertain or cannot be ascertained, it may be necessary to measure the baseline and post-installation, *in-situ* boiler efficiency as outlined in Section 4.2. Hourly baseline and post-installation measurements can be used for such retrofits if the loading and temperature have remained relatively constant and if the boiler fuel input and thermal output are being measured.

For boiler-replacement projects that have constant baseline and post-installation loading conditions and operating temperatures, obtaining utility billing data for 12 months prior to the retrofit is recommended. If electricity savings and electric demand are being evaluated, a separate demand analysis should be performed that compares the demand for a given month with the demand for that same month of the year prior to the retrofit.

Calculating Energy Savings. Energy savings from a boiler retrofit that has constant baseline and post-installation loading profiles and operating temperatures can be determined in the following fashion: First, develop a baseline model using the methods previously described. Second, calculate the retrofit energy savings by comparing the energy use predicted by the baseline model to measured post-installation data. Savings are determined to be significant if the difference between baseline and post-installation energy use is greater than model error as determined by the RMSE.

Calculating Electric Demand Reductions. Electric demand reductions from an electric boiler retrofit can be determined in the following fashion: First, develop a baseline demand model using the methods previously described. Second, calculate the post-installation electric demand savings by comparing the monthly demand predicted by the baseline model to measured post-installation demand data. Demand savings are determined to be significant if the difference between baseline and post-installation electric demand is greater than model error as determined by the RMSE.

Limitations of Calculating Savings from a Boiler Retrofit Using Utility Billing Data. Boiler retrofit savings calculations can be adversely affected by the following:

- Savings in energy consumption can be affected when the boiler operates at different baseline and post-installation setpoint conditions. This is due to the additional fuel required to produce higher temperatures.

- Savings in energy consumption can be affected if the baseline and post-installation loading on the boiler is substantially different. This can be a significant problem if the new boiler is oversized and must operate under on/off cycling conditions.

- Savings in energy consumption can also be affected by combustion settings, changes in the environment surrounding the boiler and changes in the boiler operating schedule.

All of the above operating conditions should be noted carefully during both the baseline and post-retrofit periods.

4.3.2 Whole-Facility or Main Meter Hourly Before/After Analysis.

For projects where hourly monitoring equipment has been installed at least 9-12 months prior to the retrofit, the following procedures can be used to document savings. In order for the measurements to be valid, monitoring equipment should be installed to economically capture a significant portion of the energy use of the equipment to be replaced and/or upgraded.

The equipment, where feasible, should also be installed to minimize noise that might be introduced by other (non-retrofitted) equipment. For example, if a lighting retrofit is being analyzed with a derived whole-facility "lights and receptacles" measurement, care should be taken to document the receptacle loads so that changes can be noted and adjustments made should there be significant change in the receptacle loads that might affect savings measurements, e.g., the purchase and installation of extensive 120 VAC office computer equipment.

Data Requirements for Hourly or Daily Models. In many cases whole-facility or main-meter hourly baseline and post-installation measurements can utilize the same revenue meters as the local utility to bill the owner. Such meters must be equipped or modified to provide a digital pulse (or in some cases a 4-20 mA signal) that can be recorded by the monitoring equipment. Each recorded pulse then represents a specific unit of consumption over a given time period, i.e., kWh/hour or CCF/hour. In some instances, such equipment may provide a 4-20 mA signal that can be recorded as an accumulated analog signal by the monitoring equipment.

EMCSs can be used to record energy use using the "trend" capability. However, most EMCSs use change of value (COV) data that are not immediately useful for calculating energy savings because of varying intervals. Such data will need to be converted to interval data before they are useful for energy savings calculations (Claridge et al. 1993). In almost all cases, care should be taken to accurately calibrate the "kWh/pulse" constant against a known reference, or to determine the scale and offset values to be entered into the data logger to convert the 4-20 mA signal into engineering units. Often, this can be accomplished by comparing acquired, recorded data values against similar data recorded by the utility revenue meter, provided the revenue meter has been recently calibrated according to National Institute of Standards and Technology (NIST) traceable standards. Additional material concerning calibration may be found in Section 5.0.

In most cases, hourly measurements are adequate to characterize energy and demand profiles of the equipment to be retrofitted. However, where savings changes to the electric demand plays an important factor in energy savings calculations, the minimum time step for recording data should match the utility demand time interval. For example, if the local utility is calculating peak demand using a 15-minute window, then the loggers should be set to record data every 15 minutes. In some cases, utilities use "sliding windows" to record electric demand data. This type of demand measurement requires a special data recorder that has sliding-window recording capabilities. This can also be accomplished by setting the data acquisition system to the one-minute level, recording one minute of data and then recreating the sliding 15-minute window using post-processing software. In most cases 15-minute clock measurements will suffice for sliding 15-minute data. However, care should be taken to ensure that the facility does not contain unusual combinations of equipment that are indeed generating high one-minute peak loads. After processing the data for the demand analysis, the 15-minute data can then be converted to hourly data for archiving and further analysis against hourly weather data.

In most facilities, hourly baseline and post-installation whole-facility electricity, cooling, heating and motor-control-center measurements are usually sufficient to capture lighting retrofits, HVAC system retrofits and facility envelope modifications. Such data usually need to be recorded for 9-12 months prior to the retrofit to adequately ensure that sufficient data are recorded to adjust for weather normalization measure (Haberl et al. 1995). Other data required include average hourly dry bulb temperatures and humidity data. These data can be recorded on site or obtained from a nearby National Weather Service weather station.

Developing a Baseline Energy Use Using an Hourly or Daily Inverse (Regression) Model. Inverse or regression models of a facility's hourly or daily baseline energy use are developed in the same fashion as those developed using monthly data except hourly or daily models must often incorporate a switching variable to account for differences in facility operation. At the monthly level, switching variables are usually not used because the data have been aggregated to the monthly level. However, in daily and hourly inverse models, significantly more data scatter makes the fit of the regression line less accurate (Katipamula et al. 1994). Reasons for scatter in hourly data include:

- on/off switching of HVAC systems
- schedule variations
- dynamic effects of thermal mass, etc.

- solar effects
- latent cooling loads due to dehumidification of moist air during the cooling season

In many facilities, scatter in the data can be reduced without losing significant accuracy by aggregating the data to the daily level prior to analysis.

In daily data a fair amount of this scatter can be accounted for by using the appropriate weekday/weekend model. To accomplish this, the analyst should first sort the daily data into weekday and weekend groups, perform the appropriate analysis on the separate groups, and then create a combined model that automatically switches between the weekday and weekend (or holiday) mode, depending on the day of the week. Such daily models have been shown to be capable of accounting for 90-95% of the variation in a facility's weather-dependent energy use (Reddy 1994). Appropriate models for the analysis of daily energy use include one-, two-, three-, four- and five-parameter models using a single influencing variable and, in some cases, multiparameter models.

Hourly models tend to be significantly more complex, i.e., 8,760 data points for an hourly model of a facility's energy use versus 365 data points for a daily model, or 12 data points for a monthly model. In general, such models must account for hourly scheduling differences and often need to account for additional parameters such as solar and humidity, and dynamic parameters such as thermal mass. Models that have been shown to be effective in hourly applications include: simple one-, two-, three-, four-, and five-parameter models, inverse bin models, and more advanced models such as neural network models (Kreider and Haberl 1994).

Selecting the Best Hourly or Daily Regression Model. The best hourly or daily regression model is selected in the same fashion as the monthly models. Several regression models are calculated, and the best model is selected based on the best match as measured by the R^2, coefficient of variation of the normalized annual consumption, i.e., CV(NAC), or coefficient of variation of the RMSE.

Lighting Efficiency and/or Controls Project. Savings from lighting efficiency and/or lighting controls projects can be analyzed using baseline and post-installation hourly measurements, provided the savings are greater than the uncertainty of the regression model. Existing baseline conditions should be documented according to the procedures outlined in Section 4.0.1.

Hourly measurements for at least 12 months should be taken to adequately characterize cooling and heating season performance. In some cases, nine months of data can adequately characterize performance if the period under analysis adequately reflects all normal environmental and schedule conditions. If electricity savings and electric demand are being evaluated, a separate demand analysis will need to be performed that compares the demand for a given month to the demand for the same month of the year prior to the retrofit.

Total measured energy savings should consider electricity savings from the reduced lighting load, cooling savings from reduced internal heating and increased heating (i.e., negative savings) to make up for internal heating reductions caused by lighting retrofits.

Calculating Electricity Savings. The analysis method used to determine electricity savings from a lighting retrofit depends on facility loads being monitored by the hourly data acquisition system. To achieve the highest level of accuracy, it is best to monitor the retrofit at the end-use level. However, in most situations this is not economically feasible. In most facilities, electricity savings from a lighting retrofit can be monitored adequately with hourly data if the following loads can be monitored: i) whole-facility electricity, ii) motor-control-center electricity, iii) electricity used for powering chillers (and electric heating) and iv) other electric loads that are easily identified as "non-lighting," such as exterior security lighting, electric slab heaters (for ice melting on sidewalks), etc. Assuming such channels can be monitored, a proxy lighting channel can be created by subtracting the sum of the motor-control-center (plus cooling and heating loads), from the whole-facility electric as indicated in Equation C-9.

Once such measurements have been obtained, electricity savings from a lighting retrofit can be determined in the following fashion: First, develop a baseline model using the methods previously described for each of the channels being monitored. Second, calculate electricity savings by comparing electricity use predicted by baseline parameters (projected into the post-installation period by multiplying by post-installation weather and operating conditions) to measured post-installation data. In situations where significant data are missing in the post-installation period, a post-installation model can be created to fill in missing data. Electricity savings are then calculated by comparing the energy use predicted by the baseline model to the energy use predicted by the post-

E(lights, proxy) =
E(whole-facility) –
[E(motor control center) +
E(chiller) +
E(boiler) +
E(other, non-lighting)] (Eq. C-9)

where

E(lights, proxy) = the hourly (or 15-minute) electricity use of the derived lights and receptacles load.
E(whole-facility) = the hourly (or 15-minute) electricity use of the whole facility.
E(motor control center) = the hourly (or 15-minute) electricity use of all motors in the facility.
E(chiller) = the hourly (or 15-minute) electricity use of all chillers or large cooling equipment and associated equipment.
E(boiler) = the hourly (or 15-minute) electricity use of all heating equipment and associated equipment.
E(other, non-lighting) = the hourly (or 15-minute) electricity use of all other significant non-lighting loads.

installation model. Savings are determined to be significant if the difference between baseline and post-installation energy use is greater than model error as determined by RMSE.

In cases where only whole-facility electricity is available on a 15-minute or hourly basis, savings can be calculated if at least 12 months of data are available, and the data can be statistically separated into heating season, cooling season, and non-heating/non-cooling season data. Several methods have been developed for accomplishing this separation, including weather day-types (Bou Saada and Haberl 1995a) and calibrated simulations (Bou Saada and Haberl 1995b; Akbari, et al. 1988). In general, these techniques synthesize end-use loads by breaking down 8,760 hours of use into average profiles for weekday and weekend loads that represent non-cooling/non-heating loads. Retrofit savings are calculated by comparing average profiles for baseline and post-installation, non-cooling/non-heating loads. Savings are determined to be significant if the difference between baseline and post-installation profiles is greater than model error as determined by RMSE.

Calculating Electric Demand Reductions. If end-use lighting loads are being monitored or if proxy lighting loads, i.e., lights and receptacles, are being measured, electric demand reductions can be determined if the measurements are taken at the same demand time interval used for utility billing purposes. Savings can be determined by comparing the baseline and post-installation maximum values for the appropriate demand interval. Several calculations may be required for more complex utility rate structures that can include on-peak, off-peak summer and/or winter electric demand rates.

Calculating Interactive Cooling Savings. Interactive cooling savings can be measured using hourly data if the chillers (or large cooling equipment) are being measured directly, or statistically if only whole-facility electric data are available, which includes heating, cooling and all other electrical uses.

If separate metering data are available for the cooling system, cooling reductions can be determined by developing a baseline cooling model using the methods previously described. Cooling energy savings are calculated by comparing the electricity use predicted by baseline parameters (projected into the post-installation period by multiplying by post-installation weather and operating conditions) to measured post-installation data. In situations where significant data are missing in the post-installation period, a post-installation model can be created to fill in missing data. Energy savings are then calculated by comparing energy use predicted by the baseline model to energy use predicted by the post-installation model. Savings are determined to be significant if the difference between baseline and post-installation energy use is greater than model error as calculated by the RMSE.

If cooling-energy use is part of the main meter, cooling savings will be combined with electricity reduction (due to lighting fixture retrofits) in the whole-facility statistical model. Combined electricity and cooling reductions can be determined by developing a baseline model using the methods previously described. Electricity plus cooling savings are calculated by comparing electricity use predicted by baseline parameters (projected into the post-installation period by multiplying by post-installation weather and operating conditions) to measured post-installation data. Electricity plus cooling-energy savings are then calculated by comparing energy use predicted by the baseline model to energy use predicted by the post-installation model. Savings are determined to be significant if the difference between baseline and post-installation energy use is greater than model error as determined by the RMSE.

Calculating Interactive Heating Savings. The interactive increase in heating use can be measured using hourly data if heating equipment is being measured directly, or statistically if only whole-facility electric data are available that include: heating, cooling and all other electrical uses.

If separate metering data are available for the heating system, heating reductions can be determined by developing a baseline heating model using the methods previously described. Additional heating energy is calculated by comparing the energy use predicted by the baseline energy use (projected into the post-installation period by multiplying by post-installation weather and operating conditions) to measured post-installation data. The additional energy is then calculated by comparing energy use predicted by the baseline model to energy use predicted by the post-installation model. Additional heating energy is determined to be significant if the difference between baseline and post-installation energy use is greater than model error as determined by the RMSE.

If heating-energy use is part of the main electric meter, the additional heating will be combined with the electricity reduction (due to lighting fixture retrofits) in the whole-facility statistical model. An evaluation of reduced lighting electricity and increased heating electricity can be determined by developing a baseline model using the methods previously described. Electricity savings plus additional heating are then calculated by comparing the electricity use predicted by the baseline model (projected into the post-installation period by multiplying by post-installation weather and operating conditions) to measured post-installation data. Electricity plus additional heating energy is calculated by comparing energy use predicted by the baseline model to energy use predicted by the post-installation model. Combined savings are determined to be significant if the difference between baseline and post-installation energy use is greater than model error as determined by the RMSE.

Limitations of Calculating Retrofit Savings from Lighting and/or Lighting Controls Projects Using Hourly, Whole-facility, Before/After Data. Savings from lighting efficiency and/or lighting controls projects that are calculated with hourly, whole-facility, baseline and post-installation utility billing data should be greater than the uncertainty as calculated by the CV(RMSE) and/or R^2. These savings can be adversely affected by the following factors:

- Savings measured with end-use lighting measurements are most accurate. However, such savings can be affected by lighting fixture outages and/or changes in lighting system operational patterns.

- Savings that utilize proxy lighting measurements can also be highly accurate. However, such savings can be affected by lighting fixture outages and/or changes in operational patterns. In addition, such savings can be affected by significant additions to the plug (or receptacle loads), since such loads are typically combined with lighting loads.

- Changes in hourly, whole-facility interactive cooling savings can be affected by procedures used to operate the cooling systems. In particular, the average chiller kW/ton ratio is affected by the rate of the cooling load on a particular chiller. Chillers that are loaded below 50% of their capacity tend to have significantly higher kW/ton ratios, which can increase overall electricity consumption.

- Changes in the hourly, whole-facility interactive heating savings may be affected by heating system operating procedures. Boilers or furnaces run at low loads can cycle excessively, which decreases fuel conversion efficiency.

All of the above operating conditions should be noted carefully during both the baseline and post-retrofit periods.

Constant-Load Motor Replacement Project. Constant-load motor replacement project savings can be analyzed using baseline and post-installation hourly measurements, provided the savings are greater than the uncertainty of the regression model. Existing baseline conditions should be documented according to the procedures outlined in Section 4.0.1.

Hourly measurements should be obtained for several months to adequately characterize baseline motor performance. In some cases, 12 months of data may be needed to adequately characterize performance if the facility has several occupancy schedules throughout the year. If electricity savings and electric demand are being evaluated, a separate demand analysis will need to be performed that compares demand for a given month with the demand for that same month in the year prior to the retrofit. Total measured energy savings should include end-use electricity savings from reduced motor load where possible.

In special cases where the motor is being used to deliver heating and/or cooling loads, an additional analysis may need to be performed to evaluate the impact of reduced heating and/or cooling requirements due to the change in the motor used to run the pump.

Calculating Electricity Savings. The analysis method used to determine electricity savings from a constant-load motor replacement project depends on the facility loads being monitored by the hourly data acquisition system. To achieve the highest level of accuracy, it is best to monitor the retrofit at the end-use level. However, in most situations this is not economically feasible. In some facilities, electricity savings from a constant-load motor retrofit can be adequately monitored with hourly, whole-facility electricity data if the change in baseline and post-installation electricity use is greater than the uncertainty in the whole-facility baseline model and there are no substantial changes to other electric-consuming subsystems. Caution must be taken to identify any significant additions or deletions to the whole-facility electricity use.

Retrofit savings using whole-facility hourly measurements can be determined in the following fashion: First, develop a baseline model using the methods previously described for each of the channels being monitored. Second, calculate retrofit electricity savings by comparing electricity use predicted by the baseline parameters (projected into the post-installation period by multiplying by post-installation weather and operating conditions) to measured post-installation data. In situations where significant data are missing in the post-installation period, a post-installation model can be created to fill in missing data. Electricity savings are calculated by comparing energy use predicted by the baseline model to energy use predicted by the post-installation model.

In those cases where end-use hourly electricity measurements are available, electricity savings are calculated by comparing end-use electricity use predicted by the baseline model to energy use predicted by the post-installation model.

Savings are determined to be significant if the difference between baseline and post-installation energy use is greater than model error as determined by the RMSE.

Calculating Electric Demand Reductions. If end-use motor loads are being monitored, electric demand reductions can be determined if the measurements are taken at the same demand time intervals used for utility billing purposes. Savings can be determined by comparing baseline and post-installation maximum values for the appropriate demand interval. Several calculations may be required for more complex utility rate structures which may include on-peak, off-peak summer and/or winter electric demand rates.

Calculating Cooling or Heating Savings. If the motor is being used to deliver heating and/or cooling energy to the facility from a mechanical room or central plant, there may be changes to overall heating and/or cooling energy use due to changes in motor operation. Savings can be measured using hourly data if cooling or heating equipment is being measured directly, or statistically if only whole-facility electric data are available. If separate metering data are available for the heating/cooling system, heating/cooling reductions can be determined by developing a baseline model using the methods previously described.

Limitations of Calculating Retrofit Savings from Constant-Load Motor Retrofits Using Hourly Before/After Data. Constant-load motor retrofit savings using utility billing data can be adversely affected by:

- Savings calculated using whole-facility, hourly data can be affected by additions or deletions to electric-consuming equipment in the facility. Care should be taken to document major equipment in the facility. Electricity use at the whole-facility level can also be affected by operational changes in the equipment schedules.

The above operation condition should be noted carefully during both the baseline and post-installation periods.

Variable-Speed Drive Motor Project. Savings resulting from the replacement of a constant-speed motor drive with a variable-speed motor drive can be analyzed with baseline and post-installation hourly measurements, provided the savings are greater than the uncertainty of the regression model. Existing baseline conditions should be documented according to the procedures outlined in Section 4.0.1.

Hourly measurements should be taken for several months to adequately characterize baseline motor performance. In some cases, 12 months of data may be needed to adequately characterize performance if the facility has several occupancy schedules throughout the year. If electricity savings and electric demand are being evaluated, a separate demand analysis should be performed that compares demand for a given month with the demand for that same month in the year prior to the retrofit. Total measured energy savings should include end-use electricity savings from the reduced motor load where possible.

In cases where the motor is being used to deliver heating and/or cooling loads, an additional analysis may need to be performed to evaluate the impact of reduced heating and/or cooling requirements due to the change in the motor used to run the pump.

Calculating Electricity Savings. The analysis method used to determine electricity savings from a variable-speed motor drive is the same method used to calculate a constant-speed motor retrofit using hourly data. To achieve the highest level of accuracy, it is best to monitor the retrofit at the end-use level. However, in most situations this is not economically feasible.

Savings are determined to be significant if the difference between baseline and post-installation energy use is greater than model error as determined by the RMSE.

Calculating Electric Demand Reductions. If end-use motor loads are being monitored, electric demand reductions can be determined if the measurements are taken at the same demand time intervals used for utility billing purposes. Savings can be determined by comparing baseline and post-installation maximum values for the appropriate demand interval. Several calculations may be required for more complex utility rate structures which may include on-peak, off-peak, summer and/or winter electric demand rates.

Calculating Cooling or Heating Savings. If the motor is being used to deliver heating and/or cooling energy to the facility from a mechanical room or central plant, there may be changes to the overall heating and/or cooling energy use due to changes in motor operation. Savings can be measured using the same method outlined for measuring constant-load motor retrofits with hourly data.

Limitations of Calculating Retrofit Savings from Constant-Load Motor Retrofits Using Hourly Before/After Data. Savings from a constant load motor retrofit using utility billing data can be adversely affected by:

- Savings calculated using whole-facility hourly data can be affected by additions or deletions to electric-consuming equipment in the facility. Care should be taken to document major equipment in the facility. Electricity use at the whole-facility level can also be affected by operational changes in equipment schedules.

- In certain cases, special monitoring equipment may be required to analyze variable-speed motor drive electricity use to check for power factor changes and adverse harmonics caused by an improperly installed variable-speed drive.

All of the above operating conditions should be noted carefully during both the baseline and post-installation periods.

HVAC and/or EMS Project. Savings from HVAC and/or EMCS retrofits can be analyzed using baseline and post-installation hourly measurements, provided the savings are greater than the uncertainty of the regression model. Existing baseline conditions should be documented according to the procedures outlined in Section 4.0.1.

Hourly measurements should be taken for at least 12 months to adequately characterize cooling and heating season performance. In some cases, nine months of data can adequately characterize performance if the period under analysis adequately reflects all normal environmental and schedule conditions. If electricity savings and electric demand are being evaluated, a separate demand analysis should be performed that compares the demand for a given month with the demand for that same month in the year prior to the retrofit.

Total measured energy savings should include electricity savings from reduced motor loads, as well as cooling and heating savings.

Calculating Electricity Savings. The analysis method used to determine electricity savings from HVAC and EMCS retrofits depends on facility loads being monitored by the hourly data acquisition system. To achieve the highest level of accuracy, it is best to monitor the retrofit at the end-use level. However, in most situations this is not economically feasible. In most facilities, electricity savings from HVAC and EMCS retrofits can be adequately monitored with hourly data if the following loads can be monitored: i) whole-facility electricity, ii) motor control center electricity, iii) electricity used for powering chillers (and electric heating) and iv) other electric loads that are easily identified as "non-HVAC," such as exterior security lighting, exterior electric slab heaters (for ice melting on sidewalks, etc.). Assuming such channels can be monitored, models for electricity, cooling and heating savings can be created.

Once these measurements have been obtained, electricity savings from HVAC or EMCS retrofits can be determined in the following fashion: First, develop a baseline model for electricity savings using measurements from the motor control center. Second, calculate the retrofit electricity savings by comparing electricity use predicted by baseline parameters (projected into the post-installation period by multiplying by post-installation weather and operating conditions) to measured post-installation data. In situations where significant data are missing in the post-installation period, a post-installation model can be created to fill in missing data. Electricity savings are calculated by comparing energy use predicted by the baseline model to energy use predicted by the post-installation model.

In cases where only whole-facility electricity data are available on a 15-minute or hourly basis, savings can be calculated if at least 12 months of data are available, and the data can be statistically separated into heating season, cooling season and nonheating/non-cooling season data. Several methods have been developed for accomplishing this separation including weather day-types and calibrated simulations.

Calculating Electric Demand Reductions. If end-use loads are being monitored, electric demand reductions from an HVAC or EMCS retrofit can be determined if the measurements are taken at the same demand time intervals used for utility billing purposes. Savings can be determined by comparing baseline and post-installation maximum values for the appropriate demand interval. Several calculations may be required for more complex utility rate structures which may include on-peak, off-peak, summer and/or winter electric demand rates.

Demand savings can also be calculated from whole-facility measurements. However, caution should be taken to verify that reductions are caused by the HVAC or EMCS retrofit, not from an unknown cause.

Calculating Heating and Cooling Savings. Heating and cooling savings can be measured using hourly data if heating and/or cooling equipment is being measured directly, or statistically if only whole-facility electric data are available.

If separate metering data are available for heating/cooling systems, heating/cooling reductions can be determined by developing a baseline model using the methods previously described. Heating/cooling energy savings are calculated by comparing energy use predicted by baseline parameters (projected into the post-installation period by multiplying by post-installation weather and operating conditions) to measured post-installation data. In situations where significant data are missing in the post-installation period, a post-installation model can be created to fill in missing data. Energy savings are calculated by comparing energy use predicted by the baseline model to energy use predicted by the post-installation model. Savings are determined to be significant if the difference between baseline and post-installation energy use is greater than model error as determined by the RMSE.

If heating/cooling energy use is part of the main meter, savings will be combined with other electricity reductions (due to reductions in motor loads) in the whole-facility model. A combined electricity and heating/cooling reduction can be determined by developing a baseline model using the methods previously described. Electricity plus heating/cooling savings are then calculated by comparing electricity use predicted by baseline parameters (projected into the post-installation period by multiplying by post-installation weather and operating conditions) to measured post-installation data. Electricity plus heating/cooling energy savings are calculated by comparing energy use predicted by the baseline model to energy use predicted by the post-installation model. Savings are determined to be significant if the difference between baseline and post-installation energy use is greater than model error as determined by the RMSE.

Limitations of Calculating Retrofit Savings fom HVAC or EMCS Projects Using Hourly, Whole-facility, Before/After Data. Savings resulting from HVAC or EMCS retrofits calculated with hourly, whole-facility, baseline and post-installation utility billing data should be greater than the uncertainty as calculated by the RMSE. These savings can be adversely affected by the following factors:

- Savings measured with end-use measurements are the most accurate. However, such savings can be affected by changes to the operational parameters of the HVAC or EMCS.

- Changes in hourly, whole-facility heating/cooling savings can be affected by heating/cooling system operating changes. In particular, the average chiller kW/ton ratio is affected by the rate of the cooling load on a particular chiller, and boilers or furnaces run at low loads can cycle excessively, which decreases fuel conversion efficiency.

All of the above operating conditions should be noted carefully during both the baseline and post-installation periods.

Chiller Project. Savings resulting from chiller replacements can be analyzed with baseline and post-installation hourly measurements, provided the savings are greater than the uncertainty of the regression model. Existing baseline conditions should be documented according to procedures outlined in Section 4.0.1.

Hourly measurements during the previous cooling season should be taken to adequately characterize cooling season performance. If electric demand savings are being evaluated, a separate demand analysis should be performed that compares the demand for a given month with the demand for that same month in the year prior to the retrofit.

Total measured energy savings from a chiller retrofit should include electricity savings from reduced chiller load and associated loads such as pumps, etc., that accompany the chiller where possible.

Calculating Electricity Savings. The most accurate analysis method used to determine electricity savings from chiller retrofits is the method developed by ASHRAE RP 827, which involves a calibrated thermodynamic model of the chiller similar to that described by Gordon and Ng (1994) and Anderson and Breene (1995). Such a model captures part load performance at varying chilled water and condenser temperatures, and only requires hourly baseline and post-installation measurement of chiller electricity use, thermal output, chilled water supply temperature and condenser water return temperature.

Chiller savings can also be measured with fewer channels if operating conditions in post-installation periods are exactly the same as operating conditions in the baseline period. In instances where the baseline and post-installation chilled water temperature is constant, and the baseline and post-installation condenser return temperature is constant, savings can be measured using hourly chiller thermal output and electric input measurements. In instances where temperature and chiller loading are constant, only electric measurements need be obtained.

Once such measurements have been obtained, electricity savings from the chiller retrofit can be determined by developing a baseline chiller model and comparing the electricity use predicted by baseline parameters (projected into the post-installation period by multiplying by post-installation weather and operating conditions) to measured post-installation data. In situations where significant data are missing in the post-installation period, a post-installation model can be created to fill in missing data. Electricity savings are calculated by comparing energy use predicted by the baseline model to energy use predicted by the post-installation model.

In cases where only whole-facility electricity data are available on a 15-minute or hourly basis (and baseline and post-installation chiller temperatures and loading conditions are constant) savings can be calculated if at least 12 months of data are available, and the data can be statistically separated into heating, cooling and non-heating/non-cooling season data. Several methods have been developed for accomplishing this separation including weather day-types and calibrated simulations.

Calculating Electric Demand Reductions. If end-use loads are being monitored, electric demand reductions from a chiller retrofit can be determined if measurements are taken at the same demand time intervals used for utility billing purposes. Savings can be determined by comparing baseline and post-installation maximum values for the appropriate demand interval. Several calculations may be required for more complex utility rate structures which may include on-peak, off-peak, summer and/or winter electric demand rates.

Demand savings can also be calculated from whole-facility measurements. However, caution should be taken to ascertain that reductions could only have been caused by the chiller retrofit, not by an unknown cause.

Limitations of Calculating Retrofit Savings from Chiller Projects Using Hourly, Whole-facility, Before/After Data. Savings resulting from chiller retrofits that are calculated with hourly, whole-facility, baseline and post-installation utility billing data should be greater than the uncertainty as calculated by the RMSE. These savings can be adversely affected by the following factors:

- Savings measured with electricity measurements only can be adversely affected by chiller loading, chilled water supply temperature, condenser water return temperature and other changes to operational settings that affect chiller efficiency, i.e., chilled water flow rate through the chiller, etc.

- Savings measured with electricity measurements only can also be affected by changes to the facility cooling load, such as changes to the facility envelope and/or changes to interior lighting loads, etc.

- Savings measured with chiller thermal and electric measurements can be adversely affected by chilled water supply temperature, condenser water return temperature and other changes to operational settings that affect chiller efficiency.

All of the above operating conditions should be noted carefully during both the baseline and post-installation periods.

Boiler Project. Savings resulting from boiler replacements can be analyzed with baseline and post-installation hourly measurements, provided the savings are greater than the uncertainty of the regression model. Existing baseline conditions should be documented according to the procedures outlined in Section 4.0.1.

During the previous heating season, hourly measurements should be taken to adequately characterize heating-season performance. As well, measurements for the non-heating season may be required to measure standby losses and/or non-heating-season use, i.e., domestic water heating, etc. If electric demand savings are being evaluated, a separate demand analysis will need to be performed that compares the demand for a given month with the demand for that same month in the year prior to the retrofit.

Total measured energy savings from a boiler retrofit should include electricity and thermal savings resulting from boiler replacement, as well as associated loads (pumps, etc.) that accompany the new boiler package.

Calculating Energy Savings. In a similar fashion to chillers, boiler efficiency is affected by boiler loads, control settings for combustion, fuel energy content and surrounding environmental conditions. Therefore, is it important to record enough baseline data to develop an adequate baseline model which captures the "part-load" performance at various boiler loads as well as surrounding environmental conditions. Useful information regarding boilers can be found in Dukelow (1991), Babcock and Wilcox (1992) and Dyer and Maples (1981).

In cases where the boiler load is the same for both the baseline and post-installation periods, energy savings can be calculated using hourly measurements of fuel input to the boiler. Be sure to note both baseline and post-installation operating conditions affecting boiler operation. Savings can be calculated by developing a baseline model using methods previously described, forecasting baseline use into the post-installation period and comparing post-installation baseline use to measured post-installation data.

For cases where the post-installation boiler load is different than the baseline boiler load, fuel input and thermal output should be measured so that a baseline input/output model can be developed. Be sure to note baseline operating conditions affecting boiler operation. This baseline input/output model can then be forecast into the post-installation period. Ultimately, savings can be determined by comparing energy use forecasted by the baseline input/output model to post-installation energy use. In situations where significant data are missing in the post-installation period, a post-installation model can be created to fill in missing data.

In cases where only whole-facility energy use is available on a 15-minute or hourly basis, and baseline and post-installation boiler conditions are constant, savings can be calculated if at least 12 months of data are available, and the data can be statistically separated into heating, cooling and non-heating/non-cooling season data. Several methods have been developed for accomplishing this separation, including weather day-types and calibrated simulations.

Calculating Electric Demand Reductions. If end-use loads are being monitored, electric demand reductions can be determined if the measurements are taken at the same demand time interval used for utility billing purposes. Savings can be determined by comparing baseline and post-installation maximum values for the appropriate demand interval. Several calculations may be required for more complex utility rate structures which include on-peak, off-peak, summer and/or winter electric demand rates.

Demand savings can also be calculated from whole-facility measurements. However, be sure that the reductions could only have been caused by the boiler retrofit, not by an unknown cause.

Limitations of Calculating Retrofit Savings from Boiler Retrofits Using Hourly, Whole-facility, Before/After Data. Savings resulting from boiler retrofits calculated with hourly whole-facility, baseline and post-installation utility billing data should be greater than the uncertainty as calculated by the CV(RMSE) and/or R^2. These savings can be adversely affected by the following factors:

- Savings calculated with whole-facility measurements only can be adversely affected by boiler loading, boiler supply temperature, combustion settings and other changes to operational settings affecting boiler efficiency. Savings calculated with boiler thermal and fuel input measurements can be adversely affected by boiler operating settings.

The above operating condition should be noted carefully during both the baseline and post-installation periods.

4.3.3 Expected Accuracy

Calculating energy savings using a monthly baseline regression model is expected to be accurate ±20% for those facilities that do not have significant schedule changes during the course of one year. Energy savings calculated using a daily or hourly baseline and post-installation, whole-facility models should be accurate to ±5-10% for facilities that do not have significant schedule changes during the course of one year.

4.3.4 Expected Cost

The expected cost of Option C should be 1-10% of the installed retrofit cost depending on whether utility billing methods or hourly data are used. If monthly utility billing methods are used, the expected cost should be approximately 1% of the installed retrofit cost. If hourly monitoring equipment is installed in a facility, costs can vary from 3-10%, depending on the amount of instrumentation and end-use measurements being recorded. If continuous hourly post-installation monitoring is planned, savings recording and reporting for the second and subsequent years should not exceed 1-3% of the retrofit cost.

Installing and maintaining a data logger, as well as collecting and archiving data over the life of the retrofit, significantly increases the accuracy of daily and hourly models. For most applications, whole-facility data loggers can be installed for the first year for 5% of the retrofit cost. Recording and reporting during the second and subsequent years should cost approximately 1% of the cost of the retrofit each year.

SECTION 5.0: OTHER M&V ISSUES

5.1 Reporting Format and Invoicing

Reporting is an inherent part of any M&V plan because it is the method used to track and verify project value. Report formats and invoicing procedures should be agreed upon prior to contract execution for two reasons: i) this planning helps avoid potential conflicts between the owner and contractor/ESCO and ii) this planning is necessary to accurately determine M&V costs.

5.2 M&V Professionals

A "payment based on performance" arrangement requires that both parties believe the information on which the payments are based is valid and accurate. Often, an unbiased third party trained to measure and verify projects may be helpful to both parties agreeing on measurement validity. Should conflicts arise over the course of the project payback period, this third-party professional can become an invaluable tool as an unbiased source of information, independent of the ESCO. The level of involvement of the M&V professional depends on the amount of information necessary to determine contract value.

5.3 Instrumentation and Measurement Techniques

This section provides a review of the instrumentation and techniques applicable to measurement of electricity, runtime, temperature, humidity, flow and thermal energy. Although measurements of electrical power and energy form the basis for analyzing equipment performance and energy savings calculations, additional information may be required. Runtime information is sometimes useful alone, or to substantiate electrical power and energy data. It is often desirable to adjust baseline energy use to account for indoor or outdoor temperature and relative humidity. Flow data are required to determine natural gas consumption and as part of thermal energy calculations.

5.3.1. The Measurement of Electric Parameters

While the measurement of electrical energy seems simplistic on the surface, there are numerous opportunities for error. All electrical information is derived from two types of measurements: current and voltage. Numerous manufacturers have developed equipment to gather one or both of these types of measurements. Indiscriminate selection or use of voltmeters and ammeters can lead to errors in the calculation of power and energy. Error is usually due to effects of the electrical engineering principle termed "power factor," which is defined as "the ratio of the active power to the apparent power." Often this error occurs because it was assumed that the electrical load possessed a sinusoidal waveform, when, in fact, the waveform was distorted due to the presence of harmonics.

The Nuances of Power Measurement. *Energy* is not the same as *power*. Power is an instantaneous quantity. Energy includes a time function, i.e., the length of time the power has been applied. Utility companies often bill customers based on "demand," which is defined as "the average value of power over a specified interval of time (typically 15 minutes)." Most energy efficiency projects measure electrical power in terms of kilowatts (kW) and electrical energy in terms of kilowatt-hours (kWh). These terms apply to practical units of active power (power that has the ability to perform work, e.g., to move air or pump fluids). This type of power is termed "real power" or "actual power" and is the basis upon which utilities invoice their customers. Utility revenue meters record whole-facility watt-hours from voltage and current sensors, which is the measurement that forms the basis for energy savings calculations. However, watt-hour information is not usually available for specific end uses. Therefore, it is often impossible to determine project-specific energy savings without additional metered data. Obtaining those data should follow the accepted engineering practices discussed in this section.

Electrical power also exists in the form of reactive power, i.e., power required to generate the magnetic fields required of motors, transformers and lighting ballasts. Reactive power produces no work. Instead, it builds magnetic fields that collapse upon themselves and are then built again. Although most utilities do not bill customers directly for reactive power, it does exist and has an important place in power measurement theory.

Reactive power combined vectorially with real power determines apparent (or total) power which is measured in volt-amperes. Apparent power is an important power consideration and is the cause of many savings calculation errors. Using hand-held metering equipment, it is possible to measure both the voltage supplied to a load and the current drawn by that load. The product of these two measurements is volt-amperes, a measurement that includes real power (in watts) and any associated reactive power. Apparent power (in volt-amperes) is related to, but not always equal to, real power (in watts). An adjustment factor, the "power factor," must be applied to the apparent power to obtain real power. This power factor (a number between zero and one) represents the ratio of real-to-apparent power. Electrical loads that are resistive in nature, such as incandescent lamps and heating strips, have a power factor of one (a "unity" power factor), thus the measurements of real and apparent power are the same. Electrical loads such as fluorescent lamps and motors do not have unity power factors, thus apparent power does not equal real power. Assuming the two are equal could lead to errors of as much as 40%. Wattmeters on chiller equipment typically read much lower power levels than individual current and voltage measurements due to lower power factors associated with partially loaded chillers.

Power Measurement Equipment. Real power can be measured directly using watt transducers (devices that determine power from voltage and current sensors). Devices that integrate power over time are called "watt-hour transducers." They provide real energy data and eliminate the error inherent in assuming or ignoring power factor. Standalone watt-hour transducers are available to produce pulses representative of some number of watt-hours. These pulses are typically input to a pulse-counting data logger for storage and subsequent retrieval and analysis.

An alternate technology involves combining metering and data logging functions into a single piece of hardware. This integrated metering approach incorporates virtual digital watt-hour meters into a single solid-state device capable of monitoring 8 to 32 single-phase power channels. Whereas pulse-counting technology makes kW and kWh information available to the user, the integrated approach allows access to much more information. In addition to kW and kWh, each defined power channel can record voltage, current, apparent power in kVA, kVAh and power factor. Many integrated meter/monitors have the ability to perform waveform analysis, capturing harmonic information for both voltage and current waveforms.

True RMS Metering. Until recently, most loads were linear, i.e., the nature of the load remained essentially constant regardless of the applied voltage. These linear loads resulted in smooth sinusoidal voltage and current waveforms. Conventional meters usually measure the average value of the amplitude of the waveform. Some meters are calibrated to read the equivalent RMS value, equal to 0.707 times the sinusoidal peak value. This type calibration is a true representation only when the waveform is a non-distorted sine wave.

With the advent of computers, uninterruptible power supplies and variable-speed motor drives, nonlinear waveforms are more the norm. When distortion occurs, the relationship between average readings and true RMS values changes drastically. Peak-sensing and averaging meters are inaccurate and inappropriate technologies for the measurement of distorted waveforms. Digital sampling technology is the recommended method of measuring non-sinusoidal waveforms. Solid-state digital metering equipment samples voltage and current simultaneously to produce instantaneous values which are stored in memory along with their product and individual squares. Periodically, the meter calculates RMS and average values to obtain true RMS power and energy.

True RMS power- and energy-metering technology, based on digital sampling principles, is recommended over individual voltage and current readings due to its ability to accurately measure distorted waveforms and properly record load shapes.

5.3.2 Measurement of Runtime

Measurement and verification of energy savings often involves little more than an accurate accounting of the amount of time that a piece of equipment is operated or "on." Constant-load motors and lights are examples of equipment that need not be metered with full-featured RMS power-metering equipment to establish energy consumption. Self-contained, battery-powered monitoring devices are available to record equipment runtime and, in some cases, time-of-use information. This equipment provides a reasonably priced, simple-to-install solution to energy savings calculations.

5.3.3 Measurement of Temperature

The computerized measurement of temperature has become an "off-the-shelf" technology. The most commonly used computerized temperature measurements use one of four basic methods for measuring temperatures: i) resistance temperature detectors (RTDs), ii) thermoelectric sensors (thermocouples), iii) semiconductor-type resistance thermometers (thermistors) and iv) junction semiconductor devices which are also called integrated circuit temperature (IC) sensors.

Resistance Temperature Detectors (RTDs). The most common method of measuring temperature in the energy management field is with RTDs, which are among the most accurate, reproducible, stable and sensitive thermal elements available. The theory behind an RTD is that electrical resistance in many materials changes with temperature. In some materials this change is very reproducible and, therefore, can be used as an accurate measure of the temperature. These devices are economical and readily available in configuration packages to measure indoor and outdoor air temperatures as well as fluid temperatures in chilled water or heating systems. Considering its overall performance, the most popular RTDs are 100- and 1,000-Ohm platinum devices in various packaging, including ceramic chips, flexible strips and thermowell installations.

Depending on the application specifics, two-, three- and four-wire RTDs are available. Required accuracy, distance and routing between the RTD and the data-logging device can determine the specific type of RTD for a project. Three-wire RTDs exist to compensate for applications where an RTD required a long wire lead exposed to varying ambient conditions. This is because three wires of identical length and material exhibit similar resistance-temperature characteristics and can be used to cancel the effect of the long leads in an appropriately designed bridge circuit. Two-wire RTDs must be field-calibrated to compensate for lead length and should not have lead wires exposed to conditions that vary significantly from those being measured.

Installation of RTDs is relatively straightforward with the advantage that conventional copper lead wire can be used as opposed to the more expensive thermocouple wire. Most metering equipment allows for direct connection of RTDs by providing internal signal conditioning and the ability to establish offsets and calibration coefficients.

Thermocouples. In general, thermocouples are used when reasonably accurate high-temperature data are required. In thermocouple thermometry, the magnitude of the voltage is dependent on the type of material and the temperature difference. The most commonly used thermocouple materials are: i) platinum-radium (Type S or R), ii) chromel-alumel (Type K), iii) copper-constantan (Type T) and iv) iron-constantan (Type J). The main disadvantage of thermocouples is their weak output signal, making them sensitive to electrical noise and always requiring amplifiers. Few performance contracts require the accuracy and complexities of thermocouple technology.

Thermistors. Thermistors are semiconductor temperature sensors and usually consist of an oxide of either manganese, nickel, cobalt or one of several other types of materials that is milled, mixed, pressed and sintered. One of the primary differences between thermistors and RTDs is that thermistors have a very large negative resistance change with temperature. Thermistors are not interchangeable, and their temperature-resistance relationship is very non-linear. They are fragile devices and require the use of shielded power lines, filters or DC voltage. Like thermocouples, these devices are infrequently encountered in performance contracting.

Integrated Circuit Temperature Sensors. Certain semiconductor diodes and transistors also exhibit reproducible temperature sensitivities. Such devices are usually ready-made integrated circuit (IC) sensors and can come in various shapes and sizes. These devices are occasionally found in HVAC applications where low cost and a strong linear output are required. Temperature sensors have a fairly good absolute error, but they require an external power source, are fragile and are subject to errors due to self-heating.

5.3.4 Measurement of Humidity

Accurate, affordable and reliable humidity measurement has always been a difficult and time-consuming task. Recently, such measurements have become more important in HVAC applications for purposes of control, comfort and system diagnosis. The amount of moisture in the air can be described by several interchangeable parameters including relative humidity, humidity ratio, dew-point temperature and wet-bulb temperature.

In energy performance contract work, it will occasionally be necessary to measure relative humidity, the measure of moisture concentration expressed as a percentage of the moisture at saturated conditions. In general, most measurements of humidity do not actually "measure" the humidity but rather measure the effect of moisture using an indirect measurement. Relative humidity measurements (indirect) include:

- the evaporation psychrometer
- electrical resistance or conductivity
- elongation
- capacitance-reactance
- infrared
- radio frequency
- acoustic measurements

Equipment to measure relative humidity is available from several vendors and installation is relatively straightforward.

5.3.5 Measurement of Flow

In many situations, whole-facility Btu measurements are needed for a facility or group of facilities. Most often this requires accurate measurements of liquid flow and temperature, usually at the service entrance to the facility. Even in cases where steam flow must be measured in a closed loop, it is easier (and much safer) to measure the returning liquid condensate than to measure the live steam as it enters the facility.

Choosing a flow meter for a particular application requires a knowledge of what type of fluid is being measured, how dirty or clean that fluid is, what the lowest expected flow velocities for that fluid are and what type of budget one has available. This section discusses the most common liquid-flow measurement devices that are used in conjunction with temperature measurements to determine the thermal energy in a fluid flow.

In general, flow sensors can be grouped into four different types of meters: i) differential pressure flow meters (e.g., orifice plate meter, venturi meter, pitot tube meter), ii) obstruction flow meters (e.g., variable-area meter, positive displacement meter, turbine meter, tangential paddlewheel meter, target meter, vortex meter), iii) non-interfering meters (e.g., ultrasonic meter, magnetic meter) and iv) mass flow meters (e.g., coriolis mass flow meter, angular momentum mass flow meter). While there are specific applications for each of these metering technologies, the most common flow meters found in thermal energy calculations are turbine meters and vortex meters. There is interest in non-interfering metering technology to defray the costs of shutting down pumps and cutting pipe. Each of these technologies will be addressed in this section.

Non-Pressure-Differential Obstruction Flow Meters. Several types of obstruction flow meters have been developed that are capable of providing a linear output signal over a wide range of flow rates, oftentimes without the severe pressure-loss penalty that is incurred with an orifice plate or venturi meters. In general these meters place a much smaller target, weight or spinning wheel in the flow stream that then allows the velocity of the fluid to be determined by the force on the meter body (target or variable-area meter), and by the rotational speed of the meter, (turbine, paddlewheel meters).

Turbine meters measure fluid flow by counting the rotations of a rotor placed in a flow stream. Turbine meters can be axial-type or insertion-type. Axial turbine meters usually have an axial rotor and a housing that is sized for an appropriate installation. An insertion turbine meter allows the axial turbine to be inserted into the fluid stream and uses the existing pipe as the meter body. Because the insertion turbine meter only measures the fluid velocity at a single point on the cross-sectional area of the pipe, total volumetric flow rate for the pipe can only be accurately inferred if the meter is installed per manufacturer's specifications, most importantly installation along straight sections of pipe removed from internal turbulence. This type of meter can be hot-tapped into existing pipelines through a valve system without having to shut down the system. Insertion meters can be used on pipelines above 4" with very low pressure loss. The speed of rotation of a bladed turbine, driven by the fluid, provides an output linear with flow rate. This output can usually be obtained either as a signal pulse representing a quantity of fluid flow or as an analog signal proportional to flow rate. Either output can be captured by meter/monitoring equipment to build trends.

Vortex meters utilize the same basic principle that makes telephone wires oscillate in the wind between telephone poles. This effect is due to oscillating instabilities in a low field after it splits into two flow streams around a blunt object. Vortex meters have no moving parts and are suitable for gas, steam or liquid flow measurements. They require minimal maintenance and have good accuracy and long-term repeatability. Vortex meters provide a linear digital (or analog) output signal that can be captured by meter/monitoring equipment to build trends.

Non-Interfering Flow Meters. In all of the previously mentioned meters, some interference with the flow stream was necessary to extract a measurement. Recently, a relatively new class of meters has been developed that is able to extract a measurement without placing an obstruction in the fluid stream.

Ultrasonic flow meters measure clean fluid velocities by detecting small differences in the transit time of sound waves that are shot at an angle across a fluid stream. Various designs have been developed that utilize multiple-pass, multiple-path configurations. Accurate clamp-on ultrasonic flow meters have been developed that now facilitate rapid measurement of fluid velocities in pipes of varying sizes. An accuracy from 1% of actual flow to 2% of full scale are now possible, although this technology is still quite expensive. Recently, an ultrasound meter that uses the Doppler principle in place of transit time has been developed. In such a meter a certain amount of particles and air are necessary in order for the signal to bounce off and be detected by the receiver. Doppler-effect meters are available with an accuracy between 2-5% of full scale and command prices somewhat less than the standard transit time-effect ultrasonic devices. Meter cost is independent of pipe size.

5.3.6 Measurement of Thermal Energy

The measurement of thermal energy used in a facility's heating or cooling system often requires the measurement and recording of Btus. The cooling provided by the facility chillers is recorded in Btus and is a calculated value determined by measuring the chilled water flow in gallons per minute (GPM) and the temperature differential (delta-T) between the chilled water supply and the chilled water return. A Btu meter, either a stand-alone device or a "virtual" Btu meter as part of a larger meter/recorder device, performs an internal Btu calculation in real time based on inputs from a previously described flow meter and temperature sensors. These electronic Btu meters offer an accuracy better than 1%. They are most attractive on larger or more critical installations where accuracy is a prime concern. A side benefit is the availability of real-time operating data such as flow rate, temperature (both supply and return), and Btu rate.

When measuring the narrow differential temperature (delta-T) range typical of chilled-water systems, the two temperature sensors should be matched or calibrated to the tightest tolerance possible. For the purpose of computing thermal loads in Btu per hour, it is more important that the sensors be matched or calibrated with respect to one another than for their calibration to be traceable to a standard. Attention to this detail will maximize the accuracy of the Btu computation. Suppliers of RTDs can provide sets of matched devices when ordered for this purpose. Typical purchasing specifications are for a matched set of RTD assemblies (each consisting of an RTD probe, holder, connection head with terminal strip, and a stainless steel thermowell), calibrated to indicate the same temperature within a tolerance of 0.1°F over the range 25°F to 75°F. A calibration data sheet is normally provided with each set.

Thermal energy measurements for steam can require steam flow measurements (e.g., steam flow or condensate flow), steam pressure, temperature and feedwater temperature where the energy content of the steam is then calculated using steam tables. In instances where the steam production is constant this can be reduced to measurements of steam flow or condensate flow only (i.e., assumes a constant steam temperature-pressure and feedwater temperature-pressure).

5.4 Verification of Project Maintenance

Part of every performance contract is the implied fact that the specified maintenance of the Energy Conservation Measures will be performed. Independent of whether the maintenance is performed by the vendor, the purchaser or another party, the contract measurement and verification plan should include the procedure by which the implementation of the maintenance plan is verified.

5.5 Special Notes on New Construction

New construction by definition will not have pre-retrofit information for use in calculating energy savings. Therefore, it will be necessary for both parties to agree on how the baseline energy usage will be determined. Many contracts use the Minimum Energy Standards that are in effect for the jurisdiction where the project is to be constructed. Energy savings are calculated as the difference between the "minimum standard energy performance" and the actual performance.

5.6 Minimum Energy Standards (Federal, State and Local)

One of the difficulties in determining energy savings is defining at what level the baseline energy use should be established. Many facilities where retrofits are considered have equipment that does not meet current standards for energy efficiency. Agreement must be made between the contracting parties as to whether the baseline will be established at the actual performance level of the existing equipment or whether to ignore actual conditions and use the current standard as the baseline.

5.7 Dealing with Data Collection Errors and Lost Data

No data collection is without error. Methodologies for data collection differ in degrees of difficulty and therefore in the amounts of data that are either in error or missing. Regardless of the method, two concepts should be agreed to in advance by both parties. First, a minimum level of data performance should be established. This level should be part of the overall accuracy of results calculation needed to provide the confidence levels desired by both parties. The contract should stipulate penalties for the party who fails to collect the minimum data requirement. Higher levels of data accuracy may dramatically effect the cost of verification and should be decided as part of the overall project economics. The second concept is the methodology by which data that are missing or determined to be incorrect will be interpolated for final analysis.

5.8 Commissioning Projects

Project commissioning is not only preferable, but highly recommended for performance-dependent projects. ASHRAE Standard 1-89, Guideline for Commissioning of HVAC Systems, should be consulted for recommendations concerning HVAC commissioning procedures. The properly chosen M&V option should reinforce the total commissioning process.

The commissioning process itself may be divided into pre-design, design, construction, acceptance and post-acceptance phases. Ideally, verification methods should be chosen during the design phase and implemented prior to the acceptance phase. The chosen M&V option should aid the commissioning agent (CA) in determining optimum performance and acceptability of a given retrofit project during the acceptance process.

Project commissioning also involves monitoring use, occupancy and maintenance beyond the project acceptance phase. Therefore, an additional benefit of implementing M&V protocols in conjunction with commissioning is that it allows the facility owner, CA or maintenance foreman to determine the point at which re-commissioning of a facility or project should be considered. In cases where projects are completed under a performance contract, this would allow monitoring and verification activities to proceed, possibly beyond the performance contract limit. In such cases, a qualified commissioning agent might act as the M&V agent as well. This would have three-fold benefit since it would:

1. allow for monitoring under the chosen M&V option,
2. give independent party verification of the monitoring results and
3. allow long-term monitoring of persistence of savings extending for the useful life of a project.

5.9 Interactive Effects Between Measures

Information not included in this version 1.0.; deferred until next version of the M&V protocol.

5.10 Dispute Resolution

Information not included in this Version 1.0.; deferred until next version of the M&V protocol.

5.11 Special Notes on Residential Projects

All three paths are applicable to residential projects, however there are some practical considerations and limitations to be aware of. The choice of an M&V protocol will be strongly influenced by the type of residence and the type of purchaser, in addition to the retrofit technology. Projects can be grouped into three categories:

1. a large, multifamily facility or complex (above about 20 to 40 units),

2. an individual residence or a small multifamily facility and
3. a large number of individual residences.

Large, multifamily facilities or complexes can be treated in much the same manner as commercial facilities. It is, however, crucial to recognize that less of the energy use in a residential facility is based on systematic scheduling than is the case in commercial or industrial applications. Hours of operation for interior lighting, for example, tend to fall in only two classes, always on (hallways and other common areas), and unknown (private areas). Unitary HVAC equipment is more common in residential and small commercial facilities, so the M&V plan may also need to recognize that case. In particular, there are many facilities in some areas of the country with central heat and unitary air conditioning. The design of the M&V plan must also consider whether the facility is master-metered or if each residence has its own billing meter.

For individual residences, performance contracting is rare and Option A M&V is generally the only feasible path. The submetering of loads required by Option B is usually far too costly. In addition, many of the retrofit measures affect more than one energy-using piece of equipment, such as insulation lowering both heating and cooling requirements. Application of either Option B or C to a single residence is also not accurate due to the high variability in energy use patterns over time for any individual home. In Option A, verification of installation and potential to perform is generally by nameplate only, without spot measurement. Stipulated values for hours of use are usually gross averages, sometimes reflecting regional values.

A large number of individual residences can receive efficiency improvements under a performance contract between an ESCO and a utility or a government entity. Occasionally, a utility acts as an ESCO under an agreement with its regulators, acting on behalf of all customers. The largest programs of this type provide services to tens of thousands of homes, although the techniques are applicable to groups of less than one hundred homes. All three Options can be applicable to this type of program, with Option C being highly preferable for most situations.

The stipulated values in Option A are often based on prior research using Options B or C. Option B is expensive for individual residences, relative to the value of the saved energy, but with a large group of homes sampling can be used. With a carefully stratified sample, as few as 40 points can yield 90% confidence of 10% accuracy, while 200 points can achieve 95% confidence of 5% accuracy. Metering techniques are derived from those used by utilities for appliance load research. The Electric Power Research Institute has a large body of literature on the topic.

Option C is the preferred method to measure the energy savings from a large scale residential program. Often the conservation effort is a "whole house" retrofit, including heating, cooling, water heating and lighting improvements, so the facility utility meter is the end-use meter. Sampling can be used, but most programs use a census of all treated homes. This improves accuracy without significant cost increase, since production computer programs are usually needed to process the data. For large sample sizes, smaller savings levels can be found than would be possible for single buildings.

When using option C in the residential sector, whether for single family or multifamily homes, the preferred analysis tool is the Princeton Scorekeeping Method (PRISM). PRISM is an automated process that develops pre- and post-retrofit normalized annual consumption. PRISM uses a regression of daily average use against variable-base heating and/or cooling degree days to normalize and annualize the data. The program is deterministic, so given the same data and input parameter, different analysts will get the same results. This makes the technique highly amenable for contractual use. A base period of three to five years can normally be easily assembled from utility-archived billing data. Although a single base year has been used, a longer period is preferable to mitigate unusual external events, since no normalization process is perfect. The post-period measurement can be as long as the expected life of the savings, but most projects limit the measurement to the first three to five years, to demonstrate savings and persistence. A comparison group of untreated homes is often used as a further normalization method. (The definitive reference on PRISM is the special issue of *Energy and Facilities* edited by Margaret F. Fels—Volume 9, Numbers 1 & 2, February/May 1986). Other physical models exist, usually employing a fixed-base, degree-day system to model the facility's response to weather. Econometric analysis is sometimes used, but cannot readily model physical reality and is difficult to reduce to contractual language.

5.12 Calibration of Instrumentation

For the highest-quality measurements it is recommended that the calibration procedures developed by the National Institute of Standards and Technology (NIST) be used. A list of calibration procedures is provided in the references and includes: Baker and Hurley (1984), Benedict (1984), Bevington and Robinson (1992), Bryant and O'Neal (1992), Cortina (1988), Doebelin (1990), EEI (1981), Haberl et al. (1992), Harding (1982), Huang (1991), Hyland and Hurley (1983), Hurley and Schooley (1984), Hurley (1985), ISA (1976), Kulwicki (1991), Lee (1988), Leider (1990), Miller (1989), Morrissey (1990), O'Neal et al. (1990), Ramboz and McAuliff (1983), Robinson et al. (1992), Ross (1990), Taylor (1981), Wiesman (1989), Wise and Soulen (1986) and Wise (1976).

5.13 Calculating Uncertainty

The issue of uncertainty affects the degree of confidence one has in the calculation of energy savings. In the most simple cases, a continuously running constant load can be measured (per Section 5.3.1) before and after an energy retrofit and the savings calculated directly. Uncertainty becomes an issue when multiple loads are sampled and the total load is derived from the sample data. How certain can one be that the sample truly represents the load characteristics of the whole (the population)? The largest concerns of uncertainty occur when one develops a regression equation to represent the population load characteristics. Load data are obtained upon which a regression equation is constructed. How certain that equation represents the actual load is a great concern to all involved. Inaccuracies in the load data (from a poor instrumentation plan design or inaccuracies in load-measuring equipment) can lead to uncertainties in the ability of the equation to predict true loads. A poorly constructed regression equation can provide uncertain information.

The three statistical indices used to evaluate the models are defined below (SAS 1990):

1. The coefficient of determination, R^2 (%):

$$R^2 = \left(1 - \frac{\sum_{i=1}^{n} \left(y_{pred,i} - y_{data,i}\right)^2}{\sum_{i=1}^{n} \left(\bar{y}_{data} - y_{data,i}\right)^2} \right) \times 100$$

2. The coefficient of variation CV (%):

$$CV = \frac{\sqrt{\dfrac{\sum_{i=1}^{n}\left(y_{pred,i} - y_{data,i}\right)^2}{n-p}}}{\bar{y}_{data}} \times 100$$

3. The mean bias error, MBE (%):

$$MBE = \frac{\dfrac{\sum_{i=1}^{n}\left(y_{pred,i} - y_{data,i}\right)}{n-p}}{\bar{y}_{data}} \times 100$$

where

$y_{data,i}$ is a data value of the dependent variable corresponding to a particular set of the independent variables,

$y_{pred,i}$ is a predicted dependent variable value for the same set of independent variables above,

$\bar{y}_{data}$ is the mean value of the dependent variable of the data set,

n is the number of data points in the data set,

p is the total number of regression parameters in the model (which was arbitrarily assigned as 1 for all models).

The regression equation is used in the calculation of energy savings. After the retrofit action, data are obtained on the system operation and used as input to the regression equation. The equation is used to determine the load which would have occurred had the original equipment been left in place and the system operated under the conditions currently observed. The regression equations must be sensitive to runtimes, indoor and outdoor temperatures and humidity, chilled water temperatures and possibly thermal loads removed (or added) to the system. Uncertainties in obtaining the base data upon which equation is developed or in the structure of the equation will lead to unfounded projections of savings. In general, the result of a measurement is only an approximation or estimate of the value of the specific quantity subject to measurement. It is important that the equation express not simply a physical law but a measurement process, and, in particular, it should contain all quantities that can contribute a significant uncertainty to the measurement result. If the measurement situation is especially complicated, one should consider obtaining the guidance of a statistician.

In terms of prioritization, the greatest source of error in this process is in the collection of energy information and the externalities (i.e., temperature/humidity, runtime and occupancy) which affect energy use. A poorly designed metering and instrumentation plan can result in poor quality inputs to build regression equations. Equations so constructed will be poor predictors of energy consumption.

5.13.1 Effect of Short Pre-Retrofit Data Sets

Ideally, a full year or more of energy use and weather data should be used to construct regression models. The data can then be deemed to contain the entire range of variation in both climatic conditions and in the different operating modes of the facility and of the HVAC system. However, in many cases a full year of data are not available and one is constrained to develop models using less than a full year of data.

How temperature-dependent regression models of energy use fare in such cases is discussed by Kissock et al. (1993). That study constructed temperature-dependent linear regression models of daily energy use from 1-, 3-, and 5-month data sets. Annual energy use predicted by these models was compared to the annual energy use predicted by a model based on an entire year of data. It was found that annual heating energy use can be more than 400% greater than the annual energy use predicted by models from short data sets. In addition, in the climate of central Texas, models of heating energy use have prediction errors four to five times greater than those of cooling energy models.

Two characteristics of data sets were identified which influence their ability to predict annual energy use:

- As expected, longer data sets provide a better estimate of annual energy use than shorter data sets. In the sample of facilities chosen, the average cooling prediction error of short data sets decreased from 7.3% to 3.0% and the average annual heating prediction error decreased from 27.5% to 12.9% as the length of data sets increased from one month to five months.

- More important than the length of the data set, however, was the season during which it occurred. Cooling models identified from months with above-average temperatures tend to over-predict annual energy and vice-versa. The converse seems to hold for heating models.

The best predictors of both cooling and heating annual energy use are models from data sets with mean temperatures close to the annual mean temperature. The range of variation of daily temperature values in the data set seems to be of secondary importance. One-month data sets in spring and fall, when the above condition applies, are frequently better predictors of annual energy than five-month data sets from winter and summer.

5.13.2 Uncertainty in Savings Determination

The duration of metering and monitoring must be sufficient to ensure an accurate representation of the average amount of energy used by the affected equipment both before and after project installation. The measurements should be taken at typical system outputs within a specified time period, such as one month. These measurements can then be extrapolated to determine annual and time-of-use period energy consumption.

The required length of the metering or monitoring period depends on the type of project. If, for instance, the project is a system that operated according to a well-defined schedule under a constant load, such as a constant-speed exhaust fan motor, the period required to determine annual savings could be quite short. In this case, short-term energy savings can be easily extrapolated to the entire year. However, if the project's energy use varies both across the day and across seasons, as with air-conditioning equipment, a much longer metering or monitoring period may be required to characterize the system. In this case, long-term data are used to determine annual and time-of-use period energy savings.

If the energy consumption of the metered equipment or systems varies by more than 10% from month to month, measurements must be taken at sufficient detail and over a long-enough period of time to identify and document the source of the variances. Any major energy consumption variances due to seasonal production increases or periodic fluctuations in occupancy or use must also be tracked and recorded. If these variances cannot be integrated into the regression equations for whatever reason, they must be built into the annual energy consumption figure through an agreed-upon mathematical adjustment.

In statistics, ascertaining the uncertainty of a prediction is as important as the prediction itself. Hence determining the uncertainty in the retrofit savings estimate is imperative. Model identification has direct bearing on determining the uncertainty because the same issues equally affect the nature and magnitude of errors. The uncertainty in savings can be attributed to measurement errors (both in the independent and dependent variables) and to errors in the regression model. The former are relatively well-known to engineers and the methodology of estimating their effect is adequately covered in classical engineering textbooks. Errors in regression models, on the other hand, are more complex and arise from several sources.

Model prediction errors arise due to the fact that a model is never "perfect." Invariably a certain amount of the observed variance in the response variable is unexplained by the model. This variance introduces an uncertainty in prediction. Model extrapolation errors arise when a model is used for prediction outside the region covered by the original data from which the model has been identified.

SECTION 6.0: DEFINITIONS OF TERMS

Term	Definition
Annual Energy Audit	A procedure established within the contract for determining the annual energy savings attributed to a project.
Baseline Usage (Demand & Energy)	The calculated energy usage (demand) by a piece of equipment or a site prior to the implementation of the project. Baseline physical conditions such as equipment counts, nameplate data and control strategies will typically be determined through surveys, inspections, and/or spot or short-term metering at the site.

Billing Data — Energy data collected from invoices sent to the owner from the power supplier, i.e., an electric or gas bill.

Building Automation System — An electronic computer that can be programmed to control the operations of energy-consuming equipment in a facility.

Chauffage — Type of performance contract where the ESCO assumes the operating risk and provides the owner with a guaranteed amount of energy at a guaranteed price over time.

Commissioning — The process of documenting and verifying the performance of VAC systems so that they operate in conformity with the design intent. System/equipment commissioning is expected to be completed by the ESCO. Current editions of ASHRAE's commissioning guideline GPC-1 can be the basis for commissioning activities.

Demand Reduction Estimates — Electric demand reductions (in kW) derived from sample metering and estimation equations, in accordance with the provisions of the contract's approved measurement and verification plans, and documented in regular true-up reports.

Demand Savings — [Peak-period baseline energy less peak-period, post-installation energy] divided by the number of hours in the peak period.

Demand-Side Management (DSM) — The concept of achieving overall energy use reductions through the use of conservation techniques at the end-use equipment, rather than changing or controlling the supply of the energy source.

Detailed Energy Survey	Often referred to as an energy audit. A complete inventory of the energy-consuming equipment at a given facility. This information is used in determining the scope of work for a project.
Energy Audit	Procedure to establish baseline energy use and verify achievement of energy savings.
Energy Conservation Measure (ECM)	Installation of equipment or systems, or modification of equipment or systems, for the purpose of reducing energy use and/or costs.
Energy Conservation Opportunity	An alternative to a new or existing system or component specifically intended to reduce energy consumption.
Energy Cost	The actual unit cost of power, i.e., electric cost = $/kW, $/kWh.
Energy Cost Savings	Reduction in the cost of energy expenses.
Energy Management System	Usually a computer-controlled device that is capable of sensing building conditions and making pre-set logic decisions about how energy-consuming equipment should operate.
Energy Savings	Actual reduction in energy use or demand in electrical or thermal units.
Energy Savings Estimates	Electric energy savings (in kWh) derived from sample metering and estimation equations, in accordance with the provisions of the contract's measurement and verification plans, and documented in regular true-up reports.

Energy Savings Performance Contract (ESPC)	A contract where the cost of ECM implementation is recovered through savings created by the ECMs.
Energy Services Company (ESCO)	An organization which designs, procures, finances, installs and possibly maintains one or more ECMs at an owner facility or facilities.
Error Analysis	A mathematical determination of the errors present in the representation of any savings reports.
Investment-Grade Audit	Detailed energy survey with sufficient detail to allow for the project value with respect to financing
Measurement & Verification (M&V)	The act of obtaining and verifying energy efficiency equipment performance.
Measurements, Long-Term	Measurements taken over a period of several years.
Measurements, Short-Term	Measurements taken for several hours, weeks or months.
Measurements, Spot	Measurements taken one time; snapshot measurements.
Metered Data	Data collected at a facility over time through a meter for a specific end-use, energy-using system or location.
Metering	Collection of energy consumption data over time at a facility through the use of meters.
Models, Calibrated Engineering	Simulation models that are forced to fit measured data.

Models, Regression	Inverse models that require data to extract parameters.
Models, Simulation	Use algorithms that calculate energy used based on engineering equations.
Monitoring	The collection of data at a facility over time for the purpose of savings analysis, i.e., energy consumption, temperature, humidity, hours of operation, etc.
M&V Method	A generic, not project-specific, M&V approach defined which applies one of the three M&V options to a specific ECM technology category, such as lighting efficiency retrofits, constant-load motor retrofits, or variable-load, variable-operating-hour project retrofits.
M&V Option	One of three generic M&V approaches defined for energy performance contracts.
M&V Technique	An evaluation tool for determining energy and cost savings. M&V techniques discussed in this document include engineering calculations, metering, utility billing analysis and computer simulation.
Non-Variable Loads	Power-consuming equipment that has steady, non-changing energy consumption over time.
Outsource	The concept of subcontracting an entire area of service in exchange for a fee; often referred to as "turnkey operations."
Owner	Person or persons who have possession of a facility or facilities where an ESCO provides ECM-related services.

Post-Installation Energy Use (Demand)
The calculated energy use (or demand, e.g., in kW) by a piece of equipment or a site after implementation of the project. Post-installation energy use is verified by the ESCO and the host customer that the proper equipment/systems were installed, are operating correctly and have the potential to generate the predicted savings.

Preliminary Energy Survey
A quick inventory of energy-consuming equipment often used for the first determination of whether a potential project exists for improved energy performance. Not to be used for investment decisions.

Project Pre-Installation Report
The ESCO, prior to the installation of energy-efficient equipment, will provide pre-specified documentation that verifies the proposed equipment/systems and associated energy savings, and demonstrates proper maintenance and operation to have the potential to generate the predicted savings. Documentation that provides a description and inventory of existing and proposed energy-efficiency equipment, estimates of energy savings and a site-specific M&V plan (if not included in contract).

Project Post-Installation Report
The ESCO, after the installation of energy-efficient equipment, will provide pre-specified documentation that verifies the installed equipment/systems and associated energy savings, and demonstrates proper maintenance and operation to have the potential to generate the predicted savings. Documentation that provides a description and inventory of old and installed energy-efficiency equipment, estimates of energy savings and M&V results.

RMSE	Root mean square error—see Section 5.13 Calculating Uncertainty.
Variable Loads - Accuracy	Power-consuming equipment that has a changing energy-consumption level over time.
Walk -Through Audit	See "Preliminary Energy Survey."

SECTION 7.0: REFERENCES

NOTE: *The following references are meant to provide the reader with sources of additional information. These sources consist of publications, textbooks and reports from government agencies, national laboratories, universities, professional organizations and other recognized authorities on building-energy analysis. For the most part care has been taken to cite the publication, publisher or source where the document can be obtained.*

Akbari, H., Heinemeier, K.E., LeConiac, P. and Flora, D.L. 1988. "An Algorithm to Disaggregate Commercial Whole-facility Hourly Electrical Load Into End Uses," Proceedings of the ACEEE 1988 Summer Study on Energy Efficiency in Buildings, Vol. 10, pp. 10.14-10.26.

Alexander, R. 1992. *Handbook for Electricity Metering,* 9th Edition, Edison Electric Institute, Washington, D.C., ISBN 0-931032-30-X.

Anderson, S. and Breene, L. 1995. "Field Measurement Efficiency of Commercial Electric Chillers," International District Energy Association Conference, Miami, Florida, Swiki Anderson & Associates, 1516 Shiloh Avenue, Bryan, Texas.

ASHRAE 1990. *Handbook: HVAC Applications,* Chapter 37—"Building Energy Monitoring," Atlanta, Georgia.

Babcock and Wilcox 1992. *Steam: Its Generation and Use.* Babcock and Wilcox Co., Barberton, Ohio, ISBN 0-9634570-0-4.

Baker, D. and Hurley, W. 1984. "On-Site Calibration of Flow Metering Systems Installed in Buildings," NBS Building Science Series Report No. 159, January.

Balcomb, J.D., Burch, J.D. and Subbarao 1993. "Short-Term Energy Monitoring Of Residences," ASHRAE Transactions, 99(2), pp. 935-944.

Benedict, R. 1984. *Fundamentals of Temperature, Pressure and Flow Measurement.* John Wiley and Sons, New York, New York, ISBN 0-471-89383-8.

Bevington, P.R. and Robinson, D.K. 1992. *Data Reduction and Error Analysis for the Physical Sciences.* 2nd Edition, McGraw-Hill, New York, ISBN-0-07-911243-9.

Biesemeyer, William D. and Jowett, Jeffrey 1994. "Facts and Fiction of HVAC Motor Measuring for Energy Savings," Proceedings of ACEEE 1994 Summer Study on Energy Efficiency in Buildings, Vol. 5, pp. 5.17-5.24.

Bou Saada, T.E. and Haberl, J.S. 1995a. "A Weather-Daytyping Procedure for Disaggregating Hourly End-Use Loads in an Electrically Heated and Cooled Building from

Whole-facility Hourly Data," 30th Intersociety Energy Conversion Energy Conference, July 30-August 4.

Bou Saada, T.E. and Haberl, J.S. 1995b. "An Improved Procedure for Developing Calibrated Hourly Simulated Models," International Building Performance Simulation Association, 4th International Conference, August 14-16.

Bowman, M. 1994. "Power Measurement," Technical Bulletin, Measuring & Monitoring Services, Inc., Tinton Falls, New Jersey, 07701.

Bryant, J. and O'Neal, D. 1992. "Calibration of Relative Humidity Transducers for use in the Texas LoanSTAR Program," Proceedings of the 1992 Hot and Humid Conference, Texas A&M University, Energy Systems Laboratory Report No. ESL-PA-92/02-15.

Bullock, C. 1992. *All About ESCOs*, copyright KENETECH Corporation, 500 Sansome Street, San Francisco, California, 1995 revision in progress.

Bullock, C. 1995. *Energy Services: A Competitive Opportunity for Utilities*, copyright KENETECH Corporation, 500 Sansome Street, San Francisco, California.

Bullock, C. 1994. *Project Financing for Utilities DSM Programs*, copyright KENETECH Corporation, 500 Sansome Street, San Francisco, California.

Claridge, D., Haberl, J., Bryant, J., Poyner, B. and McBride, J. 1993. "Use of Energy Management and Control Systems for Performance Monitoring of Retrofit Projects," Final Summary Report, USDOE Grant #DE-FG01-90CE21003, Submitted to the USDOE Office of Conservation and Energy, Energy Systems Laboratory Report ESL-TR-91/09/02, Texas A&M University, March.

Claridge, D., Haberl, J., Liu, M., Houcek, J. and Aather, A. 1994. "Can You Achieve 150% of Predicted Retrofit Savings? Is it Time for Recommissioning?," Proceedings of the 1994 ACEEE Summer Study, pp. 5.73-5.88, August.

Claridge, D., Haberl, J., O'Neal, D., Heffington, W., Turner, D., Tombari, C., Roberts, M. and Jaeger, S. 1991. "Improving Energy Conservation Retrofits with Measured Results," ASHRAE Journal, Vol. 33, No. 10, pp. 14-22, October.

Cortina, V. (ed.) 1988. "Precision Humidity Analysis," EG&G Environmental Equipment, 151 Bear Hill Road, Waltham, Massachusetts, (IR sensors).

DOE 1980. "Architects and Engineers Guide to Energy Conservation in Existing Buildings," United States Department of Energy Federal Energy Management Program, Washington, D.C., DOE/CS-0132, February.

Doebelin, E. 1990. *Measurement Systems*. McGraw-Hill, New York, New York, ISBN 0-07-017338-9.

Dubin, F.S. and Long, C.G. 1978. *Energy Conservation Standards for Building Design, Construction and Operation*. McGraw-Hill, New York, ISBN 0070178836.

Dubin, F.S., Mindell, H.L., Bloome, S. 1976. "How to Save Energy and Cut Costs in Existing Industrial and Commercial Buildings," Noyes Data Corp., Park Ridge, New Jersey, ISBN 0815506384.

Dukelow, S.G. 1991. "The Control of Boilers," Instrument Society of America, Research Triangle Park, North Carolina, ISBN 1-55617-330-x.

Dutt, G.S. and Harrje, D. 1988a. "Multifamily Building Energy Audit Procedure: Interim Report," The Center for Energy and Environmental Studies—Working Paper #95, Princeton University, Princeton, New Jersey, July.

Dutt, G.S. and Harrje, D. 1988b. "Multifamily Building Energy Audit Procedure: Technical Reference Manual," The Center for Energy and Environmental Studies - Working Paper #96, Princeton University, Princeton, New Jersey, July.

Dyer, D.F. and Maples, G. 1981. "Boiler Efficiency Improvement," Boiler Efficiency Institute, Department of Mechanical Engineering, Auburn University, P.O. Box 2255,

Auburn, Alabama.

EEI 1981. "Handbook for Electricity Metering," Edison Electric Institute, Washington, D.C., ISBN-0-931032-11-3.

EPA 1993. Energy Star Buildings Manual. United States Environmental Protection Agency, Global Change Division, 401 M Street, S.W., Washington, D.C., October.

Fels, M. (ed.) 1986. "Special Issue Devoted to Measuring Energy Savings, The Princeton Scorekeeping Method (PRISM)," Energy and Buildings, Vol. 9, Nos. 1 and 2.

Fels, M., Kissock, K., Marean, M.A., Reynolds, C. 1995. "Advanced PRISM User's Guide," Center for Energy and Environmental Studies, Princeton University, Princeton, New Jersey, January.

Fracastoro, G.V. and Lyberg, M.D. 1983. "Guiding Principles Concerning Design of Experiments, Instrumentation and Measuring Techniques," International Energy Agency and the Swedish Council for Building Research, Stockholm, Sweden.

Gordon, J.M., Ng, K.C. 1994. "Thermodynamic Modeling of Reciprocating Chillers," *Journal of Applied Physics*, Volume 75, No. 6, March 15, 1994, pp. 2769-2774.

Haberl, J., Athar, A., Abbas, M., Claridge, D. and MacDonald 1992. "Comparing Monitoring Protocols for Energy Retrofits," ASHRAE Transactions Symposium Paper, Vol. 98, Part 1, pp. 1081-1096, January.

Haberl, J., Claridge, D., Harrje, D. 1990. "The Design of Field Experiments and Demonstrations," Proceedings of the IEA Field Monitoring for a Purpose Workshop, Gothenburg, Sweden, April.

Haberl, J., and Komor, P. 1989. "Investigating An Analytical Basis for Improving Commercial Building Energy Audits: Results from a New Jersey Mall," Center for Energy and Environmental Studies, Report No. 264, June.

Haberl, J., Reddy, A. 1995 "Measuring Energy-Saving Retrofits: Experiences from the LoanSTAR Program," Energy Systems Laboratory, Texas Engineering Experiment Station, Texas A&M University System, in progress.

Haberl, J., Turner, W.D., Finstad, C., Scott, F. and Bryant, J. 1992. "Calibration of Flowmeters for use in HVAC Systems Monitoring," Proceedings of the 1992 ASME/JSES/KSES International Solar Energy Conference.

Hadley, D.L. and Tomich, S.D. 1986. "Multivariate Statistical Assessment or Meteorological Influences in Residence Space Heating," Proceedings of the ACEEE 1986 Summer Study on Energy Efficiency in Buildings, Vol. 9, pp. 9.132-9.145.

Harding, J. (ed). 1982. "Recent Advances in Chilled Mirror Hygrometry General Eastern Corporation Technical Bulletin, 50 Hunt Street, Watertown, Massachusetts.

Harrje, D. 1982. "Monitoring Energy Use—What is Needed?," Proceedings of the ACEEE Summer Study on Energy Efficiency in Buildings, August.

Harrje, D. 1986. "Obtaining Building Energy Data, Problems and Solutions," Proceedings of the Field Data Acquisition for Building an Equipment Energy Use Monitoring Workshop, ORNL Publication No. CONF-8510218, pp. 205-209.

Huang, P. 1991. "Humidity Measurements and Calibration Standards," ASHRAE Transactions, Vol. 97, p. 3521.

Hurley, C.W. and Schooley, J.F. 1984. "Calibration of Temperature Measurement Systems Installed in Buildings," N.B.S. Building Science Series 153, January.

Hurley, W. 1985. "Measurement of Temperature, Humidity, and Fluid Flow," Field Data Acquisition for Building and Equipment Energy Use Monitoring, ORNL Publication No. CONF-8510218, March.

Hyland, R.W. and Hurley, C.W. 1983. "General Guidelines for the On-Site Calibration of Humidity and Moisture Control Systems in Buildings," N.B.S. Building Science Series 157, September.

IES 1987. *Lighting Handbook,* Illuminating Engineering Society of North America, New York, New York.

ISA 1976. *Recommended Environments for Standards Laboratories,* Instrument Society of America—Recommended Practice, Research Triangle Park, North Carolina.

Jilar, T. 1990. "Proceedings of the Field Monitoring for a Purpose Workshop," Gothenburg, Sweden, sponsored by the Swedish Council for Building Research, Stockholm, Sweden, prepared by the Building Services Engineering Department, Chalmers University of Technology.

Katipamula, S. and Haberl, J. 1991. "A Methodology to Identify Diurnal Load Shapes for Non-Weather-Dependent Electric End Uses," Proceedings of the 1991 ASME-JSES International Solar Energy Conference, ASME, New York, New York, pp. 457-467, March.

Katipamula, S., Reddy, T.A. and Claridge, D.E. 1994. "Development & Applications of Regression Models to Predict Cooling Energy Consumption in Large Commercial Buildings," ASME International Solar Energy Conference.

Katipamula, S., Reddy, T.A. and Claridge, D. 1995. "Effect of Time Resolution on Statistical Modeling of Cooling Energy Use in Large Commercial Buildings," ASHRAE Transactions, 1995, Vol. 101, Part 2.

Kissock, K., Wu, X., Sparks, R., Claridge, D., Mahoney, J. and Haberl, J. 1994. EModel Version, 1.4d, Energy Systems Laboratory ESL-SW-94/12-01, Texas Engineering Experiment Station, Texas A&M University System, December.

Knebel, D.E. 1983. "Simplified Energy Analysis Using the Modified Bin Method," ASHRAE, Atlanta, Georgia, ISBN 0-910110-39-5.

Kreider, J. and Haberl, J. 1994. "Predicting Hourly Building Energy Usage: The Great Energy Predictor Shootout—Overview And Discussion Of Results," ASHRAE Journal, Atlanta, Georgia, ASHRAE, pp. 72-81, June.

Kulwicki, B. 1991. "Humidity Sensors," Journal of the American Ceramic Society, Vol. 74, pp. 697-707.

Landman, D.S. and Haberl J.S. 1995a. "Monthly Power Level Indices: Comparative Results of Using Different Analysis Tools In Energy Modeling," Energy Systems Laboratory Report ESL-TR-95/11-01I, Texas A&M University, in progress.

Landman, D.S. and Haberl J.S. 1995b. "Monthly Power Level Indices: A Brochure That Shows How To Calculate Monthly Power Level Indices To Compare Energy Consumption For Buildings," Energy Systems Laboratory Report ESL-TR-95/11-02I, Texas A&M University, in progress.

Lee, I. (ed.) 1988. *Review of Humidity Sensors,* Texas Instruments Incorporated, Attleboro, Massachusetts, general reference.

Leider, M. et al. 1990. *A Solid State Amperometric Humidity Sensor,* Journal of Applied Electrochemistry, Chapman and Hill: Vol. 20, pp. 964-8.

Liptak, B. 1995. *Instrument Engineers' Handbook, 3rd Edition: Process Measurement and Analysis,* Chilton Book Company, Radnor, Pennsylvania, ISBN 0-8019-8197-2.

Liu et al. 1994. "Identifying and Implementing Improved Operation and Maintenance," Proceedings of ACEEE 1994 Summer Study on Energy Efficiency in Buildings, Vol. 5, pp. 5.153-5.165.

Liu et al. 1995 "Application of Calibrated HVAC Systems to Identify Component Malfunctions and to Optimize the Operation and Control Schedules," ASME/JSME International Solar Energy Conference, pp. 209-217.

Lyberg, M.D. 1987. *Source Book for Energy Auditors,* Volumes 1 and 2, International Energy Agency and the Swedish Council for Building Research, Stockholm, Sweden.

MacDonald, J.M., Sharp, T.R. and Gettings, M.B. 1989. "A Protocol for Monitoring Energy

Efficiency Improvements in Commercial and Related Buildings," Oak Ridge National Laboratory Report, ORNL/CON-291, September.

Miller, R. 1989. *Flow Measurement Handbook,* McGraw Hill Publishing Company, New York, New York, ISBN 0-07-042046-7.

Morrissey, C.J. 1990. "Acoustic Humidity Sensor," NASA Tech Brief. Vol. 14, No. 19, April, (acoustic).

O'Neal, D., Bryant, J., Turner, W. and Glass, M. 1990. "Metering and Calibration in LoanSTAR Buildings," Proceedings of the Seventh Annual Symposium on Improving Building Systems in Hot and Humid Climates, Texas A&M University, College Station, Texas, October.

Phelan, J., Brandemuehl, M. and Krarti, M. 1995. "Methodology to Measure *In-Situ* Chiller Performance," Joint Center for Energy Management Report, University of Colorado at Boulder.

Phelan, J., Brandemuehl, M. and Krarti, M. 1996. "*In-Situ* Testing of Fan and Pump Performance for Energy Analysis," Proceedings of the ASME/JSME Solar Engineering Conference, San Antonio, Texas, March, in preparation.

Rabl, A. 1988. "Parameter Estimation in Buildings: Methods for Dynamic Analysis of Measured Energy Use," *Journal of Solar Energy Engineering,* Vol. 110, pp. 52-66.

Rabl, A. and Riahle, A. 1992. "Energy Signature Model for Commercial Buildings: Test With Measured Data and Interpretation," *Energy and Buildings,* Vol. 19, pp. 143-154.

Ramboz, J.D. and McAuliff, R.C. 1983. "A Calibration Service for Wattmeters and Watt-Hour Meters," N.B.S. Technical Note 1179.

Reddy, T.A., Kissock, J.K., Katipamula, S., Ruch, D.K. and Claridge, D.E. 1994. "An Overview of Measured Energy Retrofit Savings Methodologies Developed in the Texas LoanSTAR Program," Publication No. ESL-TR-94/03-04, Energy Systems Laboratory Technical Report, Department of Mechanical Engineering, Texas A&M University.

Reynolds and Fels. 1988. "Proceedings of ACEEE 1994 Summer Study on Energy Efficiency in Buildings," Vol. 10, pp. 10.237-10.241.

Robinson, J., Bryant, J., Haberl, J. and Turner, D. 1992. "Calibration of Tangential Paddlewheel Insertion Flowmeters," Proceedings of the 1992 Hot and Humid Conference, Texas A&M University, Energy Systems Laboratory Report No. ESL-PA-92/02-09.

Ross, I.J. and White, G.M. 1990. "Humidity," Instrumentation and Measurement for Environmental Sciences: Transactions of the ASAE, 2nd ed., p. 8-01.

Ruch, D. and Claridge, D. 1991. "A Four Parameter Change-Point Model for Predicting Energy Consumption in Commercial Buildings," Proceedings of the ASME-JSES-JSME International Solar Energy Conference, Reno, NV, March 17-22, pp. 433-440.

SMACNA 1985. *Handbook of HVAC Retrofit,* Sheet Metal and Air Conditioning Contractors National Association, Chantilly, Virginia.

Schiller, S. 1994. Power-Saving Partners Measurement and Verification Procedures Manuals, Pacific Gas and Electric Company.

Schiller, S. 1995. DSM Bidding Program Measurement and Verification Guidelines, Southern California Edison Company.

Stein B. and Reynolds, J.S. 1992. *Mechanical and Electrical Equipment for Buildings,* Eighth Edition, John Wiley and Sons, New York, New York, ISBN 0471525022.

Taylor, J. 1981. *An Introduction to Error Analysis,* University Science Books, Oxford University Press, Mill Valley, California, ISBN 0-935702-07-5.

Ternes, M.P. 1987. "Single-Family Building Retrofit Performance Monitoring Protocol: Data Specification Guideline," Oak Ridge National Laboratory, Report No. ORNL/CON-196.

Turner, W.C. 1993. *Energy Management Handbook*, Fairmont Press, Inc., Lilburn, Georgia, ISBN 0881731463.

Wexler, A. and Schiller, S. 1983. *Monitoring Methodology Handbook*, EPRI EM3003, Electric Power Research Institute.

Wiesman, S. (ed.) 1989. *Measuring Humidity in Test Chambers*, General Eastern Corporation, 50 Hunt Street, Watertown, Massachusetts.

Wise, J.A. 1976. "Liquid-In-Glass Thermometry," N.B.S. Monograph 150, January.

Wise, J.A. and Soulen, R.J. 1986. "Thermometer Calibration: A Model for State Calibration Laboratories," N.B.S. Monograph 174, January.

Witte, L.C., Schmidt, P.S. and Brown, D.R. 1988. *Industrial Energy Management Utilization*, Hemisphere Publishing Company, New York, New York, ISBN 0891163220.

7.2 BIBLIOGRAPHY

Air Conditioning and Refrigeration Center, Mechanical Engineering, University of Illinois, 140 Mechanical Engineering Building, 1206 W. Green St., Urbana, Illinois 61801, TEL: 217-333-3115, FAX: 217-244-6534.

American Council for an Energy Efficient Economy (ACEEE), 1001 Connecticut Avenue, N.W., Suite #535, Washington, D.C. 20036, TEL: 202-429-8873.

American Society of Heating, Refrigerating, and Air Conditioning Engineers (ASHRAE), Attn.: Ms. Phyllis Maurer, 1791 Tullie Circle, N.E., Atlanta, Georgia 30329-2305, TEL: 404-636-8400, FAX: 404-321-5478.

American Society of Mechanical Engineers (ASME), 22 Law Drive, Fairfield, New Jersey 07007, TEL: 800-843-2763.

Association of Energy Engineers (AEE), Dept. 192, P.O. Box 1026, Lilburn, Georgia 30226, TEL: 404-925-9558, FAX: 404-381-9865.

Boiler Efficiency Institute, Department of Mechanical Engineering, Auburn University, P.O. Box 2255, Auburn, Alabama.

Brandemuel, M. 1995. "ASHRAE RP 827: Method For Determining *In-Situ* Efficiency Tests For Pumps, Ahus, And Chillers: Progress Report," ASHRAE Research, Atlanta, Georgia.

Center for Energy and Environment (CEE), 100 North 6th Street, Suite 412A, Minneapolis, Minnesota 55403-1520, TEL: 612-348-9336.

Center for Energy and Environmental Studies (CEES), Princeton University, Engineering Quadrangle, P.O. Box CN5263, Princeton, New Jersey 08544-5263, TEL: 609-452-5445, FAX: 609-258-3661.

Edison Electric Institute (EEI), Attn: Publications, P.O. Box 2800, Kearneysville, West Virginia 25430, TEL: 800-334-5453.

Energy Systems Laboratory, Texas Engineering Experiment Station, Attn: Publications Department, College Station, Texas 77843-3581, TEL: 409-845-9213, FAX: 409-862-2457, WWW homepage: www.esl.tamu.edu.

Energy Technology Center, University of Wisconsin, 432 North Lake Street, Room 825, Madison, Wisconsin 53706-1498, TEL: 608-262-0692, FAX: 608-262-6209.

IESNA Publications, 120 Wall Street, 17th Floor, New York, New York 10005, TEL: 212-248-5000, ext. 112, FAX: 212-248-5017.

Joint Center for Energy Management (JCEM), Department of Civil, Environmental and Architectural Engineering, University of Colorado, Boulder, Colorado 80309-0428, TEL: 303-492-3915, FAX: 303-492-7317.

Lawrence Berkeley National Laboratory (LBNL), Attn: Ms. Karen Olsen, Building 90, Room

4000, 1 Cyclotron Road, Berkeley, California 94720, WWW homepage: www.eande.lbl.gov.

National Association of Energy Service Companies (NAESCO), Attn: Ms. Mary Lee Berger-Hughs, 1140 New York Avenue, N.W., Washington, D.C. 20005, TEL: 202-371-7816.

National Energy Information Center, Energy Information Administration (EIA), Forrestal Building, Room 1F-048, Washington, D.C. 20585, TEL: 202-586-8800.

National Renewable Energy Laboratory (NREL), Attn: Ms. Sally Evans, 1617 Cole Boulevard, Building 15, TEL: 303-275-4363.

National Technical Information Service (NIST), U.S. Department of Commerce, 5285 Port Royal Road, Springfield, Virginia 22161. *NOTE: This is the repository for all publications by the federal labs and contractors.*

Office of Scientific and Technical Information, P.O. Box 62, Oak Ridge, Tennessee 37831, TEL: 615-576-8401.

Superintendent of Documents, U.S. Government Printing Office, Washington, D.C. 20402, TEL: 202-783-3238, FAX: 202-275-0019.

Thumann, A. 1979. *Handbook of Energy Audits*, Fairmont Press, Atlanta, Georgia, ISBN 0915586185.

Index